Texts in Applied Mathematics

Volume 81

The mathematization of all sciences, the fading of traditional scientific boundaries, the impact of computer technology, the growing importance of computer modelling and the necessity of scientific planning all create the need both in education and research for books that are introductory to and abreast of these developments. The aim of this series is to provide such textbooks in applied mathematics for the student scientist. Books should be well illustrated and have clear exposition and sound pedagogy. Large number of examples and exercises at varying levels are recommended. TAM publishes textbooks suitable for advanced undergraduate and beginning graduate courses, and complements the Applied Mathematical Sciences (AMS) series, which focuses on advanced textbooks and research-level monographs.

Wilfredo Urbina-Romero · Ricardo Rios

An Introduction to the Modern Martingale Theory and Applications

An Analytic View

 Springer

Wilfredo Urbina-Romero
College of Sciences, Health and Pharmacy
Roosevelt University
Chicago, IL, USA

Ricardo Rios
Department of Mathematics
Universidad Central de Venezuela
Caracas, Distrito Capital, Venezuela

ISSN 0939-2475   ISSN 2196-9949 (electronic)
Texts in Applied Mathematics
ISBN 978-3-031-88902-8   ISBN 978-3-031-88903-5 (eBook)
https://doi.org/10.1007/978-3-031-88903-5

Mathematics Subject Classification: 60G40, 46B09, 60G42, 60G46

*This book is dedicated to the memory of **Sebastián Feliz Rios Ludeña** our beloved Tuchi, to our dear friend **Diómedes Bárcenas** who taught us, among many other things, about martingales and geometry of Banach spaces, and last but not least to **Richard (Dick) Gundy** a great mathematician and magician, and a very good friend.*

Foreword

This book delves into martingale theory and its extensive range of applications. Originally introduced by Doob in the 1950s, this theory was developed to generalize the sum of independent random variables. Specifically, it aimed to establish a Law of Large Numbers, a Central Limit theorem, and other key concepts under dependence conditions. These goals were achieved, and as often happens in mathematics, the initial objectives were quickly surpassed. Over time, an entire spectrum of applications emerged. Concepts such as stopping times, convergence theorems, martingale decomposition, and later the introduction of the notion of martingale transform significantly advanced the theory, forging a close connection with the field of stochastic integration. A real turning point came with Burkholder's groundbreaking article on the Martingale Transform. This work showcased the immense potential of the theory across various fields, particularly in its applications to analysis. In this book, the reader will find not only a clear and concise development of Burkholder's contributions but also those of other researchers on this important topic. Collectively, these findings are referred to in the present book as modern martingale theory.

The book guides readers through the fundamental concepts of martingale theory, presenting each topic within a coherent and carefully structured narrative. It also develops the tools of discrete stochastic calculus and explores some of its consequences, such as the discrete Itô's formula. This foundation prepares readers to approach these topics in the context of continuous-time processes, which are significantly more complex and beyond the scope of this book.

An observation that should not be overlooked concerns the applications discussed in the final chapter of the book. Among those presented, my favorites are, first, the relationship between martingale theory and symmetric semigroups of operators, and second, the connection between vector martingale theory and the geometry of Banach spaces. In the section on semigroups, the authors elegantly present G. C. Rotta's profound theorem on constructing a martingale that represents the semigroup. Using this representation, they provide a remarkably simple proof of the boundedness of the semigroup's maximal function. In relation to the geometry of Banach spaces, the final application explores the Radon-Nikodým property and its connection to the convergence theorems of vector valued martingales, defined using the Bochner integral. The entire section serves as a preamble, inviting readers to delve deeper into these fascinating topics. The text strikes a balance between mathematical rigor and accessibility, making the subject not only intellectually profound but also approachable for a wide audience.

About the Authors

In the 1970s, a group of mathematicians gathered in Venezuela, marking the return of the first generation of graduates from the Facultad de Ciencias at the Central Universidad Central de Venezuela. This faculty, established in 1958 following the fall of the dictatorship, quickly became a vibrant hub of academic activity, further enriched by the arrival of numerous researchers from across the globe. Many of these scholars came from the Southern Cone, having been exiled due to the military dictatorships that afflicted the region. During this dynamic period, two research groups stood out: the Analysis Group and the Probability and Statistics Group. A hallmark of these groups was their close collaboration and frequent exchange of ideas. The authors of this book, Ricardo Ríos and Wilfredo Urbina, emerged from this rich academic environment. Ríos, after completing his studies in Caracas, pursued a doctorate in non-parametric statistics at the University of Orsay in Paris. Urbina, initially focused on classical harmonic analysis, later shifted his attention to one of his enduring passions: harmonic analysis in Gaussian measure spaces, becoming a world specialist in this topic. Despite this shift, he retained a strong interest in the applications of probability to analysis. This book reflects the interplay between these two disciplines, offering a comprehensive treatment of modern martingale theory—a subject positioned at the crossroads of Analysis and Probability.

Overview of the Book

Chapter 1 introduces the central themes of the book with clarity and elegance. The authors begin with orthogonal systems formed by functions supported on dyadic intervals, covering classical systems such as Rademacher, Haar, and Walsh. They address fundamental problems, such as the convergence of partial sums of functions in L^p spaces, setting the stage for solutions rooted in martingale theory, which is developed in subsequent chapters. Chapters 2 and 3 focus on conditional expectation and the classical theory of martingales, respectively. The authors adopt a clear and didactic approach, ensuring that these chapters provide a robust foundation for both analysts and probabilist who may wish to revisit or deepen their understanding of these essential topics. The exposition is concise and accessible, tailored to meet the needs of a broad audience. Chapter 4 serves as the heart of the book. It presents the pioneering ideas of Burkholder and Burkholder-Gundy with precision and depth. Advanced topics are explored, including Doob's decomposition theorems and additional decompositions by Krickeberg and Gundy. The martingale convergence theorem is revisited with fresh insights, and significant results on martingale transforms are established. The chapter concludes with proofs of the strong law of large numbers for martingales, an exposition of the celebrated inequalities of Burkholder and Gundy, and a discussion of the "good lambda" inequalities, extending these results to general operators. The final chapter is filled with applications—some fully developed and others sketched to encourage further exploration by the reader.

Audience and Purpose

This book is designed to cater to both students seeking a foundational understanding and researchers aiming for deeper insights into stochastic analysis. The authors' meticulous presentation, coupled with carefully chosen examples, will inspire readers to appreciate the elegance and power of Martingale Theory.

Final Words

This book, an earlier version of which was prepared for the "Escuela Venezolana de Matemáticas" (EVM)—courses for postgraduate students and the precursor to the "Escuela de Matemática de América Latina y del Caribe" (EMALCA)—stands as a testament to the close collaboration of two researchers and lifelong friends. Born out of their formative years in Venezuela, this work is set to become a valuable reference for both students and scholars. I am confident that readers will share the same sense of enjoyment and discovery that I experienced while delving into its pages.

José Rafael León Ramos
Professor at IMERL
Universidad de la República
Montevideo, Uruguay

Former Professor at the School
of Mathematics
Universidad Central de Venezuela
Caracas, Distrito Federal, Venezuela

Preface

Martingale theory is nowadays one of the central components of probability theory. Martingales is a remarkably flexible tool used throughout probability and its applications to other areas of mathematics. They are central to modern stochastic analysis. In a way, they are a natural generalization of the study of sums of independent random variables. Although P. Lévy already discussed martingales in 1937, without explicitly naming them, it was J. L. Doob in the 1940s who not only named them but also realized their potential and did the fundamental development of the theory. His landmark 1953 book *Stochastic Processes* [69], which is still a primary reference nowadays, formalized the theory as we know it now, and it is what we call the classical martingale theory. Since then, it has been continuously studied. Several mathematicians worked on objects that turned out to be martingales before the theory was developed; for example, among others, R. E. A. C. Paley [134], J. Marcinkiewicz and A. Zygmund [116], [117] A. Khintchine and A. Kolmogorov [107], S. Kaczmarz and H. Steinhaus [106], and H. Rademacher [142]. In another direction, in 1928, R. Courant, K. Fredricks, and H. Lewy used ideas related to the notions of martingales, although without randomness, to study harmonic functions.

In 1966 D. L. Burkholder published his seminal paper *Martingale Transform* [29] starting a new era for martingale theory. From there, an intensive research developed obtaining some of the now famous results such as Burkolder's inequality, Burkholder, Davis, Gundy's inequality and their generalizations, the good-λ inequalities which constitute what we call advanced martingale theory.

Today martingale theory has become recognized as an essential tool in a diversity of topics in mathematical analysis and has a huge number of applications in several other areas. For instance, applications of martingale theory is having an increasingly important impact on Banach space theory, harmonic analysis, and in data sciences.

The book covers not only results from the classical theory for discrete martingale, which is what is contained nowadays in almost any book on probabilities or stochastic processes, but also covers in detail most of the results of the research by authors such as D. Burkholder, B. Davis and R. Gundy, completing the work on martingales by J. Doob. The book is strongly influenced by the results of

D. Burkholder which are the basis of further generalizations of martingale theory of what we call advanced martingale theory and applications. In that respect this is not an "standard" book on martingales but a unified exposition of classical and more recent results.

This book is an outgrowing of the notes written for a course that we taught in the 5th Venezuelan School of Mathematics in September 1992 at Mérida Venezuela, and later discussed in a workshop at the University of La Plata, Argentina, and has been the basis of several graduate courses at Universidad Central de Venezuela (UCV). We will consider only the martingale theory for discrete parameter, which can be considered the basic one, since many results for discrete parameter can be easily extended to the continuous case. For the continuous parameter case, there are several excellent books, such as C. Delacherie and P. A. Meyer [66], [67], I. Karatzas and E. Shreve [105], R. S. Liptser, and A. N. Shiryayev [113], and A. N. Shiryayev [148], for instance. The publication in 2010 of *Selected Works of Donald L. Burkholder* by B. Davies and R. Song as editors [65] was of incredible help in preparing these notes.

The organization of the book is as follows. Chapter 1 provides a historical introduction, exploring the Rademacher, Haar, and Walsh functions, along with several key results involving them. The main goal is to present these foundational results that later played a significant role in the development of martingale theory, discussing notions that are the origin of many ideas in Chapters 3 and 4, through their history and their simplest form of dyadic martingales, creating a bridge between harmonic analysis and probability theory. Thus it constitutes a historical prelude that could be skipped by impatient readers. The chapter draws considerable inspiration from the beautiful book of Mark Kac [103]. To make these notes as self-contained as possible in Chap. 2 we review in detail the notion of conditional probability which is crucial for the theory; we also include an appendix with the basic results of probability theory. Chapter 3 is on the classical results of the theory developed by J. L. Doob. Thus, Chaps. 2 and 3 are a recap of probability theory and basic martingale theory to prepare the reader for what follows next. Chapter 4 is the main core of the book and it is on advanced martingale theory i.e., the results obtained since D. L. Burkholder published his paper *Martingale Transform* [29]. Another important reference here is the seminal paper of D. L. Burkholder and R. Gundy *Extrapolation and interpolation and quasilinear operators on martingales* [44]. However, the proofs in that article, while containing all the basic ideas for good λ inequalities, do not use them to their full potential. We will use the formulation of good λ inequalities given by D. L. Burkholder in [30] gives alternative proofs of those results that greatly simplifies the original proofs. Thus, we present some refined forms of the Burkholder-Davis-Gundy circle of ideas, with clear proofs progressively generalizing from L^p norms and square/maximal functions to general Orlicz norms and quasilinear operators, extending Theorems 3.2 and 4.2 of [44] for general operators, using good λ inequalities, see [123]. Also, we shall study some local properties for martingales transforms. Finally, Chap. 5 is on applications of martingale theory. As we have said these days the number of applications of martingale theory is so vast that results almost impossible to review all of them in detail, thus some of them are throughly explained but for others we just provide brief overviews and further references. We hope to provide to a panoramic view of the scope of applications of martingale theory.

The book is intended for a very diverse audience, from graduate students to researchers working in a broad spectrum of areas in probability and analysis.

Finally, the first author wants to thank INRIA Saclay Centre, Marc Lavielle, and Universidad Central de Venezuela (UCV); the second author wants to thank Roosevelt University for a research leave during the spring semester of 2021 in which an important amount of the work writing these notes was done. We also want to thank Melanie Pivarski, Steve Cohen, and Joaquin Ortega, Cristina Pereyra and Alex Stokolos for important corrections to the manuscript that have improved it in important ways all the remaining errors and typos are of our entirely responsibility. Nashla Báez gave us an invaluable bibliographical support and help in the graphics as well as Hugo Villarroel. Additionally, we want to thank all the anonymous reviewers who, with their comments and corrections, greatly improved and complete the final version of the manuscript. Last but not least, we are indebebted to Donna Chernyk, editor of Springer Nature, for all her help and support. We were very fortunate to have had several conversations with Richard Gundy about these notes. Noblesse oblige, we were fortunate to have had enriching conversations on martingale theory and their applications to the geometry of Banach spaces with our beloved friend Diomedes Bárcenas and his adviser Joe Diestel, unfortunately, both long gone.

<table>
<tr><td>Caracas, Venezuela</td><td style="text-align:right">Ricardo Rios</td></tr>
<tr><td>Chicago, USA</td><td style="text-align:right">Wilfredo Urbina-Romero</td></tr>
<tr><td>January 2025</td><td></td></tr>
</table>

Contents

Introduction 1

Rademacher functions have been the subject of intense research since, in 1922, Hans Rademacher introduced them as an orthogonal system of functions in [142]. He soon wrote a sequel on the completion of this system but unfortunately he was discouraged from publishing it. Meanwhile, J. L. Walsh [169] in 1923, independently completed the system and published a paper containing much of Rademacher's work. Rademacher never published his second paper, which, for many years, was considered to be permanently lost but in 1979, the paper was miraculously unearthed. Rademacher functions combine in their nature both an analytical characteristic (see the work of Paley [134], Marcinkiewicz [115], Kaczmarz and Steinhaus [106] and Zygmund [176]) as well as a probabilistic one (see the works of Burkholder [28,29], Gundy [85,86] and Burkholder and Gundy [44]) and it is precisely in his study that these two aspects continue to be valid to this day.

From the probabilistic point of view they, and the systems derived from them, Walsh's and Haar's systems, not only have served as an inexhaustible source of inspiration for the development of contemporary martingale theory (especially in the works of D. L. Burkholder and R. Gundy among others) but also as a working tool to study more general cases, both in the field of probability theory as well as of harmonic analysis. In the present chapter, we study the properties of these orthogonal systems that later developed from the probabilistic point of view into the martingale theory that will be discussed in later chapters. This chapter is strongly influenced by the beautiful work of Mark Kac [103], which we highly recommend.

One of the goals of this chapter is to discuss several results that later are extensively developed in martingale theory establishing a bridge between probability theory and harmonic analysis.

W. Urbina-Romero and R. Rios, *An Introduction to the Modern Martingale Theory and Applications*, Texts in Applied Mathematics 81,
https://doi.org/10.1007/978-3-031-88903-5_1

1.1   Rademacher Functions

Consider the interval $[0, 1)$, the family of Borel sets in $[0,1)$, $\mathscr{B}[0, 1)$, i.e., the σ-algebra of the sets generated by the open sets of $[0, 1)$ and the Lebesgue measure, m, note that $m([0, 1)) = 1$. Let these choices be denoted by the triplet

$$([0, 1), \mathscr{B}[0, 1), m)$$

Each $t \in [0, 1)$ has a binary (or dyadic) decomposition

$$t = \frac{\varepsilon_1(t)}{2} + \frac{\varepsilon_2(t)}{2^2} + \cdots = \sum_{n=1}^{\infty} \frac{\varepsilon_n(t)}{2^n}, \quad \varepsilon_n(t) = 0 \text{ or } 1. \tag{1.1}$$

The functions $b\varepsilon_n$ are called dyadic or Steinhaus functions. To guarantee the uniqueness of these expansions, we will use the convention of considering only expansions containing an infinite number of zeros, and so the Steinhaus functions will be right-continuous and are then defined, for each $n \in \mathbb{N}$ as

$$\varepsilon_n(t) = \begin{cases} 0, & \text{if } \frac{k-1}{2^n} \leqslant t < \frac{k}{2^n} \text{ if k is odd, } \ 1 \leqslant k \leqslant 2^n \\ 1, & \text{otherwise.} \end{cases}$$

See (Fig. 1.1). The Rademacher functions can be defined in several ways. First as

$$r_n(t) = 1 - 2\varepsilon_n(t) = \begin{cases} 1, & \text{if } \frac{k-1}{2^n} \leqslant t < \frac{k}{2^n} \text{ if } k \text{ is odd, } \ 1 \leqslant k \leqslant 2^n \\ -1, & \text{otherwise.} \end{cases} \tag{1.2}$$

See (Fig. 1.2). Equivalently, define for each $n \in \mathbb{N}$

$$r_n(t) = \sum_{k=1}^{2^n} (-1)^{k+1} \chi_{I_n^k}(t) \tag{1.3}$$

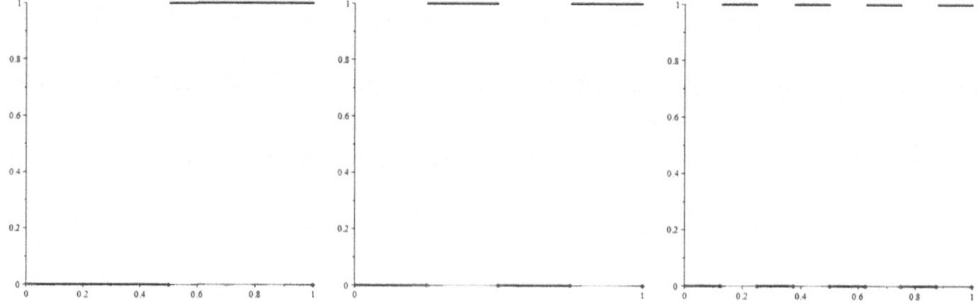

Fig. 1.1 The first three Steinhaus functions $\varepsilon_1, \varepsilon_2, \varepsilon_3$

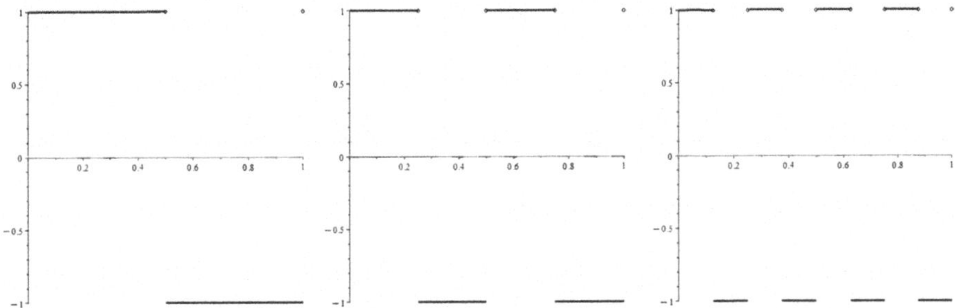

Fig. 1.2 The first three Rademacher functions r_1, r_2, r_3

where $I_n^k = \left[\frac{k-1}{2^n}, \frac{k}{2^n}\right)$, $1 \leqslant k \leqslant 2^n$, the dyadic interval and $\chi_A(t) = \begin{cases} 1, & \text{if } t \in A \\ 0 & \text{if } t \notin A \end{cases}$

is the characteristic function of the set A.

The Rademacher functions can be also represented as

$$r_n(t) = \text{sign}\,\sin(2^n \pi t) \tag{1.4}$$

since

$$r_1(t) = \text{sign}\,\sin(2\pi t) \quad \text{and} \quad r_n(t) = r_1(2^{n-1}(t)), \quad \text{for } n > 1,$$

and $\sin(2^n \pi t)$ changes sign precisely on the intervals $I_n^k = \left[\frac{k-1}{2^n}, \frac{k}{2^n}\right]$.

Another alternative definition of the Rademacher functions, defined not only on $[0, 1)$, but on all $\mathbb{R}$ periodically with period 1, i.e., $r_n(t + 1) = r_n(t)$, $t \in [0, 1)$ is the following. Define

$$r_1(t) = \begin{cases} 1 & \text{if } 0 \leqslant t < 1/2 \\ -1 & \text{if } 1/2 \leqslant t < 1 \end{cases} = \chi_{[0,\frac{1}{2})} - \chi_{[\frac{1}{2},1)}, \tag{1.5}$$

extends r_1 periodically as $r_1(t + 1) = r_1(t)$, and then

$$r_n(t) = r_1\left(2^{n-1}t\right), \quad n > 1 \tag{1.6}$$

Note that from (1.1)

$$\sum_{n=1}^{\infty} \frac{r_n(t)}{2^n} = \sum_{n=1}^{\infty} \frac{1 - 2\varepsilon_n(t)}{2^n} = \sum_{n=1}^{\infty} \frac{1}{2^n} - 2\sum_{n=1}^{\infty} \frac{\varepsilon_n(t)}{2^n} = 1 - 2t$$

On the other hand, observe that $\{r_n\}$ is an orthonormal system in $L^2([0, 1), m)$ since for all n

$$\int_0^1 [r_n(t)]^2 \, dt = \int_0^1 dt = 1 \tag{1.7}$$

and if $n \neq m$, assuming that $m > n$, then

$$\int_0^1 r_n(t)r_m(t)dt = \sum_{k=1}^{2^n} \int_{(k-1)/2^n}^{k/2^n} r_n(t)r_m(t)dt = \sum_{k=1}^{2^n}(-1)^k \int_{(k-1)/2^n}^{k/2^n} r_m(t)dt = 0$$

$$(1.8)$$

since on

$$I_n^k = \left[\frac{k-1}{2^n}, \frac{k}{2^n}\right) = \left[\frac{2^{m-n}(k-1)}{2^m}, \frac{2^{m-n}k}{2^m}\right) = \bigcup_{i=1}^{2^{m-n}} \left[\frac{2^{m-n}(k-1)+i}{2^m}, \frac{2^{m-n}k+i}{2^m}\right)$$

the function r_n changes sign an even number of times and therefore its integral is equal to zero. Using an analogous argument one can prove that if $n_1 < n_2 < \cdots < n_k$, then

$$\int_0^1 r_{n_1}(t)r_{n_2}(t) \cdots r_{n_k}(t)dt = 0 = \int_0^1 r_{n_1}(t)dt \cdots \int_0^1 r_{n_k}(t)dt \qquad (1.9)$$

It is important to emphasize that $\{r_n\}$ is not a complete orthonormal system, because if we take the constant function 1 (i.e., $f(t) = 1$ for all $t \in [0, 1)$), then for all n

$$\int_0^1 f(t)r_n(t)dt = \int_0^1 r_n(t)dt = 0 \qquad (1.10)$$

thus, f is orthogonal to the space generated by $\{r_n\}$ but it is not zero. Moreover, by (1.8) it is clear that by taking f as any product of Rademacher functions $r_{n_1}r_{n_2} \cdots r_{n_k}$ then f is orthogonal to r_n if $n \leqslant n_1 \leqslant n_2 \cdots \leqslant n_k$. Furthermore, using (1.13) we will see that f is orthogonal to any r_n.

However, as we will see later, from the Rademacher functions it is possible to define complete orthonormal systems in $L^2([0, 1), m)$, such as the Haar functions and the Walsh functions that will be discussed in the next section.

From a probabilistic point of view, the dyadic or Steinhaus functions have an important property. First, note that for all $n \in \mathbb{N}$

$$m\{t \in [0, 1) : \varepsilon_n(t) = 0\} = m\{t \in [0, 1) : \varepsilon_n(t) = 1\} = 1/2$$

and if $a_1, \ldots, a_n$ are numbers equal to 0 or 1, then, we have:

$$m\{t \in [0, 1) : \varepsilon_1(t) = a_1, \ldots, \varepsilon_n(t) = a_n\}$$
$$= m\left\{t \in [0, 1) : \frac{a_1}{2} + \frac{a_2}{2^2} + \cdots + \frac{a_n}{2^n} \leqslant t < \frac{a_1}{2} + \frac{a_2}{2^2} + \cdots + \frac{a_n}{2^n} + \frac{1}{2^n}\right\}$$
$$= m\left[\frac{a_1}{2} + \frac{a_2}{2^2} + \cdots + \frac{a_n}{2^n}, \frac{a_1}{2^n} + \cdots + \frac{a_n}{2^n} + \frac{1}{2^n}\right) = \frac{1}{2^n}.$$

Thus

$$m\{t \in [0, 1) : \varepsilon_1(t) = a_1, \ldots, \varepsilon_n(t) = a_n\} = \prod_{i=1}^{n} m\{t \in [0, 1) : \varepsilon_i(t) = a_i\}.$$

$$(1.11)$$

Since the Rademacher functions are defined in terms of the Steinhaus functions, it is not surprising that they have an analogous property to (1.11): if $\delta_1, \ldots, \delta_n$ is a sequence of $+1$ and -1, then

$$m\{t \in [0, 1) : r_1(t) = \delta_1, \ldots, r_n(t) = \delta_n\} = \prod_{i=1}^{n} m\{t \in [0, 1) : r_i(t) = \delta_i\}$$

(1.12)

That is to say, in probabilistic terms, Rademacher and Steinhaus functions are independent.[1] The relation (1.8) is then a particular case of the following relation:

$$\int_0^1 r_{n_1}^{\alpha_1}(t) r_{n_2}^{\alpha_2}(t) \cdots r_{n_k}^{\alpha_k}(t) dt = \int_0^1 r_{n_1}^{\alpha_1}(t) dt \cdots \int_0^1 r_{n_k}^{\alpha_k}(t) dt \qquad (1.13)$$

for $\alpha_1, \ldots, \alpha_k \in \mathbb{N}$, and this is a consequence of independence (1.12). To prove (1.13) it suffices to note that the product $r_{n_1}^{\alpha_1} \cdots r_{n_k}^{\alpha_k}(t)$ is a step function and then

$$
\begin{aligned}
\int_0^1 r_{n_1}^{\alpha_1}(t) r_{n_2}^{\alpha_2}(t) \cdots r_{n_k}^{\alpha_k}(t) dt &= \sum_{\delta_1,\ldots,\delta_k \in \{-1,1\}} \delta_1^{\alpha_{n_1}} \cdots \delta_k^{\alpha_{n_k}} \, m\{t \in [0, 1) : r_1(t) \\
&= \delta_1, \ldots, r_n(t) = \delta_n\} \\
&= \sum_{\delta_1,\ldots,\delta_k \in \{-1,1\}} \prod_{i=1}^{k} \delta_i^{\alpha_i} m\left\{t \in [0, 1) : r_{n_i}(t) = \delta_i\right\} \\
&= \prod_{i=1}^{k} \sum_{\delta_i \in \{-1,1\}} \delta_i^{\alpha_i} m\left\{t \in [0, 1) : r_{n_i}(t) = \delta_i\right\} \\
&= \prod_{i=1}^{k} \int_0^1 r_{n_i}^{\alpha_i}(t) dt.
\end{aligned}
$$

Using the linearity of the integral, (1.13) can be generalized to

$$\int_0^1 p_1\left(r_{n_1}(t)\right) \cdots p_k\left(r_{n_k}(t)\right) dt = \int_0^1 p_1\left(r_{n_1}(t)\right) dt \cdots \int_0^1 p_k\left(r_{n_k}(t)\right) dt$$

where $p_1, \ldots, p_k$ are polynomials. However, this relation can be obtained directly in general form since if $f_1, \ldots, f_k$ are integrable functions on $[0, 1)$ using the same type of argument as in (1.8) we have that

$$\int_0^1 f_1\left(r_{n_1}(t)\right) \cdots f_k\left(r_{n_k}(t)\right) dt = \int_0^1 f_1\left(r_{n_1}(t)\right) dt \cdots \int_0^1 f_k\left(r_{n_k}(t)\right) dt \quad (1.14)$$

---

[1] See Appendix.

Now let us prove that the Rademacher functions form a symmetric sequence in the following sense: Given a measurable function $F : \mathbb{R}^k \to \mathbb{R}$ and a numeric sequence

$$\alpha = (\alpha_1, \ldots, \alpha_k) \in \{-1, 1\}^k,$$

then the functions

$$F\left(r_{n_1}(t), r_{n_2}(t), \ldots, r_{n_k}(t)\right) \text{ and } F\left(\alpha_1 r_{n_1}(t), \alpha_2 r_{n_2}(t), \ldots, \alpha_k r_{n_k}(t)\right)$$

are equidistributed, i.e.,

$$m\left\{t \in [0, 1) : F\left(r_{n_1}(t), r_{n_2}(t), \ldots, r_{n_k}(t)\right) \in A\right\}$$
$$= m\left\{t \in [0, 1) : F\left(\alpha_1 r_1(t), \alpha_2 r_{n_2}(t), \ldots, \alpha_k r_{n_k}(t)\right) \in A\right\}$$

for any $A \in \mathscr{B}[0, 1)^k$.

In order to prove this, it is enough to consider $F = \chi_E$, $E \in \mathscr{B}[0, 1)^k$ and then prove that

$$\mu(E) = m\left\{t \in [0, 1) : (r_{n_1}(t), \ldots, r_{n_k}(t)) \in E\right\}$$
$$= m\left\{t \in [0, 1) : (\alpha_1 r_{n_1}(t), \ldots, \alpha_n r_{n_k}(t)) \in E\right\} = \mu_\alpha(E).$$

It is clear that μ and μ_α are supported on the same set $S = \{s \in \mathbb{R}^k : |s_k| = 1, i = 1, \ldots k$. Moreover, for each $s \in S$ there exists an interval $I_k^j = \left[\frac{j-1}{2^k}, \frac{j}{2^k}\right)$ for some $j = j(s) \in \{1, 2, \ldots, 2^k\}$ such that $(r_{n_1}(t), \ldots, r_{n_k}(t)) = s$ for any $t \in I_k^j$, as there is a bijective map between S and the intervals I_k^j. Then, for any $s \in S$

$$\mu(\{s\}) = m\left\{t \in [0, 1) : (r_{n_1}(t), \ldots, r_{n_k}(t)) = s\right\} = m(I_k^j) = 2^{-k}.$$

On the other hand,

$$(r_{n_1}(t), \ldots, r_{n_k}(t)) = s \text{ if and only if } (\alpha_1 r_{n_1}(t), \ldots, \alpha_k r_{n_k}(t)) = s_\alpha$$

where $s_\alpha = (\alpha_1 s_1, \ldots, \alpha_k s_k)$ and therefore $\mu(\{s\}) = 2^{-k} = \mu_\alpha(\{s_\alpha\})$, for any $s \in S$. Therefore, for these discrete measures, we can conclude $\mu = \mu_\alpha$.

From the symmetry property, we obtain another proof of (1.14) since in particular, we have proved that the joint distribution of $\{r_{m_1}(t), r_{m_2}(t), \ldots, r_{m_n}(t)\}$ is the product of n probability measures in $\mathbb{R}$, each of which is equal to $\frac{1}{2}(\delta_1 + \delta_{-1})$, where δ_a is the Dirac's delta at a.

We want to consider now Rademacher's series $\sum_{n=1}^{\infty} c_n r_n(t)$ where $\{c_n\}$ can be either real or complex (in general, in our exposition we will only consider the real case). Note that since $\{r_n\}$ is not a complete system, then Parseval's identity is not satisfied, since the two conditions are equivalent. However the identity

$$\left(\sum_{n=1}^{\infty}|c_n|^2\right)^{1/2} = \left(\int_0^1\left|\sum_{n=1}^{\infty}c_n r_n(t)\right|^2 dt\right)^{1/2}$$

holds for any $\{c_n\} \in \ell^2$ since it only requires the orthonormality. Note that for fixed $t \in [0, 1)$, $\{r_n(t)\}$ is a sequence of ones (1) and negative ones (-1) that can be interpreted as a "random" choice of signs or as the fortune of a gambler in a fair game.

A very natural question is the a.e. convergence of Rademacher series, a problem proposed in 1922 by H. Steinhaus (and independently by Wiener [171]). We have the following theorem.

Theorem 1.1 (Rademacher, Kolmogorov and Khintchine) *Given a Rademacher series $\sum_{k=1}^{\infty} c_k r_k$ it converges almost everywhere if $\{c_k\} \in \ell_2$, i.e., if $\sum_{k=1}^{\infty}|c_k|^2 < \infty$.*

Furthermore, if $\sum_{k=1}^{\infty}|c_k|^2 = \infty$ then the series diverges a.e.

This result, as we will see in Chap. 5, will be generalized using martingale theory, see Theorem 5.8, see also Theorem B.2.

Proof To prove the first statement of the theorem we use two important theorems in analysis: the Riesz-Fischer theorem and the Lebesgue differentiation theorem.

Since Rademacher functions are an orthonormal system in $L^2([0, 1], m)$ given $\{c_k\}$ such that $\sum_{k=1}^{\infty}|c_k|^2 < \infty$, using the Riesz-Fischer theorem, there exists a function $f \in L^2([0, 1], m)$ such that

$$\lim_{N\to\infty}\int_E\left(f(t) - \sum_{k=1}^{N}c_k r_k(t)\right)^2 dt = 0.$$

Let x be a Lebesgue point of f (i.e., a point of the set of measure one, where the Lebesgue differentiation theorem holds) and take a dyadic approximation of it $a_m = \frac{k_m}{2^m} < x < \frac{k_m+1}{2^m} = b_m$ (excluding the countable set of the dyadic rationals). Then, by the Cauchy-Schwarz inequality we have

$$\left|\int_{a_m}^{b_m}\left(f(t) - \sum_{k=1}^{N}c_k r_k(t)\right) dt\right| \leqslant (b_m - a_m)^{1/2}$$

$$\left(\int_0^1\left(f(t) - \sum_{k=1}^{N}c_k r_k(t)\right)^2 dt\right)^{1/2} \overset{N\to\infty}{\to} 0.$$

Thus,

$$\int_{a_m}^{b_m} f(t)dt = \sum_{k=1}^{\infty} c_k \int_{a_m}^{b_m} r_k(t)dt,$$

but as $\int_{a_m}^{b_m} r_k(t)dt = 0$, if $k > m$ and $\int_{a_m}^{b_m} r_k(t)dt = (b_m - a_m) r_k(x)$ if $k \leqslant m$, then

$$\frac{1}{b_m - a_m} \int_{a_m}^{b_m} f(t)dt = \sum_{k=1}^{\infty} c_k r_k(x),$$

and therefore the series $\sum_{k=1}^{\infty} c_k r_k(x)$ converge a.e. to $f(x)$ by Lebesgue's theorem.   □

For the converse, we need the following lemma which is a particular version of what in probability theory is known as a 0–1 law, see Theorem 5.3.

Theorem 1.2 *The measure of the convergence set of a Rademacher series is either zero or one i.e.,*

$$m\left\{ t \in [0, 1) : \sum_{k=1}^{\infty} c_k r_k(t) < \infty \right\} = 0 \ or \ 1.$$

Proof For the proof, we will use the alternative definition of the Rademacher functions, $r_k(t) = r_1\left(2^{k-1}t\right)$. Let E be the set of convergence of the series $\sum_{k=1}^{\infty} c_k r_k(t)$. Then if $t \in E$ we have that for all n, $t + \frac{1}{2^n} \in E$, since replacing t by $t + \frac{1}{2^n}$ only a finite number of terms change and that cannot change the convergence of the series. Then χ_E, the characteristic function of the set E, is periodic with an arbitrary small period, and that implies that it is constant a.e. since taking any bounded measurable function f we have that if $(k - 1)/2^n < x < k/2^n$ for all n

$$\int_0^1 f(t)dt = \sum_{k=1}^{2^n} \int_{(k-1)/2^n}^{k/2^n} f(t)dt = 2^n \int_{(k-1)/2^n}^{k/2^n} f(t)dt \to f(x) \text{ if } n \to \infty,$$

holds a.e. by Lebesgue's differentiation theorem.

In the general case, for an arbitrary measurable function f, consider the bounded function $e^{if(t)}$. Therefore χ_E is constant and equal to 0 or 1. Thus

$$m(E) = \int_E dt = 0 \text{ or } 1.$$ □

Let us finish the proof of the converse in Theorem 1.1.

Proof Let us assume that $c_k \to 0$ as $k \to \infty$ (otherwise the proof would be immediate) Assume $m(E) > 0$, the by Theorem 1.2 $m(E) = 1$, then the Rademacher series converges a.e. and then there exists a function g such that

$$\lim_{N \to \infty} \sum_{k=1}^{N} c_k r_k(t) = g(t) \quad \text{a.e.}$$

By continuity of the exponential function, we have

$$\lim_{n \to \infty} \exp\left[i\xi \sum_{k=1}^{n} c_k r_k(t) \right] = e^{i\xi g(t)} a.e.$$

and by Lebesgue's dominated convergence theorem, if $N \to \infty$

$$\int_0^1 \exp\left[i\xi \sum_{k=1}^{N} c_k r_k(t) \right] dt = \int_0^1 e^{i\xi g(t)} dt.$$

Now,

$$\int_0^1 \exp\left[i\xi \sum_{k=1}^{N} c_k r_k(t) \right] dt = \prod_{k=1}^{N} \int_0^1 \exp\left[i\xi c_k r_k(t) \right] dt,$$

since by an analogous argument as the one given for the proof of the independence of the Rademacher functions (1.13)

$$\int_0^1 \exp\left[i \sum_{k=1}^{N} c_k r_k(t) \right] dt = \sum_{\delta_1,\dots,\delta_N \in \{-1,1\}} \exp\left(\sum_{k=1}^{N} c_k \delta_k \right)$$
$$\times m \left\{ t \in [0,1] : r_1(t) = \delta_1, \dots, r_N(t) = \delta_N \right\}$$
$$= \sum_{\delta_1,\dots,\delta_N \in \{-1,1\}} \prod_{k=1}^{N} e^{i c_k \delta_k} m \left\{ t \in [0,1] : r_k(t) = \delta_n \right\}$$
$$= \prod_{k=1}^{N} \sum_{\delta_k \in \{-1,1\}} e^{i c_k \delta_k} m \left\{ t \in [0,1] : r_k(t) = \delta_k \right\}$$
$$= \prod_{k=1}^{N} \int_0^1 e^{i c_k r_k(t)} dt.$$

On the other hand,

$$\int_0^1 \exp\left[i\xi c_k r_k(t)\right] dt = \sum_{j=1}^{2^n} \int_{(j-1)/2^n}^{j/2^n} \exp\left[i\xi c_k r_k(t)\right] dt$$

$$= \sum_{j=1}^{2n} \int_{(j-1)/2^n}^{j/2^n} \exp\left[i\xi c_k (-1)^{j-1}\right] dt$$

$$= \frac{1}{2^n} \sum_{j \text{ even}} \exp\left[-i\xi c_k\right] + \frac{1}{2^n} \sum_{j \text{ odd}} \exp\left[i\xi c_k\right]$$

$$= \frac{2^n}{2} \frac{1}{2^n} \exp\left[-i\xi c_k\right] + \frac{2^n}{2} \frac{1}{2^n} \exp\left[i\xi c_k\right]$$

$$= \frac{e^{i\xi c_k} + e^{-i\xi c_k}}{2} = \cos\left(\xi c_k\right).$$

Hence

$$\int_0^1 \exp\left[i\xi \sum_1^N c_k r_k(t)\right] dt = \prod_{k=1}^N \cos\left(\xi c_k\right).$$

but, since $\sum_{k=1}^\infty |c_k|^2 = \infty$ and $c_k \to 0$ we get

$$\lim_{N \to \infty} \prod_{k=1}^N \cos\left(\xi c_k\right) = 0,$$

since, using the Taylor's expansion of cosine, the inequality $\ln(1-x) < -x$, $0 < x < 1$ and taking $0 < a < 1/2$,

$$\ln\left(\prod_{k=1}^N \cos\left(\xi c_k\right)\right) = \sum_{k=1}^N \ln\left(\cos\left(\xi c_k\right)\right) = \sum_{k=1}^N \ln\left(1 - \xi^2 c_k^2/2 + O(\xi^4 c_k^4)\right)$$

$$\sim \sum_{k=1}^N \left[\left(-\xi^2 c_k^2/2\right) + \xi^2 c_k^2 a\right) - \xi^2 c_k^2 a\right]$$

$$= (-1/2 + a)\,\xi^2 \sum_{k=1}^N c_k^2 - \xi^2 a \sum_{k=1}^N c_k^2 \overset{N \to \infty}{\longrightarrow} -\infty.$$

Therefore,

$$\int_0^1 e^{i\xi g(t)} dt = 0 \text{ for any } \xi \neq 0.$$

Now, taking $\xi_n = \frac{1}{n}$ then $\lim_{n\to\infty} \frac{1}{n} g(t) = 0$ a.e. and $\lim_{n\to\infty} e^{i\xi_n g(t)} = 1$. a.e. thus by Lebesgue's dominated convergence theorem

$$\lim_{n\to\infty} \int_0^1 e^{i\xi_n g(t)} dt = 1;$$

but this implies $0 = 1$ a contradiction. Therefore, $m(E)$ cannot have a positive measure, so the series must diverge a.e. $\qquad\square$

Remark 1.1

(i) Observe that as a consequence of Theorem 1.1, a Rademacher series $\sum_{k=1}^{\infty} c_k r_k$ converges almost everywhere if and only if $\sum_{k=1}^{\infty} |c_k|^2 < \infty$.

(ii) It can be proved that when $\sum_{k=1}^{\infty} |c_k|^2 = \infty$, the series diverges no matter which summation method is used, see Zygmund [176, Theorem 8.2].

(iii) Moreover, in 1923 H. Rademacher and Menshov proved a general result, giving a sufficient condition for a series of orthogonal functions on an interval to converge almost everywhere: If $\{\phi_k\}$ is an orthonormal system and $\sum_k c_k \phi_k$ is the orthogonal expansion of a function f i.e., $c_k = \int f\overline{\phi_k}$ with $\sum_k |c_k|^2 < \infty$, then if $\sum_k |c_k|^2 (\log k)^2 < \infty$ then $\sum_k c_k \phi_k(x)$ converges a.e.

(iv) Theorem 1.1 can be considered a precedent result to Kolmogorov's three-series theorem, see Theorem B.2.

On the other hand, we have the following inequality, of great importance for its various applications. As we will see, it is a source of inspiration and object of study in various contexts, see for instance, Stein [152] or Strook [155] for applications from analysis and probability theory.

Theorem 1.3 (Khintchine's inequality) *Given $\{c_n\} \in \ell_2$, the series $\sum_{n=1}^{\infty} c_n r_n(t)$ belongs to $L^p[0, 1)$ for all $p > 0$, more precisely, there exist positive constants A_p, B_p that depend only on p, such that.*

$$A_p \left(\sum_{n=1}^{\infty} |c_n|^2 \right)^{1/2} \leqslant \left(\int_0^1 \left| \sum_{n=1}^{\infty} c_n r_n(t) \right|^p dt \right)^{1/p} \leqslant B_p \left(\sum_{n=1}^{\infty} |c_n|^2 \right)^2. \qquad (1.15)$$

In particular, for all $N \in \mathbb{N}$

$$A_p \left(\sum_{n=1}^{N} |c_n|^2 \right)^{1/2} \leqslant \left(\int_0^1 \left| \sum_{n=1}^{N} c_n r_n(t) \right|^p dt \right)^{1/p} \leqslant B_p \left(\sum_{n=1}^{N} |c_n|^2 \right)^{1/2}. \qquad (1.16)$$

Proof In the case $p = 2$ is trivial by the orthogonality of the functions $\{r_n\}$,

$$\left(\sum_{n=1}^{\infty} |c_n|^2 \right)^{1/2} = \left(\int_0^1 \left| \sum_{n=1}^{\infty} c_n r_n(t) \right|^2 dt \right)^{1/2}.$$

Now, let us consider first the right-hand side inequality. We need to estimate the distribution function of $\sum_{k=1} c_k r_k(t)$,

$$\lambda(\alpha) = m \left\{ t \in [0, 1] : \left| \sum_{k=1}^{\infty} c_k r_k(t) \right| > \alpha \right\}.$$

In order to do that, observe that by an analogous argument as in (1.15), we have

$$\int_0^1 \exp\left(c_n r_n(t)\right) dt = \sum_{k=1}^{2^n} \int_{(k-1)/2^n}^{k/2^n} \exp\left(c_n r_n(t)\right) dt = \sum_{k=1}^{2^n} \int_{(k-1)/2^n}^{k/2^n} \exp\left(c_n (-1)^{k-1}\right) dt$$

$$= \frac{1}{2^n} \sum_{k \text{ even}} e^{c_n} + \frac{1}{2^n} \sum_{k \text{ odd}} e^{-c_n} = \frac{2^n}{2} \frac{1}{2^n} e^{c_n} + \frac{2^n}{2} \frac{1}{2^n} e^{-c_n} = \frac{e^{c_n} + e^{-c_n}}{2}$$

$$= \cosh\left(c_n\right).$$

Now, by independence, and using the inequality $\cosh x \leqslant e^{x^2}$, for any $x \in \mathbb{R}$, we obtain

$$\int_0^1 \exp\left[\xi \sum_{n=1}^{\infty} c_n r_n(t) \right] dt = \prod_{n=1}^{\infty} \int_0^1 \exp\left(\xi c_n r_n(t)\right) dt$$

$$= \prod_{k=1}^{\infty} \cosh\left(\xi c_n\right) \leqslant \exp\left(\xi^2 \sum_{n=1}^{\infty} c_n^2 \right).$$

Then,

$$\exp\left[\xi \left| \sum_{n=1}^{\infty} c_n r_n(t) \right| \right] \leqslant \exp\left[\xi \left(\sum_{n=1}^{\infty} c_n r_n(t) \right) \right] + \exp\left[-\xi \left(\sum_{n=1}^{\infty} c_n r_n(t) \right) \right].$$

Without loss of generality we may assume $\sum_{n=1}^{\infty} c_n^2 = 1$, then by the previous inequality

$$\int_0^1 \exp\left[\xi \left| \sum_{n=1}^{\infty} c_n r_n(t) \right| \right] dt \leqslant 2 e^{\xi^2}.$$

Thus, using Tchebychev's inequality, for the function $e^{(\alpha/2)x}$, we get

$$\lambda(\alpha) = \inf\left\{ t \in [0, 1] : \left| \sum_{n=1}^{\infty} c_n r_n(t) \right| > \alpha \right\} \leqslant e^{-(\alpha/2)\alpha} \int_0^1 \exp\left[\frac{\alpha}{2} \left| \sum_{n=1}^{\infty} c_n r_n(t) \right| \right] dt$$

$$\leqslant e^{-\alpha^2/2} 2 e^{(\alpha/2)^2} = 2 e^{-\alpha^2/4}.$$

Having this estimate of the distribution function and using Fubini's theorem we obtain,

$$
\left(\int_0^1 \left|\sum_{n=1}^\infty c_n r_n(t)\right|^p dt\right)^{1/p} = \left(p\int_0^1\left(\int_0^{\left|\sum_n c_n r_n(t)\right|}\xi^{p-1}d\xi\right)dt\right)^{1/p}
$$

$$
= \left(p\int_0^1\left(\int_0^\infty \chi_{\{\xi:0\leqslant\xi<|\sum_n c_n r_n(t)|\}}(\xi)\xi^{p-1}d\xi\right)dt\right)^{1/p}
$$

$$
= \left(p\int_0^\infty\left(\int_0^1 \chi_{\{t\in[0,1]:|\sum_n c_n r_n(t)|>\xi\}}(t)\,dt\right)\xi^{p-1}d\xi\right)^{1/p}
$$

$$
= \left(p\int_0^\infty \xi^{p-1}\lambda(\xi)d\xi\right)^{1/p} \leqslant \left(2p\int_0^\infty \xi^{p-1}e^{-\xi^2/4}d\xi\right)^{1/p}
$$

$$
= \left(2^p\,p\,\Gamma\left(\tfrac{p}{2}\right)\right)^{1/p} = 2p^{1/p}\left(\Gamma\left(\tfrac{p}{2}\right)\right)^{1/p} = B_p,
$$

and then, in general, we have

$$
\left(\int_0^1 \left|\sum_{n=1}^\infty c_n r_n(t)\right|^p dt\right)^{1/p} \leqslant B_p\left(\sum_{n=1}^\infty |c_n|^2\right)^{1/2} < \infty.
$$

The left-hand side of the inequality now follows from the right-hand side. For $p > 2$, we have by Jensen's inequality (since $p/2 > 1$)

$$
\left(\sum_{n=1}^\infty |c_n|^2\right)^{p/2} = \left(\int_0^1\left|\sum_{n=1}^\infty c_n r_n(t)\right|^2 dt\right)^{p/2} \leqslant \left(\int_0^1\left|\sum_{n=1}^\infty c_n r_n(t)\right|^p dt\right).
$$

For $p < 2$, by the Cauchy-Schwarz inequality

$$
\sum_{n=1}^\infty |c_n|^2 = \int_0^1\left|\sum_{n=1}^\infty c_n r_n(t)\right|^2 dt = \int_0^1\left|\sum_{n=1}^\infty c_n r_n(t)\right|^{p/2+(2-p/2)} dt
$$

$$
\leqslant \left(\int_0^1\left|\sum_{n=1}^\infty c_n r_n(t)\right|^p dt\right)^{1/2}\left(\int_0^1\left|\sum_{n=1}^\infty c_n r_n(t)\right|^{4-p} dt\right)^{1/2}
$$

$$
\leqslant \left(\int_0^1\left|\sum_{n=1}^\infty c_n r_n(t)\right|^p dt\right)^{1/2}\left(B_{4-p}\left(\sum_{n=1}^\infty |c_n|^2\right)^{2-p/2}\right)^{1/2},
$$

and since all the norms are finite, we obtain the left inequality with $A_p = B_{4-p}^{-1/2}$. $\square$

Remark 1.2

- Other proofs of Khintchine's inequality are simpler and/or more direct than the one given, for instance, the proof is immediate using Gaussian systems, but they are beyond the techniques discussed here, see Stroock [155, Chap. 2]. Nevertheless, the one given is particularly interesting since it shows the topics (among others, Marcinkiewicz's interpolation argument) involved in this inequality. In Chap. 4 a more probabilistic proof will be discussed.
- Khintchine's inequality gives us an indirect proof that the Rademacher functions do not form a complete system, since that would imply that in the vector subspace generated by Rademacher functions the norms of $L^2[0, 1)$ and $L^p[0, 1)$ would be equivalent. In any case, Khintchine inequality shows that in that subspace all the L^p norms are equivalent.
- Note also that by this inequality the closed subspace in $L^p[0, 1)$ generated by Rademacher functions is isomorphic to $L^2[0, 1)$.
- Although the exact values of A_p and B_p have no importance in the applications of the Khintchine's inequality, the problem of computing the best constants has from time to time gained some interest. It is elementary to prove that $B_p = 1$ for $0 < p \leqslant 2$, and $A_p = 1$ for $2 \leqslant p < \infty$. In 1961 Steckin [154] proved that $B_{2n} = ((2n - 1)!!)^{1/2n}$, $n \in \mathbb{N}$. This result was recently extended by Young [175] who computed B_p for $p \in (3, \infty)$. In 1978, Szarek [156] proved that $A_1 = 1/\sqrt{2}$. This was a long outstanding conjecture due to Littlewood. Finally, Haagerup in [94] introduced a new method that allows to compute the best constants in the remaining cases $\left(A_p \text{ for } 0 < p < 2, p \neq 1 \text{ and } B_p \text{ for } 2 < p < 3\right)$, and to give a simple proof of Szarek's result $A_1 = 1/\sqrt{2}$,

$$A_p = \begin{cases} 2^{1/2-1/p} & 0 < p \leqslant p_0 \\ 2^{1/2}(\Gamma((p + 1)/2)/\sqrt{\pi})^{1/p} & p_0 < p < 2 \end{cases}$$

and

$$B_p = 2^{1/2}(\Gamma((p + 1)/2)/\pi)^{1/p} \quad 2 < p < \infty$$

where p_0 is the solution of the equation $\Gamma((p + 1)/2) = \sqrt{\pi}/2$ in the interval $(1, 2)$. It is not hard to prove that A_p cannot exceed the number indicated and that B_p cannot be less than $2^{1/2}(\Gamma((p + 1)/2)/\sqrt{\pi})^{1/p}$, so the problem is to give sufficiently good lower estimates for A_p and sufficiently good upper estimates for B_p. Consider first the case $0 < p < 2$.

- The elements of the proof of Khintchine's inequality, can be consider as an antecedent of the now standard techniques in the so called Chernoff-Cramer method for the proof of exponential and concentration inequalities, see for instance Bernstein's inequality Theorem 4.27.

As a consequence of Khintchine's inequality we have that, for all $p > 0$, given a Rademacher series $\sum_{n=1}^{\infty} c_n r_n(t)$ and considering any $p > 0$ and any change of sign of the coefficients c_k, i.e., we consider $\varepsilon_n \in \{-1, 1\}$ and $\sum_{n=1}^{\infty} \varepsilon_n c_n r_n(t)$, then, for all N

$$A_p \left(\sum_{n=1}^{N} |c_n|^2 \right)^{1/2} \leq \left(\int_0^1 \left| \sum_{n=1}^{N} c_n r_n(t) \right|^p dt \right)^{1/p} \leq B_p \left(\sum_{n=1}^{N} |c_n|^2, \right)^{1/2}$$

and therefore by $(1.16)^2$

$$\left\| \sum_{n=1}^{N} \varepsilon_n c_n r_n \right\|_p \approx \left(\sum_{n=1}^{N} |c_n|^2 \right)^{1/2} \approx \left\| \sum_{n=1}^{N} c_n r_n \right\|_p .$$

Then, we can conclude.

Proposition 1.1 *Let $\{c_n\} \in \ell^2$ i.e., $\sum_{n=1}^{\infty} |c_n|^2 < \infty$, then for $p > 0$ and for any change of sign $\varepsilon_n \in \{-1, 1\}$, we have for $p > 0$ and for all n*

$$\left\| \sum_{n=1}^{N} \varepsilon_n c_n r_n \right\|_p \approx \left\| \sum_{n=1}^{N} c_n r_n \right\|_p .$$

In particular, there exists a constant C_p such that

$$\left\| \sum_{n=1}^{N} \varepsilon_n c_n r_n \right\|_p \leq C_p \left\| \sum_{n=1}^{N} c_n r_n \right\|_p .$$

The last inequality foreshadows the important notion of *martingale transforms* that will be studied in detail in Chap. 4. Additionally, in the same direction, the following basic contraction principle is also important to be mentioned.

Proposition 1.2 (Kahane's contraction principle) *Let $\{r_n\}_{n=1}^{N}$ be a sequence of Rademacher functions. For all sequences $\{c_n\}_{n=1}^{N}$ in $\mathbb{R}$ and all $1 \leq p \leq \infty$ we have*

$$\left\| \sum_{n=1}^{N} c_n r_n \right\|_p \leq \max_{1 \leq n \leq N} |c_n| \left\| \sum_{n=1}^{N} r_n \right\|_p . \tag{1.17}$$

The inequality (1.17) is a particular version of Kahane's contraction principle, which holds even in Banach spaces, see [101, Chap. 3].

Finally, we will discuss some limit results for Rademacher functions. These limit results are central in probability theory (law of large numbers and the central limit

---

[2] $x \approx y$ means that there exists constants A, B such that $Ax \leq y \leq Bx$.

theorem) see Theorems B.1, B.3 and B.4. The sum $\sum_{n=1}^{N} r_n(t)$, can be interpreted as the fortune of a gambler in time N and then, a natural question is what is the probability that it is above a given number α_N, that is:

$$m\left\{t \in [0, 1) : \sum_{n=1}^{N} r_n(t) > \alpha_N\right\}.$$

If we are unlikely to get rich by gambling, this means that for α_N "sufficiently large" such a measure is sufficiently small, and that is what we are going to prove.

Theorem 1.4 (Weak law of large numbers) *For all $\varepsilon > 0$*

$$\lim_{N \to \infty} m\left\{t \in [0, 1) : \left|\sum_{n=1}^{N} r_n(t)\right| > \varepsilon N\right\} = 0 \tag{1.18}$$

Proof

$$\int_0^1 \left(\sum_{n=1}^{N} r_n(t)\right)^2 dt \geqslant \int_{\left\{t \in [0,1) : \left|\sum_{n=1}^{N} r_n(t)\right| > \varepsilon N\right\}} \left(\sum_{n=1}^{N} r_n(t)\right)^2 dt$$

$$\geqslant \varepsilon^2 N^2 m\left\{t \in [0, 1) : \left|\sum_{n=1}^{N} r_n(t)\right| > \varepsilon N\right\}$$

but, due to the independence of the $\{r_n\}$

$$\int_0^1 \left(\sum_{n=1}^{N} r_n(t)\right)^2 dt = \sum_{n=1}^{N} \int_0^1 r_n^2(t)dt + \sum_{n \neq j} \int_0^1 r_n(t)r_j(t)dt$$

$$= \sum_{n=1}^{N} \int_0^1 r_n^2(t)dt + \sum_{n \neq j} \int_0^1 r_n(t)dt \int_0^1 r_j(t)dt = N$$

Therefore

$$m\left\{t \in [0, 1) : \left|\sum_{n=1}^{N} r_n(t)\right| > \varepsilon N\right\} \leqslant 1/\varepsilon^2 N \to 0,$$

as $N \to \infty$. $\qquad\qquad\Box$

One can also prove a strong law of large numbers (which of course, in particular, implies the weak law),

Theorem 1.5 (Borel)

$$\lim_{N\to\infty}\frac{1}{N}\sum_{n=1}^{N}c_n r_n(t)=0,\ a.e. \tag{1.19}$$

Proof

$$\int_0^1\left|\sum_{n=1}^{N}r_n(t)\right|^4 dt=\sum_{n=1}^{N}\int_0^1 (r_n(t))^4\,dt+6\sum_{1\leqslant j<k\leqslant N}r_j^2(t)r_k^2(t)dt$$
$$=N+3\left(N^2-N\right)=3N^2-2N,$$

since

$$\int_0^1\left(\sum_{n=1}^{N}r_n(t)\right)^4 dt=\int_0^1\left[\sum_{n=1}^{N}r_n(t)^2+2\sum_{1\leqslant i<j\leqslant N}r_i(t)r_j(t)\right]^2 dt$$
$$=\int_0^1\Bigg[\sum_{n=1}^{N}r_n(t)^4+2\sum_{1\leqslant i<j\leqslant N}r_i^2(t)r_j^2(t)$$
$$+4\left(\sum_{k=1}^{N}\sum_{1\leqslant i<j\leqslant N}r_k^2(t)r_i(t)r_j(t)\right)$$
$$+4\sum_{1\leqslant j<k<l<m\leqslant N}r_j(t)r_k(t)r_l(t)r_m(t)\Bigg]dt$$
$$=\sum_{n=1}^{N}\int_0^1 r_n(t)^4 dt+6\sum_{1\leqslant j<k\leqslant N}\int_0^1 r_j^2(t)r_k^2(t)dt.$$

Then, using Tchebychev's inequality

$$m\left\{t\in[0,1]:\left|\sum_{n=1}^{N}r_n(t)\right|>N\varepsilon\right\}\leqslant\frac{1}{\varepsilon^4 N^4}\int_{n=1}^{N}\left[\sum_{n=1}^{N}r_n(t)\right]^4 dt=\frac{3N^2-2N}{\varepsilon^4 N^4}$$
$$=\frac{1}{\varepsilon^4}\left(\frac{3}{N^2}-\frac{2}{N^3}\right).$$

Thus

$$\sum_{k=1}^{\infty}P\left\{t\in[0,1]:\left|\sum_{n=1}^{N}r_n(t)>N\varepsilon\right\}<\infty,\right.$$

and then by Borel Cantelli's lemma

$$\frac{1}{N}\sum_{n=1}^{N}r_n(t)\to 0\quad a.e.$$
$$\square$$

Using the relation $r_n(t) = 1 - 2\varepsilon_n(t)$, with ε_k being the Steinhaus functions, the statement of Borel's theorem can be rewritten as $\lim_{N\to\infty} \frac{\varepsilon_1(t)+\cdots+\varepsilon_N(t)}{N} = \frac{1}{2}$ for a.e. $t \in [0, 1]$.

Observe that

$$\frac{\varepsilon_1(t) + \cdots + \varepsilon_N(t)}{N},$$

is the relative frequency of 1s which means that almost every number t has the same number of zeros and ones. Finally one has a more precise result about the rate of convergence of $\sum_{n=1}^{N} r_n(t)$ than the law of large numbers and it is none other than the Central Limit Theorem, which we will discuss now, see [18].

Theorem 1.6 (Central Limit Theorem)

$$\lim_{N\to\infty} m\left\{ t \in [0, 1] : a < \frac{\sum_{n=1}^{N} r_n(t)}{\sqrt{N}} < b \right\} = \frac{1}{\sqrt{2\pi}} \int_a^b e^{-y^2/2} dy. \qquad (1.20)$$

Proof We will give only a sketch of the proof due to Markov. Let $\chi_{(a,b)}$ be the indicator function of the interval (a, b), then its Fourier transform is

$$\widehat{\chi_{(a,b)}}(\xi) = \int_a^b e^{-2\pi i \xi x} dx = -\frac{1}{2\pi} \frac{e^{-2\pi i \xi b} - e^{-2\pi i \xi a}}{i\xi}.$$

Using the inversion formula for the Fourier transform we get, for $x \neq a, b$,

$$\chi_{(a,b)}(x) = -\frac{1}{2\pi} \int_{-\infty}^{\infty} \frac{e^{-2\pi \xi b} - e^{-2\pi i \xi a}}{i\xi} e^{2\pi i \xi x} d\xi.$$

Hence,

$$m\left\{ t \in [0, 1] : a < \frac{\sum_{k=1}^{N} r_k(t)}{\sqrt{N}} < b \right\} = \int_0^1 \chi(a, b)\left(\frac{\sum_{k=1}^{N} r_k(t)}{\sqrt{N}} \right) dt$$

$$= -\int_0^1 \frac{1}{2\pi} \int_{-\infty}^{\infty} \frac{e^{-2\pi i \xi b} - e^{-2\pi i \xi a}}{i\xi}$$

$$\times \exp\left(-i\xi \frac{\sum_{n=1}^{N} r_n(t)}{\sqrt{N}} \right) d\xi dt$$

$$= -\frac{1}{2\pi} \int_{-\infty}^{\infty} \frac{e^{-2\pi i \xi b} - e^{-2\pi i \xi a}}{i\xi}$$

$$\times \left(\int_0^1 \exp\left(-i\xi \frac{\sum_{n=1}^{N} r_n(t)}{\sqrt{N}} \right) dt \right) d\xi$$

$$= -\frac{1}{2\pi} \int_{-\infty}^{\infty} \frac{e^{-2\pi i \xi b} - e^{-2\pi i \xi a}}{i\xi} \left(\cos\frac{\xi}{\sqrt{N}} \right)^N d\xi,$$

by Fubini's theorem. Now, $\lim_{N \to \infty} \left(\cos \frac{\xi}{\sqrt{N}} \right)^N = e^{-\xi^2/2}$, since

$$\lim_{N \to \infty} \cos(\frac{\xi}{\sqrt{N}}) = \lim_{N \to \infty} \left(1 - \sin^2 \left(\frac{\xi}{\sqrt{N}} \right) \right)^{N/2} = \lim_{N \to \infty} \left(1 - \frac{\xi^2}{N} \right)^{N/2} = e^{-\xi^2/2},$$

as $\sin^2(\frac{\xi}{\sqrt{N}}) \sim \left(\frac{\xi^2}{N} \right)$, for N big enough. Then, if we interchange the limit with the integral, we would get

$$\lim_{N \to \infty} m \left\{ t \in [0, 1] : a < \frac{\sum_{n=1}^N r_n(t)}{\sqrt{N}} < b \right\} = -\frac{1}{2\pi} \int_{-\infty}^{\infty} \frac{e^{-2\pi i \xi b} - e^{-2\pi i \xi a}}{i\xi} e^{-\xi^2/2} d\xi$$

$$= \frac{1}{\sqrt{2\pi}} \int_a^b e^{-\xi^2/2} d\xi. \qquad \Box$$

As we said this just an sckech of the proof, the problem is that the limits of integration are $-\infty$ and ∞ and the function $\frac{e^{-2\pi i \xi b} - e^{-2\pi i \xi a}}{i\xi}$ is not absolutely integrable, so a more careful argument is needed. For a complete formal proof see [78]. $\qquad \Box$

1.2 Systems Derived from Rademacher's Functions

Using the Rademacher functions we can define two complete orthonormal systems in $L^2([0, 1), m)$.

1.2.1 The Walsh System

The *Walsh system*, also called the *Walsh-Paley system*, consists of products of Rademacher functions and can be defined as follows.

Let $\psi_0(t) = 1$, for all $t \in [0, 1]$ and for $n \geq 1$, if $n = 2^{n_1} + 2^{n_2} + \cdots + 2^{n_k}$ with $0 \leq n_1 < n_2 < \cdots < n_k$

$$\psi_n(t) = r_{n_1}(t) r_{n_2}(t) \cdots r_{n_k}(t), \ n \geq 1$$

See (Fig. 1.3). The Walsh functions can be grouped by levels as follows:

$$\psi_0^{(0)} = \psi_0, \text{ for all } t \in [0, 1]; \ \psi_1^{(1)} = \psi_1, \psi_1^{(2)} = \psi_2, \psi_2^{(2)} = \psi_3$$

and in general for $n > 1$

$$\psi_k^{(n)} = \psi_N \text{ with } N = 2^{n-1} + (k - 1), \ 1 \leq k \leq 2^{n-1}.$$

Note that Walsh system contains the Rademacher functions, since

$$\psi_{2^n - 1} = r_n$$

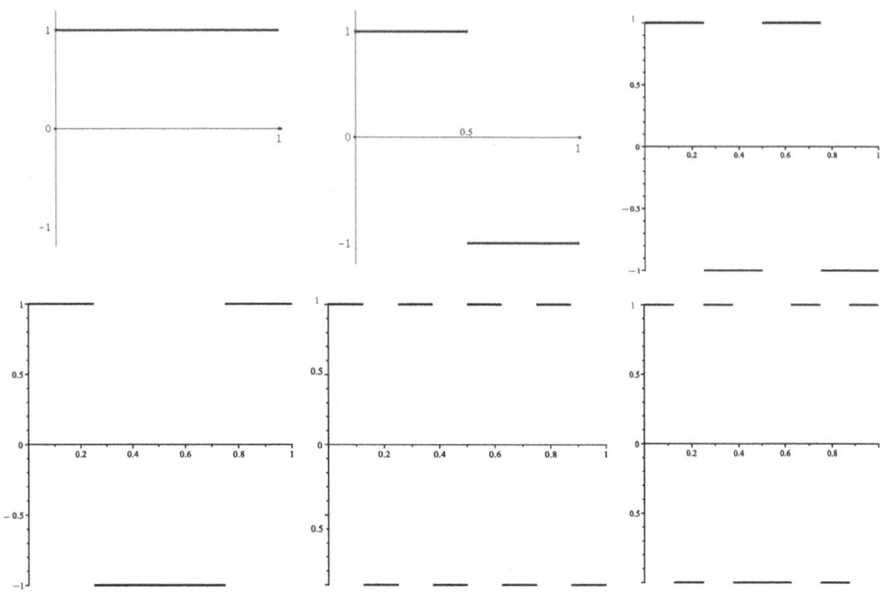

Fig. 1.3 The first six Walsh functions $\psi_0^{(0)}, \psi_1^{(1)}, \psi_1^{(2)}, \psi_2^{(2)}, \psi_1^{(3)}, \psi_2^{(3)}$

Note that the Walsh functions are closed under the operation of finite products. The Walsh system is a completion of the Rademacher functions, since, intuitively, as we discussed before, we are adding enough orthogonal functions to complete it, this was formally proved by Paley [134].

Theorem 1.7 (Paley) *The Walsh system $\{\psi_n\}$ is a complete system on $L^2([0, 1), m)$.*

Proof First of all, by the independence property of the Rademacher functions, the integral

$$\int_0^1 r_{n_1}^{\alpha_1}(t) r_{n_2}^{\alpha_2}(t) \cdots r_{n_k}^{\alpha_k}(t) dt$$

vanishes unless $\alpha_1 \cdots \alpha_k$ are all even in which case the integral is equal to one and that implies that the family $\{\psi_n\}$ is also orthonormal. Let us prove that it is also complete. Observe that all the functions $\psi_m(t)$, $0 \leqslant m \leqslant 2^n - 1$ are all constant over dyadic intervals

$$I_n^k = \left[\frac{k-1}{2^n}, \frac{k}{2^n}\right), \quad k = 1, 2, \ldots, 2^n.$$

Let $f \in L^2[0, 1]$ such that $\int_0^1 f(t)\varphi_m(t) dt = 0$, $0 \leqslant m \leqslant 2^n - 1$. Then,

$$\int_0^1 f(t)\psi_m(t) dt = \sum_{k=1}^{2n} \int_{\frac{k-1}{2^n}}^{k/2^n} f(t)\psi_m\left(\frac{k-1}{2^n}\right) dt = \sum_{k=0}^{2n} \psi_m\left(\frac{k-1}{2^n}\right) \int_{\frac{k-1}{2^n}}^{k/2^n} f(t) dt = 0$$

By the orthogonality of $\{\psi_m\}$ we get that they are linearly independent and therefore the vectors $\left\{\psi_m\left(\frac{k-1}{2^n}\right)\right\}_{m=0}^{2n-1}$ are also linearly independent, then

$$\int_{k-1/2^n}^{k/2^n} f(t)dt = 0, k = 1, 2, \ldots, 2^n, n = 1, 2, \ldots$$

Hence, since we are assuming that $\int_0^1 f(t)\psi_m(t)dt = 0$ for all m, defining

$$F(t) = \int_0^t f(u)du$$

we get that, $F\left(\frac{k-1}{2^n}\right) = F\left(\frac{k}{2^n}\right) k = 1, 2, \ldots 2^n, n = 1, 2, \ldots$ thus $F(t)$ is constant and therefore $f(t)$ is zero a.e. $\qquad\square$

Depending on the enumeration of the Walsh functions, three orthonormal systems can be obtained: the Walsh-Paley system, the original Walsh system and the Walsh-Kaczmarz system, thus each one is a rearrangement of the other.

In addition, Walsh and Paley observed many similarities between the behavior of the Walsh functions and trigonometric polynomials. In fact, many of the theorems for trigonometric series have exact analogs for Walsh series, the latter being simpler because of their dyadic structure; for more about this see Schipp, Wade, and Simon's book [147].

Each function $f \in L^1([0, 1), m)$ can be (formally) expanded using the Walsh functions,

$$f(t) \sim \sum_{k=1}^{\infty} c_k \psi_k(t), \text{ where } c_n = \int_0^1 f(t)\psi_n(t)dt$$

which is called the *Walsh-Fourier series* of f, denoted by $W(f)$ and its partial sums will be denoted by

$$W_N(f, t) = \sum_{n=0}^{N-1} c_n \psi_n(t).$$

Richard Gundy [86] noted that the sequence of dyadic partial sums $W_{2^N}(f)$ of a Walsh expansion can be rewritten as a part depending on the "present" $r_n(t)$ and a coefficient $a_n(t)$ that is a function of the "past", $a_n(t) = f_n(r_i(t), r_2(t), \ldots, r_{n-1}(t))$, and

$$W_{2^{N+1}}(f) - W_{2^N}(f) = \sum_{k=2^N}^{2^{N+1}-1} c_k \psi_k(t) = a_N(t)r_N(t).$$

Thus the Walsh expansions can be considered as generalized Rademacher series (where the coefficient also depends on t) or in the terminology introduced by Burkholder [29], they are *transforms* of Rademacher series, $\sum_{n=0}^{\infty} a_n r_n(t)$. This will be explained in detail in Chap. 3. Moreover, $\sum_{n=1}^{N} c_n(t) r_n(t)$ is the best example of what we might call a stochastic integral in the discrete case. For problems of a.e. convergence of these series see Gundy [86] and Hunt [98]. As we will see later, this problem can also be studied in the general context of martingale convergence.

Finally, the Vilenkin systems are a generalization of the Walsh system and they were introduced by Vilenkin in 1947 [166]. Since that time several authors have been dealing with the Vilenkin systems, see for instance Schipp, Wade, and Simon [147]. The extension of the results from Walsh system to Vilenkin systems is usually non-trivial. For more about Vilenkin systems see Weisz [173], Chaps. 1 and 2, see also Exercise 8.

1.2.2　The Haar System

The *Haar functions* were introduced by Alfred Haar in his doctoral dissertation in 1909 and are defined, by level, as follows:

$$\chi_0(t) = \chi_0^{(0)}(t) = 1$$
$$\chi_1(t) = \chi_1^{(1)}(t) = r_1(t)$$

and if $N = 2^{n-1} + (k-1)$, $1 \leqslant k \leqslant 2^{n-1}$.

$$\chi_N(t) = \chi_n^{(k)}(t) = \begin{cases} 2(n-1)/2 r_n(t), & \text{if } \frac{(2k-2)}{2^n} \leqslant t < \frac{k}{2^{n-1}} \\ 0 & \text{otherwise.} \end{cases}$$

Using a notation in wavelet theory, we have that the Haar system can be defined as follows:

$$\chi_N(t) = \chi_n^{(k)}(t) = 2^{(n-1)/2} r_1\left(2^{n-1} t - (k-1)\right) = 2^{(n-1)/2} r_1\left(2^{n-1} t - (k-1)\right).$$

See (Fig. 1.4). Similar to the case of the Walsh functions, it can be proved that $\{\chi_n\}$ is also a complete orthonormal system, see Exercise 5.

Now, given $f \in L^1([0,1), m)$ the Haar expansion of f, which we will call the *Haar-Fourier series* of f, is given by

$$d_0 + \sum_{n=1}^{\infty} \sum_{k=1}^{2^{n-1}} d_k^{(n)} \chi_k^{(n)}(t) \quad \text{where } d_k^{(n)} = \int_0^1 f(t) \chi_k^{(n)}(t)\, dt.$$

Sometimes, for simplicity, the Haar-Fourier expansion of f is written as

$$\sum_{n,k} c_k^{(n)} \chi_k^{(n)}(t),$$

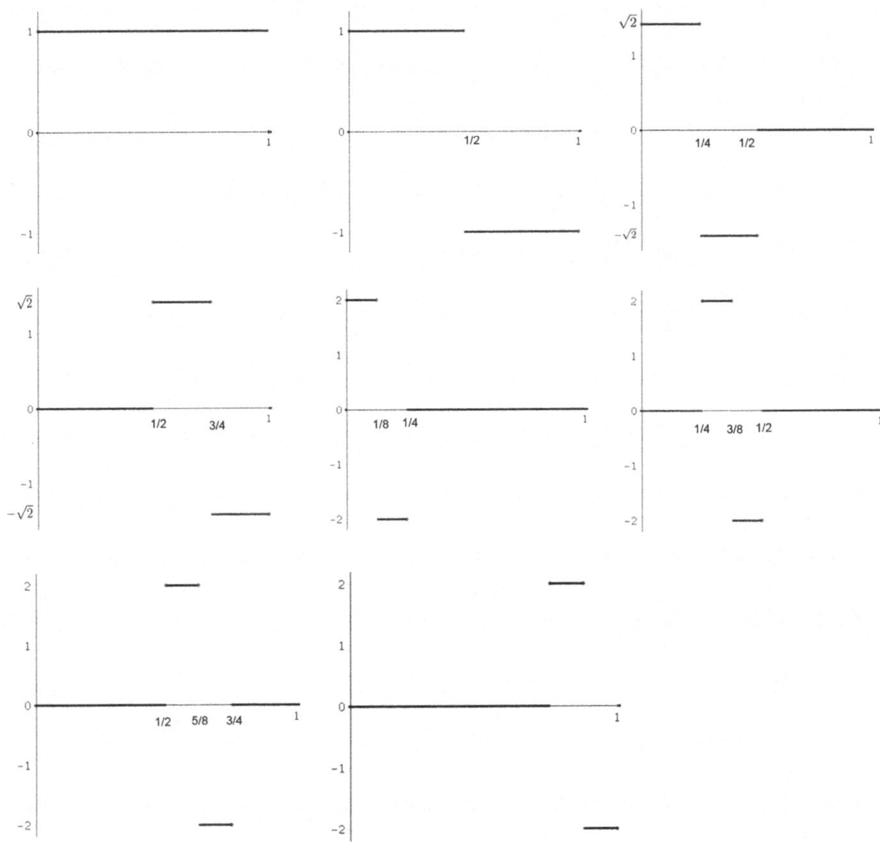

Fig. 1.4 The first eight Haar functions $\chi_0^{(0)}$, $\chi_1^{(1)}$, $\chi_1^{(2)}$, $\chi_2^{(1)}$, $\chi_2^{(2)}$, $\chi_3^{(1)}$, $\chi_3^{(2)}$, $\chi_3^{(3)}$

although the variation of the indices should be understood as in the previous expression. Its partial sums are defined as

$$\sum_{n=0}^{N-1}\sum_{k=1}^{2^{n-1}} d_k^{(n)}\chi_k^{(n)}(t).$$

The Haar-Fourier series of a function f is denoted as Hf and its partial sums by $H_N f$.

There is a close relationship between Walsh functions and Haar functions. More precisely, using the level notation (with two parameters for both levels), we have that $\{\psi_K^{(n)}\}$ can be expressed in terms of $\{\chi_k^{(n)}\}$ and vice versa as follows manner

$$\psi_0^{(0)}(t) = \chi_0^{(0)}(t), \; \psi_1^{(1)}(t) = \chi_1^{(1)}(t)$$
$$\psi_1^{(2)}(t) = \frac{1}{\sqrt{2}}\chi_1^{(2)}(t) + \frac{1}{\sqrt{2}}\chi_2^{(2)}(t)$$
$$\psi_2^{(2)}(t) = \frac{1}{\sqrt{2}}\chi_1^{(2)}(t) - \frac{1}{\sqrt{2}}\chi_2^{(2)}(t)$$
$$\psi_k^{(3)}(t) = \frac{1}{2}\chi_1^{(3)}(t) \pm \frac{1}{2}\chi_2^{(3)}(t) \pm \frac{1}{2}\chi_3^{(3)}(t) \pm \frac{1}{2}\chi_4^{(3)}(t)$$

In general, we have

$$\psi_k^{(n)}(t) = \frac{1}{\sqrt{2^{n-1}}}\chi_1^{(n)}(t) \pm \frac{1}{\sqrt{2^{n-1}}}\chi_2^{(n)}(t) \pm \cdots \pm \frac{1}{\sqrt{2^{n-1}}}\chi_{2^{n-1}}^{(n)}(t) \qquad (1.21)$$

the signs are chosen in such a way that the coefficient matrix $H = \left(b_{k,j}^{(n)}\right)$ be a symmetric, orthogonal matrix and therefore it is its own inverse, with determinant -1. This implies that the Haar functions can be expressed in terms of the Walsh functions using identical relations, i.e.,

$$\chi_k^{(n)}(t) = \frac{1}{\sqrt{2^n}}\psi_1^{(n)}(t) \pm \frac{1}{\sqrt{2^n}}\psi_2^{(n)}(t) \pm \cdots \pm \frac{1}{\sqrt{2^n}}\psi_{2^{n-1}}^{(n)}(t) \qquad (1.22)$$

In modern terminology, the Walsh and Haar systems are said to be the *Hadamard transform* of each other, see Schipp, Wade and Simon [147].

Walsh proved that the 2^n-th partial sums of the Walsh and Haar series, $W_{2^n} f$ and $H_{2^n} f$ are equal for all n. This can be easily proved using the fact that the Walsh and Haar functions can be represented in terms of the other. This is obvious for $n = 0$ and $n = 1$ since $\psi_0^{(0)}(t) = \chi_0^{(0)}(t) = 1$ and $\psi_1^{(1)}(t) = \chi_1^{(1)}(t) = r_1(t)$, for all $t \in [0, 1]$, so $W_1 f = H_1 f$. Now, for the general case using the representation of Walsh function by levels, we have,

$$W_{2^N} f - W_{2^{N-1}} f = \sum_{n=2^{N-1}}^{2^N-1} \sum_{k=1}^{2^{n-1}} c_k^{(n)} \psi_k^{(n)}(t)$$

and now using the representation of the Walsh functions in terms of the Haar functions

$$W_{2^N} f - W_{2^{N-1}} f = \sum_{n=2^{N-1}}^{2^N-1} \sum_{k=1}^{2^{n-1}} c_k^{(n)} \sum_{i=1}^{2^{n-1}} b_{k,i}^{(n)} \chi_k^{(n)}(t)$$

$$= \sum_{n=2^{N-1}}^{2^N-1} \sum_{i=1}^{2n-1} \left(\sum_{k=1}^{2n-1} c_k^{(n)} \, b_{k,i}^{(n)}\right) \chi_k^{(n)}(t).$$

Now, as $c_k^{(n)} = \int_0^1 f(t)\psi_k^{(n)}(t)dt$, we have

$$\sum_{k=1}^{2^n-1} c_k^{(n)} b_{k,i}^{(n)} = \sum_{k=1}^{2n-1} \left(\int_0^1 f(t)\psi_k^{(n)}(t)\,dt\right) b_{k,i}^{(n)}$$

$$= \int_0^1 f(t) \left(\sum_{k=1}^{2^n-1} b_{k,i}^{(n)} \psi_k^{(n)}(t)\right) dt = \int_0^1 f(t)\chi_k^{(n)}(t)dt = d_k^{(n)}.$$

Therefore,

$$W_{2^N} f - W_{2^{N-1}} f = \sum_{n=2^{N-1}}^{2^N-1} \sum_{k=1}^{2^n-1} d_k^{(n)} \psi_k^{(n)}(t) = H_{2^N} f - H_{2^{N-1}} f.$$

For more details see Kaczmarz and Steinhaus [106] and Schipp, et al. [147].

1.3 L^p Inequalities for Walsh and Haar Expansions

The Walsh and Haar functions have inequalities which are analogous to the Khintchine's inequality for the Rademacher functions. We will discuss them in this section and other important L^p inequalities.

Given a function $f \in L^1([0, 1))$ with Haar-Fourier expansion

$$\sum_{n=0}^{\infty} \sum_{k=1}^{2^n-1} d_k^{(n)} \chi_k^{(n)}(t),$$

let us define

$$S(f) = \left[\sum_{n=0}^{\infty} \sum_{k=1}^{2^n-1} \left(d_k^{(n)} \chi_k^{(n)}(t)\right)^2\right]^{1/2}$$

the *quadratic function* of associated to f. Paley [134] proved the following inequality, there exist constants c_p and C_p (independent of f) such that

$$c_p \|S(f)\|_p \leq \|f\|_p \leq C_p \|S(f)\|_p, \quad 1 < p < \infty. \tag{1.23}$$

Actually, Paley proved this for the Walsh series. The version for the Haar series was obtained by Marcinkiewicz [115] and is due precisely to the relations (1.21) and (1.22), that imply the equality of the partial sums of these expansions over the dyadic. Observe that in the case of Rademacher functions, the corresponding square function is simply $\left(\sum_{n=0}^{\infty} \sum_{n=1}^{\infty} |c_n|^2\right)^{1/2}$. Thus, this inequality is the natural extension of Khintchine's inequality for the case of the Walsh and Haar functions; an important difference is that it is only true in the range $1 < p < \infty$, thus is a missing range $0 < p < 1$.

The quadratic function has a key role in this inequality, as we shall see, this notion is of great importance also in more general contexts both in probability and analysis. For an interesting account of the development of this concept see the beautiful memoir by Stein [151].

Another very important operator is the *the maximal function* of the Haar expansions for a function f

$$f^*(t) = \sup_N \left| \sum_{n=0}^{N-1} \sum_{k=1}^{2^{n-1}} d_k^{(n)} \chi_k^{(n)}(t) \right|.$$

G. H. Hardy and J. E. Littlewood obtained the now famous *Hardy-Littlewood maximal inequality* [151]: there exists a constant c_p (independent of f) such that

$$c_p \left\| f^* \right\|_p \leqslant \| f \|_p \leqslant \left\| f^* \right\|_p , \quad 1 < p < \infty. \tag{1.24}$$

From the above two inequalities we then obtain the following inequality: There exist constants c_p and C_p (independent of f) such that

$$c_p \| S(f) \|_p \leqslant \left\| f^* \right\|_p \leqslant C_p \| S(f) \|_p, \quad 1 < p < \infty. \tag{1.25}$$

The inequality analogous to (1.15), obtained by Paley [134], and usually called Paley's inequalities, is the following: given $d_n^{(k)} \in \mathbb{R}$, $\varepsilon_n^{(k)} \in \{-1, 1\}$ then there exists a constant C_p, independent of the $d_n^{(k)}$ and the $\varepsilon_n^{(k)}$, such that

$$\left\| \sum_{n,k} \varepsilon_n^{(k)} d_n^{(k)} \chi_n^{(k)} \right\|_p \leqslant C_p \left\| \sum_{n,k} d_n^{(k)} \chi_n^{(k)} \right\|_p , 1 < p < \infty. \tag{1.26}$$

It should be noted that, precisely because of Khintchine's inequality, the two Paley's inequalities (1.23) and (1.26) are equivalent, see Exercise 6.

As we will see below, many of these ideas inspired generalizations or even, as noted by Gundy [85] some of the original proofs can be adapted in more general contexts.

On the other hand, a classical result of Schauder [149] states that the Haar system forms a basis of $L^p[0, 1)$, $1 < p < \infty$. That is, for every $f \in L^p[0, 1)$ there is a unique sequence $\left\{ d_k^{(n)} \right\}$ of real numbers satisfying

$$\left\| f - \sum_{n=0}^{N-1} \sum_{k=1}^{2^{n-1}} d_k^{(n)} \chi_k^{(n)} \right\|_p \to 0, \text{ as } N \to \infty.$$

Let β_p be the unconditional constant i.e., the smallest extended real number β with the following property: if N is a non-negative integer and $\{d_k^{(n)} : n = 0, \ldots N - 1; k = 1, \ldots, 2^{n-1}\}$ are real numbers such that $\left\| \sum_{n=0}^{N-1} \sum_{k=1}^{2^{n-1}} d_n^{(k)} \chi_n^{(k)} \right\|_p \leqslant 1$, then

$$\left\| \sum_{n=0}^{N-1} \sum_{k=1}^{2^{n-1}} \varepsilon_n^{(k)} d_n^{(k)} \chi_n^{(k)} \right\|_p \leqslant \beta, \tag{1.27}$$

for all choices of signs $\varepsilon_n^{(k)}$. Using Paley's inequality (1.26), Marcinkiewicz [115] proved that $\beta_p < \infty$ if and only if $1 < p < \infty$. D. Burkholder determined the precise value of β_p, as we are going to see in Chap. 4, see also [36], to be

$$\beta_p = p^* - 1, \quad 1 < p < \infty, \tag{1.28}$$

where $p^* = \max\{p, p/(p-1)\}$.

From here two areas developed: in harmonic analysis, the Littlewood-Paley-Stein theory which leads, among other things, to the Littlewood-Paley theorems and inequalities and Marcinkiewicz multiplier theorem, and in probability, martingale theory which we are going to study in detail in the following chapters and which is the main goal of the book.

Finally, it is worth mentioning that, from the analytical point of view, in the development of the wavelet theory one returns to Haar's functions to take up again the Paley's idea of finding orthonormal systems whose expansions give better information on the function than those given by the trigonometric series. Moreover, the recent development of powerful dyadic methods in harmonic analysis has clear roots in the study of these systems as we will discuss in Chap. 5.

1.4 Exercises

(1) Prove the following identity using Rademacher functions

$$\frac{\sin x}{x} = \prod_{k=1}^{\infty} \cos \frac{x}{2^k}$$

In particular, if $x = \frac{\pi}{2}$ we get the famous *Vieta's formula*

$$\frac{2}{\pi} = \prod_{k=1}^{\infty} \cos \frac{\pi}{2^{k+1}} = \frac{\sqrt{2}}{2} \cdot \frac{\sqrt{2+\sqrt{2}}}{2} \cdot \frac{\sqrt{2+\sqrt{2+\sqrt{2}}}}{2} \cdots .$$

[HINT: $\int_0^1 e^{ix(1-2t)} dt = \frac{\sin x}{x}$ and $\lim_{n\to\infty} \sum_{k=1}^{n} \frac{r_k(t)}{2^k} = 1 - 2t$.]

(2) Consider the ternary expansions for $t \in [0, 1]$ i.e.,

$$t = \frac{\eta_1(t)}{3} + \frac{\eta_2(t)}{3^2} + \cdots = \sum_{n=1}^{\infty} \frac{\eta_n(t)}{3^n}, \quad \eta_k(t) = 0, 1 \text{ or } 2.$$

3-(i) Prove that the functions $\{\eta_n\}$ are independent.

3-(ii) Prove that

$$\frac{\sin x}{x} = \prod_{k=1}^{\infty} \frac{1 + 2 \cos \frac{2x}{3^k}}{3}$$

and get a general version of it.

(3) Prove the inequality

$$A\sqrt{n} < \int_0^1 \left| \sum_{k=1}^{n} r_k(t) \right| dt \leqslant \sqrt{n}.$$

[HINT: use the formula $|z| = \frac{1}{\pi} \int_{-\infty}^{\infty} \frac{1 - \cos zx}{x^2} dx$ in order to prove that

$$\int_0^1 \left| \sum_{k=1}^{n} r_k(t) \right| dt = \frac{1}{\pi} \int_{-\infty}^{\infty} \frac{1 - \cos^n x}{x^2} dx > \frac{1}{\pi} \int_{-\sqrt{n}}^{\sqrt{n}} \frac{1 - \cos^n x}{x^2} dx$$

and then prove the left-hand side inequality, with $A = \frac{1}{\pi} \int_{-1}^{1} \frac{1 - \exp(-x^2/2)}{x^2} dx$. For the right-hand side inequality use the Cauchy-Schwarz inequality and the orthogonality of $\{r_n\}$].

(4) Let $\sum_{n=1}^{\infty} c_n^2 = \infty$, $c_n \to 0$ and consider the series

$$\sum_{n=1}^{\infty} c_n \sin\left(2\pi \, 2^{n-1} t\right).$$

Prove that the limit

$$\lim_{N \to \infty} \int_0^1 \left(\frac{\sum_{n=1}^{\infty} c_n \sin\left(2\pi \, 2^{n-1} t\right)}{\sqrt{\sum_{n=1}^{N} c_n^2}} \right)^4 dt,$$

exists and find its value.

[HINT: Remember that $r_n(t) = \text{sign}\left(\sin\left(2\pi \left(2^{n-1} t\right)\right)\right)$.]

(5) Prove that the Haar functions $\{\chi_n^{(k)}\}$ are also a complete orthonormal system on $L^2([0, 1])$.

(6) Prove, using Khintchine's inequality, (1.3) that the two Paley's inequalities (1.23) and (1.26) are equivalent.

(7) Prove that

$$f(t) = \sum_{n=1}^{\infty} c_n r_n(t), \quad \sum_{n=1}^{\infty} |c_n|^2 < \infty,$$

cannot be constant on a set of positive measure unless all but a finite number of coefficients are equal to 0.

(8) *Vilenkin systems.* Let $\{p_n, n \in \mathbb{N}\}$ be a sequence of natural number, such that $p_n \geqslant 2$ and let $P_0 = 1$ and

$$P_{n+1} := \prod_{k=0}^{n} p_k, \quad n \in \mathbb{N}.$$

Every point $x \in [0, 1)$ can be written in the following way:

$$x = \sum_{k=0}^{\infty} \frac{x_k}{P_{k+1}}, \quad 0 \leqslant x_k < p_k, x_k \in \mathbb{N}.$$

In case there are two different forms, we choose the one for which $\lim_{k \to \infty} x_k = 0$. With this convention the construction of x is unique.

The functions

$$r_n(x) := \exp \frac{2\pi i x_n}{p_n} \quad n \in \mathbb{N},$$

are called *generalized Rademacher functions*. Prove that these functions are independent.

The product system generated by these functions is a *one-dimensional Vilenkin system*:

$$w_n(x) := \prod_{k=0}^{\infty} r_k(x)^{n_k}$$

where $n = \sum_{k=0}^{\infty} n_k P_k, 0 \leqslant n_k < p_k$ and $n_k \in \mathbb{N}$.

Probability and Conditional Expectation

2

2.1 Probability and Conditional Expectation

As we have mentioned in the introduction, we are assuming that the reader is familiar with the basic notions of probability. Nevertheless, due to the crucial importance of the notions of conditional probability and conditional expectation for the study of martingale theory, we will give a brief review of them in this chapter.

Definition 2.1 Given $(\Omega, \mathscr{F}, \mathbf{P})$ a probability space remember that two events $A, B \in \mathscr{F}$ are *independent* if and only if

$$\mathbf{P}(A \cap B) = \mathbf{P}(A)\mathbf{P}(B). \tag{2.1}$$

If there is no independence, we would like to somehow measure the level of dependency among them, and from that idea, we get the notion of conditional probability.

Definition 2.2 If $\mathbf{P}(B) > 0$ we define the *conditional probability* of A given B as

$$\mathbf{P}(A \mid B) = \frac{\mathbf{P}(A \cap B)}{\mathbf{P}(B)}. \tag{2.2}$$

Intuitively, this means that if one knows that A has occurred and we want to calculate the probability that B occurs, we must restrict ourselves to the event A.

Analogously, if $\mathbf{P}(A) > 0$ we define the conditional probability of B given A, as

$$\mathbf{P}(B \mid A) = \frac{\mathbf{P}(A \cap B)}{\mathbf{P}(A)}.$$

© The Author(s), under exclusive license to Springer Nature Switzerland AG 2025
W. Urbina-Romero and R. Rios, *An Introduction to the Modern Martingale Theory and Applications*, Texts in Applied Mathematics 81,
https://doi.org/10.1007/978-3-031-88903-5_2

Alternatively, the conditional probability of A, given B, can be defined a any $\alpha \in \mathbb{R}$ solution of the equation

$$\mathbf{P}(A \cap B) = \gamma \mathbf{P}(B),$$

if $\mathbf{P}(B) \neq 0$ γ is unique.

Note that in the case of independent events

$$\mathbf{P}(A \mid B) = \frac{\mathbf{P}(A \cap B)}{\mathbf{P}(B)} = \frac{\mathbf{P}(A)\mathbf{P}(B)}{\mathbf{P}(B)} = \mathbf{P}(A)$$

and this means that the information that event B has occurred does not affect the probability of A.

We will illustrate these basic concepts with the following examples:

Example 2.1

(i) Suppose we roll a fair die, i.e., a balanced die such that the probability of each face is equal to $1/6$. Let $A = \{$ the face of the die is odd $\} = \{1, 3, 5\}$ then $\mathbf{P}(A) = \frac{1}{6} + \frac{1}{6} + \frac{1}{6} = \frac{1}{2}$. Now, let $B = \{1\}$, then, the probability of B is $\mathbf{P}(B) = 1/6$. On the other hand, the probability of B given A i.e., knowing that an odd number has come out? The probability of B given the information A is

$$\mathbf{P}(B \mid A) = \frac{\mathbf{P}(A \cap B)}{P(A)} = \frac{1/6}{1/2} = \frac{1}{3}.$$

Intuitively, this is clear since if it is known that $1, 3$ or 5 has been obtained, the probability of each must be $1/3$.

(ii) Suppose we toss a fair coin and let A be the event "the first toss is head" and let B be the event "the second toss is a tail". In this case, the sample space is $\Omega = \{(H, H), (H, T), (T, H), (T, T)\}$, thus $A = \{(H, H), (H, T)\}$ and $B = \{(H, T), (T, T)\}$ and $A \cap B = \{(H, T)\}$, and therefore

$$\mathbf{P}(B \mid A) = \frac{\mathbf{P}(A \cap B)}{\mathbf{P}(A)} = \frac{\mathbf{P}(\{(H, T)\})}{\mathbf{P}(\{(H, H), (H, T)\})} = \frac{1/4}{1/2} = \frac{1}{2} = \mathbf{P}(B).$$

This coincides with the intuition that the result of the first toss does not influence the outcome of the second toss. This is the essence of the fact that the events A and B are independent.

Besides, the conditional probability $\mathbf{P}(\cdot \mid A)$ is also a probability measure on Ω. Thus we go from $(\Omega, \mathscr{F}, \mathbf{P})$ to a new probability space $(\Omega, \mathscr{F}, \mathbf{P}(\cdot \mid A))$, but observe that $\mathbf{P}(\cdot \mid A)$ is concentrated in A; in the sense that if $B \cap A = \emptyset$ then $\mathbf{P}(B \mid A) = 0$. Thus $\mathbf{P}(\cdot \mid A)$ is the probability, known information that event A occurred.

Therefore, given an event A such that $P(A) > 0$, we can define the conditional probability given A, $\mathbf{P}(\cdot \mid A)$. From there we can also define the *conditional expectation* given A of an integrable random variable ξ, i.e., $\mathbf{E}|\xi| < \infty$, denoted as $\mathbf{E}[\xi \mid A]$ as

$$\mathbf{E}[\xi \mid A] = \int_A \xi \, d\mathbf{P}(\cdot \mid A). \tag{2.3}$$

This is the expected value (average or mean value) of the random variable ξ known that event A has occurred.

Note that so far, we have assumed $\mathbf{P}(A) > 0$. The problem now is how the notions of conditional probability and conditional expectation can be generalized. The generalization will be, not for events, but for σ-algebras. The generalization will be done backward, so we will start the discussion by first giving the generalization of the conditional expectation and from there getting the generalization of conditional probability.

Definition 2.3 Given $(\Omega, \mathscr{F}, \mathbf{P})$ a probability space, $\mathscr{G}$ a sub σ-algebra of $\mathscr{F}$ and $\xi : \Omega \to \mathbb{R}$ a random variable, such that $\mathbf{E}|\xi| < \infty$, we define the *conditional expectation* of ξ given $\mathscr{G}$, which we denote as $\mathbf{E}[\xi \mid \mathscr{G}]$, as a random variable such that

(i) $\mathbf{E}[\xi \mid \mathscr{G}]$ is $\mathscr{G}/\mathscr{B}(\mathbb{R})$ measurable.
(ii) For all $A \in \mathscr{G}$

$$\int_A \xi \, d\mathbf{P} = \int_A \mathbf{E}[\xi \mid \mathscr{G}] d\mathbf{P}. \tag{2.4}$$

To see why this random variable exists, assume, without loss of generality, that $\xi \geqslant 0$, and define the measure ν by

$$\nu(A) = \int_A \xi \, d\mathbf{P},$$

for all $A \in \mathscr{G}$. Then ν is a measure defined on $\mathscr{G}$ which is absolutely continuous with respect to the probability $\mathbf{P}$, $\nu \ll \mathbf{P}$. Then by the Radon-Nikodým theorem, see [159, Chap. 6], there exists the Radon-Nikodým derivative $\frac{d\nu}{d\mathbf{P}}$, which is $\mathscr{G}$-measurable and such that

$$\nu(A) = \int_A \left(\frac{d\nu}{d\mathbf{P}} \right) d\mathbf{P}.$$

Therefore,

$$\frac{d\nu}{d\mathbf{P}} = \mathbf{E}[\xi \mid \mathscr{G}] \text{ a.e.}$$

In this general case, the *conditional probability* of $B \in \mathscr{F}$ given $\mathscr{G}$ is defined from the conditional expectation as follows:

$$\mathbf{P}(B \mid \mathscr{G}) = \mathbf{E}[\chi_B \mid \mathscr{G}],$$

where χ_B is the indicator function of B. Therefore, $\mathbf{P}(B \mid \mathscr{G})$ is a random variable, such that:

(i) It is $\mathcal{G}$-measurable.

(ii) For all $A \in \mathcal{G}$

$$\int_A \mathbf{P}(B \mid \mathcal{G})d\mathbf{P} = \mathbf{P}(A \cap B).$$

Remark 2.1

(i) First, let us see that this definition is, indeed, a generalization of the definition
   of conditional probability given an event A such that $\mathbf{P}(A) > 0$.
   Let us consider the σ-algebra generated by A, $\sigma(A) = \{\emptyset, A, A^c, \Omega\}$. The conditional probability of B given $\sigma(A)$, $\mathbf{P}(B \mid \sigma(A))$, verifies the following conditions:

(i-1) $\mathbf{P}(B \mid \sigma(A))$ is $\sigma(A)$-measurable.

(i-2) For all $C \in \sigma(A)$

$$\int_C \mathbf{P}(B \mid \sigma(A))d\mathbf{P} = \mathbf{P}(C \cap B).$$

Now, as $\mathbf{P}(B \mid \sigma(A)) : \Omega \to \mathbb{R}$ is, by definition, $\sigma(A)$-measurable, then it has
to be constant over A and A^c (why?), say $\mathbf{P}(B \mid \sigma(A))(\omega) = \gamma_0$, if $\omega \in A$ and
$\mathbf{P}(B \mid \sigma(A))(\omega) = \gamma_1$ if $\omega \in A^c$. Let us see which are the values of γ_0 and γ_1.

$$\gamma_0 \mathbf{P}(A) = \int_A \mathbf{P}(B \mid \sigma(A))d\mathbf{P} = \mathbf{P}(A \cap B)$$

and

$$\gamma_1 \mathbf{P}(A^c) = \int_{A^c} \mathbf{P}(B \mid \sigma(A))d\mathbf{P} = \mathbf{P}\left(A^c \cap B\right).$$

Since $\mathbf{P}(A) > 0$ then $\gamma_0 = \frac{\mathbf{P}(A \cap B)}{\mathbf{P}(A)}$ and $\gamma_1 = \frac{\mathbf{P}(A \cap B)}{\mathbf{P}(A^c)}$.
Therefore,

$$\mathbf{P}(B \mid \sigma(A))(\omega) = \begin{cases} \frac{\mathbf{P}(A \cap B)}{\mathbf{P}(A)} & \text{if } \omega \in A \\ \frac{\mathbf{P}(A \cap B)}{\mathbf{P}(A^c)} & \text{if } \omega \in A^c \end{cases} = \begin{cases} \mathbf{P}(B \mid A) & \text{if } \omega \in A \\ \mathbf{P}\left(B \mid A^c\right) & \text{if } \omega \in A^c. \end{cases}$$

Thus, the abstract definition recovers the initial definition of conditional expectation given in (2.2).

(ii) Consider now a family of events $\{B_i\} \subset \mathcal{G}$ which are a partition of Ω, that is to
   say $B_i \cap B_j = \emptyset$ if $i \neq j$ and $\Omega = \bigcup_i B_i$. Given $\xi : \Omega \to \mathbb{R}$ a random variable,
   we want to define $\mathbf{E}[\xi \mid \mathcal{G}]$, where $\mathcal{G} = \sigma(\{B_i\})$. Observe that by construction

$$\mathbf{E}[\xi] = \sum_i \int_{B_i} \xi \, d\mathbf{P}.$$

According to Definition 2.3 $\mathbf{E}[\xi \mid \mathscr{G}]$ must be $\mathscr{G}$-measurable and verify (2.4)

$$\int_A \xi\, d\mathbf{P} = \int_A \mathbf{E}[\xi \mid \mathscr{G}]\, d\mathbf{P}.$$

Since $\mathbf{E}[\xi \mid \mathscr{G}]$ must be $\mathscr{G}$-measurable it must be constant over each B_i, let us say equal to α_i, then

$$\alpha_i \mathbf{P}(B_i) = \int_{B_i} \xi\, d\mathbf{P} = \int_{B_i} \mathbf{E}[\xi \mid \mathscr{G}]\, d\mathbf{P}.$$

If $\mathbf{P}(B_i) > 0$ then

$$\alpha_i = \frac{1}{\mathbf{P}(B_i)} \int_{B_i} \xi\, d\mathbf{P}.$$

If $\mathbf{P}(B_i) = 0$, $E[\xi \mid \mathscr{G}]$ must be constant on B_i but the value is arbitrary. Then

$$\mathbf{E}[\xi \mid \mathscr{G}] = \sum_i \frac{1}{\mathbf{P}(B_i)} \left(\int_{B_i} \xi\, d\mathbf{P} \right) \chi_{B_i}. \tag{2.5}$$

Example (i) is the particular case of this example for the trivial partition $\{A, A^c\}$.

(iii) As an application of the previous example, let $\Omega = [0, 1]$, $\mathscr{F} = \mathscr{B}([0, 1))$, the Borel σ-algebra, with Lebesgue's measure as a probability measure on Ω. Define

$$\mathscr{F}_n^d = \sigma \left(\left\{ \left(\frac{k}{2^n}, \frac{k+1}{2^n} \right) : k = 0, 1, 2, \ldots, 2^n - 1 \right\} \right),$$

for each $n \in \mathbb{N}$. Given $f : [0, 1] \to \mathbb{R}$ measurable and integrable, then.

$$E[f \mid \mathscr{F}_n^d](x) = 2^n \int_{k/2^n}^{(k+1)/2^n} f(t)\, dt,$$

if $x \in \left[\frac{k}{2^n}, \frac{k+1}{2^n} \right]$ a.e. and therefore

$$\mathbf{E}\left[f \mid \mathscr{F}_n^d \right] = \sum_{k=1}^{2^n - 1} 2^n \left(\int_{k/2^n}^{(k+1)/2^n} f(t)\, dt \right) \chi_{\left[\frac{k}{2^n}, \frac{k+1}{2^n} \right]}$$

$\{\mathscr{F}_n\}$ is called the dyadic filtration, the Walsh and Haar expansions are refer precisely to that filtration. We will study it in more detail later, in Chap. 5, Sect. 5.2.3.

(iv) The condition (2.4) in the definition of conditional expectation can be replaced by one of the following conditions:

- For all $A \in \mathscr{G}_0$

$$\int_A \xi \, d\mathbf{P} = \int_A \mathbf{E}[\xi \mid \mathscr{G}] \, d\mathbf{P}, \tag{2.6}$$

where $\sigma(\mathscr{G}_0) = \mathscr{G}$.

- For any $\mathscr{G}$-measurable random variables ζ, such that $\xi\zeta \in L^1$

$$\int_\Omega \xi\zeta \, d\mathbf{P} = \int_\Omega \mathbf{E}[\xi \mid \mathscr{G}]\zeta \, d\mathbf{P}. \tag{2.7}$$

(v) Given two random variables ξ, ζ we define the conditional expectation of ξ given ζ, $\mathbf{E}[\xi \mid \zeta]$ as $\mathbf{E}[\xi \mid \sigma(\zeta)]$, where $\sigma(\zeta)$ is the σ-algebra generated by ζ, i.e.,

$$\sigma(\zeta) = \{\zeta^{-1}(B) : B \in \mathscr{B}(\mathbb{R})\}$$

and in general, given ξ, ζ_1, ζ_2, $\ldots$, ζ_n, random variables, we define

$$\mathbf{E}[\xi \mid \zeta_1, \zeta_2, \ldots, \zeta_n] = E[\xi \mid \sigma(\zeta_1, \zeta_2, \ldots, \zeta_n)]$$

where $\sigma(\zeta_1, \zeta_2, \ldots, \zeta_n)$ is the σ-algebra generated by ζ_1, ζ_2, $\ldots$, ζ_n.

Now, let us discuss the main properties of the conditional expectation.

Theorem 2.1 *Given ξ, ζ integrable random variables and a sub-σ-algebra $\mathscr{G}$ of $\mathscr{F}$.*

(i) $\mathbf{E}[\alpha\xi + \beta\zeta \mid \mathscr{G}] = \alpha\mathbf{E}[\xi \mid \mathscr{G}] + \beta\mathbf{E}[\zeta \mid \mathscr{G}]$ a. e. for any $\alpha, \beta \in \mathbb{R}$.
(ii) If $\xi \leqslant \zeta$ then $\mathbf{E}[\xi \mid \mathscr{G}] \leqslant \mathbf{E}[\zeta \mid \mathscr{G}]$.
(iii) If ξ is $\mathscr{G}$-mesurable, then $\mathbf{E}[\xi \mid \mathscr{G}] = \xi$ a.e.
(iv) If $\mathscr{G}_1 \subset \mathscr{G}_2$ are σ-algebras, then

$$\mathbf{E}[\mathbf{E}[\xi \mid \mathscr{G}_2] \mid \mathscr{G}_1] = \mathbf{E}[\mathbf{E}[\xi \mid \mathscr{G}_1] \mid \mathscr{G}_2] = \mathbf{E}[\xi \mid \mathscr{G}_1].$$

(v) If ξ is independent of $\mathscr{G}$ then $\mathbf{E}[\xi \mid \mathscr{G}] = \mathbf{E}[\xi]$ a.e.
(vi) If $\xi_n \to \xi \in L^1(\Omega)$ monotonically, then $\mathbf{E}[\xi_n \mid \mathscr{G}] \to \mathbf{E}[\xi \mid \mathscr{G}]$ a.e.
(vii) If $|\xi_n| \leqslant \zeta \in L^1(\Omega)$ and $\xi_n \to \xi$ a.e. then $\mathbf{E}[\xi_n \mid \mathscr{G}] \to \mathbf{E}[\xi \mid \mathscr{G}]$ a.e.
(viii) If ζ is integrable and $\mathscr{G}$-mesurable, then $\mathbf{E}[\xi\zeta \mid \mathscr{G}] = \zeta\mathbf{E}[\xi \mid \mathscr{G}]$ a.e.
(ix) If $\xi, \zeta \in L^2(\Omega)$, then

$$\int_\Omega \mathbf{E}[\xi \mid \mathscr{G}]\zeta \, d\mathbf{P} = \int_\Omega \xi\mathbf{E}[\zeta \mid \mathscr{G}] \, d\mathbf{P}.$$

(x) If $\xi \in L^p(\Omega)$, $1 \leqslant p \leqslant \infty$, then

$$\|\mathbf{E}[\xi \mid \mathscr{G}]\|_p \leqslant \|\xi\|_p.$$

Property (iii) tell us that $\mathbf{E}[\cdot \mid \mathcal{G}]$ is a projection onto the space of $\mathcal{G}$-measurable random variables and property ix) says that $\mathbf{E}[\cdot \mid \mathcal{G}]$ is a self-adjoint operator on $L^2(\Omega)$. Finally, property x) says that $\mathbf{E}[\cdot \mid \mathcal{G}]$ is a contraction on $L^p(\Omega)$.

Proof

(i) As $\alpha\mathbf{E}[\xi \mid \mathcal{G}] + \beta\mathbf{E}[\zeta \mid \mathcal{G}]$ is $\mathcal{G}$-measurable and for any $A \in \mathcal{G}$, and

$$\int_A (\alpha\mathbf{E}[\xi \mid \mathcal{G}] + \beta\mathbf{E}[\zeta \mid \mathcal{G}])d\mathbf{P} = \alpha\int_A \mathbf{E}[\xi \mid \mathcal{G}]d\mathbf{P} + \beta\int_A \mathbf{E}[\zeta \mid \mathcal{G}]d\mathbf{P}$$

$$= \alpha\int_A \xi d\mathbf{P} + \beta\int_A \zeta d\mathbf{P} = \int_A (\alpha\xi + \beta\zeta)d\mathbf{P}$$

$$= \int_A \mathbf{E}[\alpha\xi + \beta\zeta \mid \mathcal{G}]d\mathbf{P},$$

we get

$$\alpha\mathbf{E}[\xi \mid \mathcal{G}] + \beta\mathbf{E}[\zeta \mid \mathcal{G}] = \mathbf{E}[\alpha\xi + \beta\zeta \mid \mathcal{G}], \text{ a.e.}$$

(ii) If $\xi \leqslant \zeta$, by the monotonicity property of the Lebesgue integral we have that for any $A \in \mathcal{G}$

$$\int_A \mathbf{E}[\xi \mid \mathcal{G}]d\mathbf{P} = \int_A \xi d\mathbf{P} \leqslant \int_A \zeta d\mathbf{P} = \int_A \mathbf{E}[\zeta \mid \mathcal{G}]d\mathbf{P}$$

Thus,

$$\mathbf{E}[\xi \mid \mathcal{G}] \leqslant \mathbf{E}[\zeta \mid \mathcal{G}], \text{ a.e.}$$

In particular, this implies that the conditional expectation is a positive operator, i.e., if $\xi \geqslant 0$, then $\mathbf{E}[\xi \mid \mathcal{G}] \geqslant 0$.

(iii) If ξ is $\mathcal{G}$-mesurable then trivially it satisfies (2.4). Thus this property is immediate from the definition of conditional expectation and allows us to interpret it as a projection onto the set of all $\mathcal{G}$-measurable random variables.

(iv) Let $A \in \mathcal{G}_1$ then by definition

$$\int_A \mathbf{E}[\xi \mid \mathcal{G}_1]d\mathbf{P} = \int_A \xi d\mathbf{P},$$

but as $\mathcal{G}_1 \subseteq \mathcal{G}_2$, then $A \in \mathcal{G}_2$ and again by definition

$$\int_A \mathbf{E}\left[\mathbf{E}[\xi \mid \mathcal{G}_1] \mid \mathcal{G}_2\right]d\mathbf{P} = \int_A \mathbf{E}[\xi \mid \mathcal{G}_1]d\mathbf{P}.$$

Therefore,

$$\mathbf{E}\left[\mathbf{E}[\xi \mid \mathcal{G}_1] \mid \mathcal{G}_2\right] = \mathbf{E}[\xi \mid \mathcal{G}_1] \text{ a.e.}$$

But also, for $A \in \mathcal{G}_1$

$$\int_A \mathbf{E}\left[\mathbf{E}\left[\xi \mid \mathcal{G}_2\right] \mid \mathcal{G}_1\right] d\mathbf{P} = \int_A \mathbf{E}[\xi \mid \mathcal{G}_2] d\mathbf{P} = \int_A \xi \, d\mathbf{P},$$

and therefore

$$\int_A \mathbf{E}\left[\mathbf{E}\left[\xi \mid \mathcal{G}_2\right] \mid \mathcal{G}_1\right] d\mathbf{P} = \int_A \xi \, d\mathbf{P} = \int_A \mathbf{E}[\xi \mid \mathcal{G}_1] d\mathbf{P},$$

thus

$$\mathbf{E}\left[\mathbf{E}\left[\xi \mid \mathcal{G}_2\right] \mid \mathcal{G}_1\right] = \mathbf{E}\left[\xi \mid \mathcal{G}_1\right] \text{ a.e.}$$

In particular, taking the trivial σ-algebra $\mathcal{G}_0 = \{\phi, \Omega\}$ we have that, as $\mathbf{E}[\xi \mid \mathcal{G}_0]$ must be $\mathcal{G}_0$-measurable, then it is constant on Ω, thus

$$\mathbf{E}[\xi] = \int_\Omega \xi \, d\mathbf{P} = \int_\Omega \mathbf{E}\left[\xi \mid \mathcal{G}_0\right] d\mathbf{P} = \mathbf{E}\left[\xi \mid \mathcal{G}_0\right].$$

Therefore, the (absolute) expectation is the conditional expectation (with respect to the trivial σ-algebra), and then,

$$\mathbf{E}[\mathbf{E}[\xi \mid \mathcal{G}]] = \mathbf{E}[\xi] \text{ a.e.}$$

On the other hand,

$$\mathbf{E}[\mathbf{E}[\xi \mid \mathcal{G}] \mid \mathcal{G}] = \mathbf{E}[\xi \mid \mathcal{G}] \text{ a.e.}$$

so, the conditional expectation is a projection onto the space of all $\mathcal{G}$-measurable random variables.

(v) If ξ is independent of $\mathcal{G}$, for any $A \in \mathcal{G}$,

$$\int_A \xi \, d\mathbf{P} = \int_\Omega \xi \chi_A d\mathbf{P} = \mathbf{P}(A) \int_\Omega \xi \, d\mathbf{P} = \mathbf{P}(A)\mathbf{E}[\xi] = \int_A \mathbf{E}[\xi] d\mathbf{P}.$$

Now, by definition

$$\int_A \mathbf{E}[\xi \mid \mathcal{G}] d\mathbf{P} = \int_A \xi \, d\mathbf{P} = \int_A \mathbf{E}[\xi] d\mathbf{P},$$

hence

$$\mathbf{E}[\xi \mid \mathcal{G}] = \mathbf{E}[\xi] \text{ a.e.}$$

(vi) Assume $\xi_n \uparrow \xi$, then applying twice the monotone convergence theorem, for any $A \in \mathcal{G}$,

$$\int_A \mathbf{E}[\xi \mid \mathcal{G}]d\mathbf{P} = \int_A \xi\,d\mathbf{P} = \int_A \lim_{n\to\infty} \xi_n\,d\mathbf{P} = \lim_{n\to\infty} \int_A \xi_n\,d\mathbf{P}$$
$$= \lim_{n\to\infty} \int_A \mathbf{E}[\xi_n \mid \mathcal{G}]d\mathbf{P} = \int_A \lim_{n\to\infty} \mathbf{E}[\xi_n \mid \mathcal{G}]d\mathbf{P},$$

as from (iii) we know that the sequence $\{\mathbf{E}[\xi_n \mid \mathcal{G}]\}$ is monotone increasing. Thus,

$$\mathbf{E}[\xi \mid \mathcal{G}] = \lim_{n\to\infty} \mathbf{E}[\xi_n \mid \mathcal{G}] \quad \text{a.e.}$$

(vii) The argument is similar to (vi). Since $|\xi_n| \leqslant \zeta \in L^1(\Omega)$ by (iii)

$$|\mathbf{E}[\xi_n \mid \mathcal{G}]| \leqslant \mathbf{E}[\zeta \mid \mathcal{G}] \in L^1(\Omega)$$

and then,

$$\int_A \mathbf{E}[\xi \mid \mathcal{G}]d\mathbf{P} = \int_A \xi\,d\mathbf{P} = \int_A \lim_{n\to\infty} \xi_n\,d\mathbf{P} = \lim_{n\to\infty} \int_A \xi_n\,d\mathbf{P}$$
$$= \lim_{n\to\infty} \int_A \mathbf{E}[\xi_n \mid \mathcal{G}]d\mathbf{P} = \int_A \lim_{n\to\infty} \mathbf{E}[\xi_n \mid \mathcal{G}]d\mathbf{P}.$$

(viii) Let ζ is integrable and $\mathcal{G}$-measurable, observe that $\zeta\mathbf{E}[\xi \mid \mathcal{G}]$ is $\mathcal{G}$-measurable and if $\zeta = \chi_B$, $B \in \mathcal{G}$ then, for any $A \in \mathcal{G}$ then

$$\int_A \zeta\mathbf{E}[\xi \mid \mathcal{G}]d\mathbf{P} = \int_{A\cap B} \mathbf{E}[\xi \mid \mathcal{G}]d\mathbf{P} = \int_{A\cap B} \xi\,d\mathbf{P} = \int_A \chi_B\xi\,d\mathbf{P} = \int_A \xi\zeta\,d\mathbf{P}.$$

Hence,

$$\zeta\mathbf{E}[\xi \mid \mathcal{G}] = \mathbf{E}[\xi\zeta \mid \mathcal{G}] \quad \text{a.e.}$$

Now, by the linearity of the conditional expectation, property (i), the result is still true if ζ is a simple function, and the general case can be obtained by taking limits.

(ix) If $\xi, \zeta \in L^2(\Omega)$ then by (viii) and the definition of conditional probability, for any $A \in \mathcal{G}$

$$\int_A \mathbf{E}[\xi \mid \mathcal{G}]\zeta\,d\mathbf{P} = \int_A \mathbf{E}[\mathbf{E}[\xi \mid \mathcal{G}]\zeta \mid \mathcal{G}]d\mathbf{P} = \int_A \mathbf{E}[\xi \mid \mathcal{G}]\mathbf{E}[\zeta \mid \mathcal{G}]d\mathbf{P}$$
$$= \int_A \mathbf{E}[\xi\mathbf{E}[\zeta \mid \mathcal{G}] \mid \mathcal{G}]d\mathbf{P} = \int_A \xi\mathbf{E}[\zeta \mid \mathcal{G}]d\mathbf{P}.$$

(x) Using the definition of the L^p norm by duality, if $\frac{1}{p} + \frac{1}{q} = 1$

$$
\begin{aligned}
\|\mathbf{E}[\xi \mid \mathscr{G}]\|_p &= \sup \left\{ \left| \int_\Omega \mathbf{E}[\xi \mid \mathscr{G}]\zeta \, d\mathbf{P} \right| : \zeta \ \text{is} \ \mathscr{G} - \text{measurable}, \ \zeta\|_q \leqslant 1 \right\} \\
&= \sup \left\{ \left| \int_\Omega \mathbf{E}[\xi \zeta \mid \mathscr{G}] d\mathbf{P} \right| : \zeta \ \text{is} \ \mathscr{G} - \text{measurable}, \ \|\zeta\|_q \leqslant 1 \right\} \\
&= \sup \left\{ |\mathbf{E}[\xi \zeta]| : \zeta \ \text{is} \ \mathscr{G} - \text{measurable}, \ \|\zeta\|_q \leqslant 1 \right\} \\
&\leqslant \sup \left\{ |\mathbf{E}[\xi \zeta]| : \zeta \ \text{is} \ \mathscr{F} - \text{measurable}, \ \|\zeta\|_q \leqslant 1 \right\} = \|\xi\|_p.
\end{aligned}
$$

This implies that the conditional expectation is a contraction on $L^p(\Omega)$ the space of p-integrable random variables in $(\Omega, \mathscr{F}, \mathbf{P})$, for $1 \leqslant p \leqslant \infty$. $\square$

Observe that two sub-σ-algebras $\mathscr{F}_1$ and $\mathscr{F}_2$ are independent if and only if any pair of finite collections in $\mathscr{F}_1$ and in $\mathscr{F}_2$ are independent of each other.

As expected, the conditional expectation "inherits" certain inequalities of the expectation, in particular, we have

Proposition 2.1 (Jensen's Inequality) *Let ϕ be a convex function such that if ξ is an integrable random variable, $\phi(\xi)$ is also integrable, then*

$$
\phi(\mathbf{E}[\xi \mid \mathscr{G}]) \leqslant \mathbf{E}[\phi(\xi) \mid \mathscr{G}].
$$

Proof Since ϕ is convex, then is easy to see that ϕ can be written as

$$
\phi(x) = \sup\{g(x) : g \ \text{afine such that} \ g \leqslant \phi\},
$$

see for instance Durret [74, Chap. 1].

Let, $L = \{g : g \ \text{affine}, \ g \leqslant \phi\}$ then if $g \in L$ by linearity and property (ii) above

$$
g(\mathbf{E}[\xi \mid \mathscr{G}]) = \mathbf{E}[g(\xi) \mid \mathscr{G}] \leqslant E[\phi(\xi) \mid \mathscr{G}]
$$

thus, taking supremum on $g \in L$ on the left-hand side, we get

$$
\phi(\mathbf{E}[\xi \mid \mathscr{G}]) = \sup_{g \in L} g(\mathbf{E}[\xi \mid \mathscr{G}]) \leqslant \mathbf{E}[\phi(\xi) \mid \mathscr{G}]. \qquad \square
$$

2.2   Exercises

(1) Show that if ξ, ζ are independent and integrable random variables then

$$
\mathbf{E}[\xi \mid \zeta] = \mathbf{E}[\xi]
$$

and that in turn this implies that

$$\mathbf{E}[\xi\zeta] = \mathbf{E}[\xi]\mathbf{E}[\zeta].$$

Give counterexamples, to prove that the reciprocal implications are false.

[HINT: take ξ with normal standard distribution $N(0, 1)$ and consider ξ^2.]

(2) (2-i) Show that for a non-negative random variable ξ,

$$\mathbf{E}[\xi \mid \mathscr{G}] = \int_0^\infty \mathbf{P}\{\xi > t \mid \mathscr{G}\}dt \quad \text{a.e.}$$

(2-ii) Prove the generalized *Markov's inequality*

$$P\{|\xi| \geqslant \alpha \mid \mathscr{G}\} \leqslant \alpha^{-k} E\left[|\xi|^k \mid \mathscr{G}\right], \ \alpha > 0.$$

(2-iii) Analogously, write and prove the generalized *Tchebychev's and Hölder's inequalities*.

(3) (3-i) If $\mathscr{G}_1 \subset \mathscr{G}_2$ and $\mathbf{E}[\xi^2] < \infty$,

$$\mathbf{E}\left[(\xi - \mathbf{E}[\xi \mid \mathscr{G}_2])^2\right] \leqslant \mathbf{E}\left[(\xi - \mathbf{E}[\xi \mid \mathscr{G}_1])^2\right].$$

Thus, the spread of ξ around its conditional mean gets smaller as the σ-algebra gets bigger.

(3-ii) Let $\mathrm{Var}[\xi \mid \mathscr{G}] = E\left[(\xi - \mathbf{E}[\xi \mid \mathscr{G}])^2 \mid \mathscr{G}\right]$. Prove the following *paralelogram law* for $\mathrm{Var}[\cdot \mid \mathscr{G}]$

$$\mathrm{Var}[\xi] = \mathbf{E}[\mathrm{Var}[\xi \mid \mathscr{G}]] + \mathrm{Var}[\mathbf{E}[\xi \mid \mathscr{G}]].$$

(4) Let $L^2(\Omega)$ be the Hilbert space of integrable square random variables on $(\Omega, \mathscr{F}, \mathbf{P})$. For $\mathscr{G}$ a sub-σ-algebra of $\mathscr{F}$, let $M_{\mathscr{G}}$ be the subspace of random variables in $L^2(\Omega)$ which are $\mathscr{G}$-measurable. Prove that the operator $P_{\mathscr{G}}$ defined for $\xi \in L^2(\Omega)$ as $P_{\mathscr{G}} = \mathbf{E}[\xi \mid \mathscr{G}]$ is an orthogonal projection on $M_{\mathscr{G}}$. With this result, an alternative definition of conditional probability and conditional expectation can be given, i.e., given $\mathscr{G}$ a sub-σ-algebra of $\mathscr{F}$, let $P_{\mathscr{G}}$ be the orthogonal projection onto $M_{\mathscr{G}}$. Prove that for $\xi \in L^2(\Omega)$, $P_{\mathscr{G}}(X)$ has the properties required for the conditional expectation $\mathbf{E}[\xi \mid \mathscr{G}]$. Use this to define $\mathbf{E}[\xi \mid \mathscr{G}]$ for $\xi \in L^2(\Omega)$ and extend it to all $\xi \in L^1(\Omega)$. Define also conditional probability.

(5) In the above exercise let $\mathscr{G} = \sigma(\zeta)$ for a random variable $\zeta \in L^2(\Omega)$. Let S_ζ be the one-dimensional space generated by ζ. Prove that S_ζ can be much smaller than $M_{\sigma(\zeta)}$, so that $E[\xi \mid \zeta]$ for $\xi \in L^2(\Omega)$ is not the projection of ξ onto S_ζ.

(6) Conditional expectation can be considered also on σ-finite measure spaces. Verify in detail that the following definition makes perfect sense. Given a σ-finite measure space $(M, \mathscr{M}, \mu)$, and a sub-σ-algebra $\mathscr{G} \subset \mathscr{M}$, then, if f is a locally integrable function on $(M, \mathscr{M}, \mu)$ then there is a unique $\mathscr{G}$-measurable function $E[f|\mathscr{G}]$ such that for all $A \in \mathscr{G}$ we have

$$\int_A f(x)\mu(dx) = \int_A E[f|\mathscr{G}](x)\mu(dx).$$

(7) Because most projections are not conditional expectations, for the most part, partial sums of Fourier series cannot be interpreted as conditional expectations. Nevertheless, there are special cases in which such an interpretation is possible. Take $\Omega = [0, 1)$, $\mathscr{F} = \mathscr{B}_{(0,1)}$, and let m be the restriction of Lebesgue measure to $[0, 1)$. Next, for $n \in \mathbb{N}$, take $\mathscr{F}_n$ to be the σ-algebra generated by those $f \in C([0, 1); \mathbb{C})$ that are periodic with period 2^{-n}. Finally, set $\mathbf{e}_k(x) = \exp(2\pi kix)$ for $k \in \mathbb{Z}$, and use elementary Fourier analysis to show that, for each $n \in \mathbb{N}$, $\{\mathbf{e}_{k2^n} : k \in \mathbb{Z}\}$ is an orthonormal basis for $L^2([0, 1), m, \mathbb{C})$. In particular, conclude that, for every $f \in L^2([0, 1), m; \mathbb{C})$,

$$\mathbf{E}[f \mid \mathscr{F}_n] = \mathbf{E}[f] + \sum_{k \in \mathbb{Z}} \langle f, \mathbf{e}_{k2^n} \rangle_{L^2([0,1);\mathbb{C})} \mathbf{e}_{k2^n},$$

where the convergence is in $L^2([0, 1], m; \mathbb{C})$.

Classical Martingale Theory

3

In this chapter, we are going to study the basic material of martingale theory for discrete parameter. Although martingales were already discussed by P. Lévy in 1937, without explicitly naming them, it was J. L. Doob in the 1940s, who not only named them but realized their potential and did the fundamental development of the theory. As we discussed in Chap. 1, several mathematicians worked on objects that turned out to be martingales before the theory was developed; for example, R. E. A. C. Paley [134] and J. Marcinkiewicz proved inequality (1.26) for the Haar and Walsh systems which are special cases of the results of D. Burkholder in his famous 1966 paper [29]. In a different direction, in 1928 R. Courant, K. Friedrichs and H. Levy [60] used ideas related to the notions of martingales, although without randomness, to study harmonic functions, in the papers which introduced the finite element method for the numerical approximation of solutions of partial differential equations, obtaining the now famous Courant–Friedrichs–Lewy.

As we have already said, the basic reference for the theory is J. L. Doob's book [69], which is what is contained nowadays in any book on probabilities or stochastic processes at the graduate level. However, our exposition is strongly influenced by the works of Burkholder [29, 30] as we shall see, these are the basis of the generalizations of this theory, which will be discussed in the next chapter. In addition, we will give some results that extend the classical results.

3.1   Definition of Martingale and Related Concepts

Let $(\Omega, \mathscr{F}, \mathbf{P})$ be a probability space and consider $\{\mathscr{F}_n\}$ a filtration, i.e., a sequence of increasing sub σ-algebras, $\mathscr{F}_n \subseteq \mathscr{F}_{n+1}$, of the σ-algebra $\mathscr{F}$, which we will call a filtration in Ω.

© The Author(s), under exclusive license to Springer Nature Switzerland AG 2025
W. Urbina-Romero and R. Rios, *An Introduction to the Modern Martingale Theory and Applications*, Texts in Applied Mathematics 81,
https://doi.org/10.1007/978-3-031-88903-5_3

Definition 3.1 A stochastic process $X = \{X_n\}_n$ is a $\{\mathscr{F}_n\}$-martingale ($\{\mathscr{F}_n\}$-supermartingale, $\{\mathscr{F}_n\}$-submartingale) if:

(i) For each n, $X_n \in L^1(\Omega)$, i.e., if $\mathbf{E}[|X_n|] < \infty$.
(ii) $\{X_n\}$ is $\{\mathscr{F}_n\}$-adapted, i.e., for each n, X_n is $\mathscr{F}_n$-measurable.
(iii) For each $n, m \in \mathbb{N}$, if $m > n$, then

$$\mathbf{E}[X_m \mid \mathscr{F}_n] = X_n \quad \text{a.e.} \quad (\leqslant, \geqslant \text{ respectively.}) \tag{3.1}$$

Remark 3.1 For this definition we have the following observations.

(1) Clearly $\{X_n\}$ is a $\{\mathscr{F}_n\}$-martingale if and only if it is a $\{\mathscr{F}_n\}$-supermartingale and a $\{\mathscr{F}_n\}$-submartingale.
(2) Condition (3.1) is the key property of the definition of these processes and it is equivalent, by the definition of conditional expectation, to the following: for $m \geqslant n$ and every $\mathscr{A} \in \mathscr{F}_n$

$$\int_A X_m d\mathbf{P} = \int_A X_n d\mathbf{P}. \quad (\leqslant, \geqslant \text{ respectively.}) \tag{3.2}$$

Condition (3.2), in turn, is equivalent to requesting that for every ξ $\mathscr{F}_n$-measurable random variable,

$$\int_\Omega \xi X_m d\mathbf{P} = \int_\Omega \xi X_n d\mathbf{P} \quad (\leqslant, \geqslant \text{ respectively}).$$

(3) Condition (iii) of Definition 3.1 is equivalent, by induction, to the following condition:
(iii-1) For any $n \in \mathbb{N}$,

$$\mathbf{E}\left[X_{n+1} \mid \mathscr{F}_n\right] = X_n \quad \text{a.e.}(\leqslant, \geqslant \text{ respectively}). \tag{3.3}$$

(4) If we define $\Delta X_1 = X_1$ and for $n > 1$, $\Delta X_n = X_n - X_{n-1}$ (or, in other words, we define $\Delta X_n = X_n - X_{n-1}$ for all n assuming that $X_0 = 0$) and calling $\Delta X = \{\Delta X_n\}$ the sequence of differences of the process X then (iii) is also equivalent to ask for the following condition:
(iii-2) For any $n \in \mathbb{N}$

$$\mathbf{E}\left[\Delta X_{n+1} \mid \mathscr{F}_n\right] = 0 \quad \text{a.e.}(\leqslant, \geqslant \text{ respectively}). \tag{3.4}$$

and by induction, if $m > n$

$$\mathbf{E}[\Delta X_m \mid \mathscr{F}_n] = 0 \quad \text{a.e.}(\leqslant, \geqslant \text{ respectively}). \tag{3.5}$$

Additionally, condition (iii-2) is sometimes expressed in "analytical" terms as

$$\int_{\Omega} \Psi(X_1, \cdots, X_{n-1}) \Delta X_n \, d\mathbf{P} = 0, \quad (\leqslant, \geqslant \text{ respectively}) \tag{3.6}$$

for any bounded and measurable function $\Psi : \mathbb{R}^{n-1} \to \mathbb{R}$.

For martingales, condition (3.4) is known as the orthogonality property of the the increments. Nevertheless, there are other stochastic processes with orthogonal increments that are not martingales.

Observe that (3.5) implies the actual orthogonality of the increments (in the analytical sense), since if $m > n$ then

$$\mathbf{E}[\Delta X_n \Delta X_m] = \mathbf{E}[\mathbf{E}[\Delta X_n \Delta X_m \mid \mathscr{F}_n]]$$
$$= \mathbf{E}[\Delta X_n \mathbf{E}[\Delta X_m \mid \mathscr{F}_n]] = \mathbf{E}[\Delta X_n \cdot 0] = 0.$$

Additionally, for $n \geqslant 1$ by the martingale property

$$\mathbf{E}\left[(\Delta X_n)^2\right] = \mathbf{E}\left[\mathbf{E}\left[(\Delta X_n)^2 \mid \mathscr{F}_{n-1}\right]\right] = \mathbf{E}\left[\mathbf{E}[X_n^2 - 2X_n X_{n-1} + X_{n-1}^2 \mid \mathscr{F}_{n-1}]\right]$$
$$= \mathbf{E}\left[\mathbf{E}[X_n^2 \mid \mathscr{F}_{n-1}]\right] - 2\mathbf{E}\left[\mathbf{E}[X_n X_{n-1} \mid \mathscr{F}_{n-1}]\right] + \mathbf{E}\left[\mathbf{E}[X_{n-1}^2 \mid \mathscr{F}_{n-1}]\right]$$
$$= \mathbf{E}\left[X_n^2\right] - 2\mathbf{E}\left[X_{n-1}^2\right] + \mathbf{E}\left[X_{n-1}^2\right] = \mathbf{E}\left[X_n^2\right] - \mathbf{E}\left[X_{n-1}^2\right] = \mathbf{E}\left[(\Delta X_n^2)\right]. \tag{3.7}$$

From the definition of the increments ΔX_n it is clear that

$$X_n = \sum_{k=1}^{n} \Delta X_k, \tag{3.8}$$

and therefore martingales are sums of orthogonal random variables in the sense of (3.4). Consequently, one can interpret X_n as the resultant of ΔX_k favorable events in a fair game. Moreover, by (3.8), we can conclude that it is equivalent to knowing a martingale X or its sequence of increments $\{\Delta X_k\}$.

(5) If $\{X_n\}$ is a stochastic process adapted to $\{\mathscr{F}_n\}$ and integrable; the prediction at time $n - 1$ of the increment at the next stage is:

$$\tilde{X}_n - \tilde{X}_{n-1} = E\left[X_n \mid \mathscr{F}_{n-1}\right] - X_{n-1}$$

and the *compensator* of $\{X_n\}$ is the sequence $\{\tilde{X}_n\}$ defined by this equation, with $\tilde{X}_0 = 0$. The sequence $\{X_n\}$ will tend to increase (resp. decrease) if its *compensator* is increasing (resp. decreasing). The compensator, like the conditional expectation, is defined a.e.

(6) The definition of martingale has several possible generalizations.

- First, we have generalizations concerning the index set: we can consider a subset or all $\mathbb{Z}$, see Chow & Teicher [56]. Moreover, we can consider any partially ordered set to be an index set.
- Second, generalizations referring to the values they take, e.g., that they take vector values or even Banach space values (see, Neveu [130] and Diestel & Uhl [68]). We will discuss martingales with Banach space values in Chap. 5.
- Third, they are not defined on a probability space but on a space of σ–finite measure, which we will study in a later chapter (see, Dellacherie & Meyer [61]).
- Finally, it is possible to study generalized martingales $X = \{X_n\}_n$, which are those processes such that X_n is not integrable but that the conditional expectation of X_n is not integrable (in a generalized sense) $\mathbf{E}[X_n \mid \mathscr{F}_m]$ are defined for any m and n, and verify (3.1) (see Shiryayev [148], Chap. VII).

A process $X = \{X_n\}_n$ is simply called a martingale (supermartingale, submartingale) if it is a $\{\mathscr{F}_n\}$-martingale ($\{\mathscr{F}_n\}$-supermartingale, $\{\mathscr{F}_n\}$-submartingale) where $\mathscr{F}_n = \sigma(X_1, \ldots, X_n)$, is called the natural σ-algebra of X and is denoted $\mathscr{F}_n^X$. In that case, condition (ii) is trivial, and condition (iii) can be written as:

(iii-3) If $m > n$, then

$$\mathbf{E}[X_m \mid X_1, \ldots, X_n] = X_n \text{ a.e.} \quad (\leqslant, \geqslant \text{ respectively}).$$

Definition 3.2 We denote by $\mathscr{M} = \mathscr{M}(\Omega, \mathscr{F}, \mathbf{P})$, the set of all $\{\mathscr{F}_n\}$-martingales defined on $(\Omega, \mathscr{F}, \mathbf{P})$. It is easy to see that $\mathscr{M}$ is closed under the operations of addition and product by a scalar, due to the properties of conditional expectation, and therefore $\mathscr{M}$ is a real vector space with those operations, in other words, if $\{X_n\}$ and $\{Y_n\}$ are $\{\mathscr{F}_n\}$-martingales then for any $\alpha, \beta \in \mathbb{R}$ if

$$Z_n = \alpha X_n + \beta Y_n$$

then $\{Z_n\}$ is also a $\{\mathscr{F}_n\}$-martingale.

Example 3.1 (1) The typical example of a martingale, are the sums of centered independent random variables, $\{\xi_n\}$ i.e., $\mathbf{E}[\xi_n] = 0$. In that case conditions (i) and (ii) are immediate and we will check (iii)': since

$$\mathscr{F}_n = \sigma\left(\xi_1, \xi_1 + \xi_2, \ldots, \sum_{k=1}^{n} \xi_k\right) = \sigma(\xi_1, \ldots, \xi_n), \tag{3.9}$$

and for the independence of $\{\xi_n\}$, we get

$$\mathbf{E}\left[\sum_{k=1}^{n+1}\xi_k \mid \mathscr{F}_n\right] = \mathbf{E}\left[\sum_{k=1}^{n}\zeta_k + \xi_{n+1} \mid \mathscr{F}_n\right]$$

$$= \mathbf{E}\left[\sum_{k=1}^{n}\xi_k \mid \mathscr{F}_n\right] + \mathbf{E}\left[\xi_{n+1} \mid \mathscr{F}_n\right]$$

$$= \sum_{k=1}^{n}\xi_n + \mathbf{E}\left[\zeta_{n+1}\right] = \sum_{k=1}^{n}\xi_n.$$

In particular, the symmetric random walk and the partial sums of Rademacher series, $\sum_{k=1}^{\infty} c_k r_k(t)$, are martingales. Sums of independent random variables $\{\xi_n\}$ such that $\mathbf{E}[\xi_n] \geqslant 0$ or $\mathbf{E}[\zeta_n] \leqslant 0$ are examples of submartingales or supermartingales, respectively.

Moreover, if $\{\xi_n\}$ are independent random variables with $\mathbf{E}[\xi_n] = \mu$, then $\{\sum_{k=1}^{n}\xi_k - n\mu\}$ is also a martingale.

(2) Let $X \in L^1(\Omega)$, if we define $X_n = \mathbf{E}[X \mid \mathscr{F}_n]$, then $\{X_n\}$ is a $\{\mathscr{F}_n\}$-martingale. Condition (i) is obtained immediately because

$$\mathbf{E}[|X_n|] = \mathbf{E}[|\mathbf{E}[X \mid \mathscr{F}_n]|] \leqslant \mathbf{E}[\mathbf{E}[|X| \mid \mathscr{F}_n]] = \mathbf{E}[|X|] < \infty.$$

Condition (ii) is a consequence of the definition of conditional expectation and finally condition (iii) is obtained by the properties of the conditional expectation: if $m > n$,

$$\mathbf{E}[X_m \mid \mathscr{F}_n] = \mathbf{E}[\mathbf{E}[X \mid \mathscr{F}_m] \mid \mathscr{F}_n] = \mathbf{E}[X \mid \mathscr{F}_n] = X_n.$$

This martingales are sometimes called Doob's martingales. As we will see in the development of the theory, many interesting martingales are of this type. In particular, we have the following concrete example:

(3) Let us consider the probability space $([0, 1), \mathscr{B}([0, 1)), m)$ and $\{\chi_N(t)\}$ the Haar functions (indexed with only one parameter). Let $\{\mathscr{F}_N\}$ be the filtration defined as

$$\mathscr{F}_N = \sigma(\{\chi_1, \chi_2, \ldots, \chi_N\}).$$

By the properties of conditional expectation (see Observation 2.1 (ii) of conditional expectation) and the structure of Haar functions we have that, given $f \in L^1([0, 1), \mathscr{B}, m)$

$$\mathbf{E}[f \mid \mathscr{F}_N](t) = \sum_{k=1}^{N} d_k \chi_k(t) \text{ with } d_k = \int_0^1 f(t)\chi_k(t)dt.$$

That is, the partial sums $\left\{\sum_{k=1}^{N} d_k \chi_k(t)\right\}$ are the Haar-Fourier expansion and this is also a $\{\mathscr{F}_N\}$-martingale.

Proposition 3.1 *If* $\{X_n\}$ *is a* $\{\mathscr{F}_n\}$-*martingale* *(*$\{\mathscr{F}_n\}$-*supermartingale,* $\{\mathscr{F}_n\}$-*submartingale) then for any* n,

$$\mathbf{E}\left[X_{n+1}\right] = \mathbf{E}\left[X_n\right] \ \left(\mathbf{E}\left[X_{n+1}\right] \leqslant \mathbf{E}\left[X_n\right], \mathbf{E}\left[X_{n+1}\right] \geqslant \mathbf{E}\left[X_n\right], respectively\right).$$

That is, in the case of a martingale, the sequence of expected values $\{\mathbf{E}\left[X_n\right]\}$ is constant, in the case of a supermartingale it is monotonically decreasing, and in the case of a submartingale it is monotonically increasing.

Proof It is enough to prove condition (3.3). Using the properties of conditional expectation,

$$\mathbf{E}\left[X_{n+1}\right] = \mathbf{E}\left[\mathbf{E}\left[X_{n+1} \mid \mathscr{F}_n\right]\right] = \mathbf{E}\left[X_n\right] \quad (\leqslant, \geqslant \text{ respectively}). \qquad \square$$

As we will see below, a generalization of this result is equivalent to condition (iii) of Definition 3.1.

Proposition 3.2 *If* $\{X_n\}$ *is a* $\{\mathscr{F}_n\}$-*martingale and* ϕ *is a convex function such that for any* n, $\phi(X_n) \in L^1(\Omega)$, *then* $\{\phi(X_n)\}$ *is a* $\{\mathscr{F}_n\}$-*submartingale.*

Proof Conditions (i) and (ii) are immediate by the hypothesis and condition (iii) is obtained by Jensen's inequality for conditional expectation, Proposition 2.1, if $m > n$

$$\mathbf{E}\left[\phi(X_m) \mid \mathscr{F}_n\right] \geqslant \phi\left(\mathbf{E}\left[X_m \mid \mathscr{F}_n\right]\right) = \phi(X_n). \qquad \square$$

As a consequence of this result, we have, in particular, that if $\{X_n\}$ is a $\{\mathscr{F}_n\}$-martingale then $\{|X_n|^p\}$, $1 \leqslant p < \infty$, $\{\exp(X_n)\}$, and $\{\exp(|X_n|)\}$ are $\{\mathscr{F}_n\}$-submartingales.

Definition 3.3 A random process $Y = \{Y_n\}_n$ is a multiplier sequence if $Y = \{Y_n\}$ is $\{\mathscr{F}_n\}$-predictable (i.e., Y_n is $\mathscr{F}_{n-1}$-measurable for each n) and uniformly bounded (i.e., there exists $M > 0$ such that $|Y_n| < M$ for any n).

Usually, we will assume that the multiplier sequence $\{Y_n\}$ is uniformly bounded by one.

Definition 3.4 Let $Y = \{Y_n\}_n$ be a multiplier sequence, then given $X = \{X_n\}_n$ a $\{\mathscr{F}_n\}$-martingale, we define the stochastic processes $(Y \circ X) = \{(Y \circ X)_n\}_n$, which we will call the transform of X by Y, as

$$(Y \circ X)_n = \sum_{k=1}^{n} Y_k \Delta X_k = \sum_{k=1}^{n} Y_n \left(X_k - X_{k-1} \right), \tag{3.10}$$

for any $n \in \mathbb{N}$. Note that $(Y \circ X)_n = (Y \circ X)_{n-1} + Y_n \Delta X_n$, and therefore

$$\Delta(Y \circ X)_n = (Y \circ X)_n - (Y \circ X)_{n-1} = Y_n \Delta X_n.$$

Proposition 3.3 *Let $X = \{X_n\}_n$ be a $\{\mathscr{F}_n\}$-martingale and let $Y = \{Y_n\}_n$ be a multiplier sequence, then the transform of X by Y, $(Y \circ X) = \{(Y \circ X)_n\}_n$ is a $\{\mathscr{F}_n\}$-martingale which is from now on called the martingale transform of X by Y.*

Proof As $\{X_n\}_n$ is a martingale and since $\{Y_n\}_n$ is uniformly bounded, say $|Y_n| \leqslant M$, for each $n \in \mathbb{N}$, then, for any n,

$$\mathbf{E}[|(Y \circ X)_n|] \leqslant \mathbf{E}[\sum_{k=1}^{n} |Y_k||\Delta X_k|] \leqslant M\mathbf{E}[\sum_{k=1}^{n} |\Delta X_k|] \leqslant 2nM\mathbf{E}[|X_n|] < \infty,$$

for any n and $(Y \circ X)_n$ is $\{\mathscr{F}_n\}$-measurable because $\{Y_n\}$ is $\{\mathscr{F}_n\}$-predictable. Finally for $m \geqslant n$

$$
\begin{aligned}
\mathbf{E}\left[(Y \circ X)_m \mid \mathscr{F}_n\right] &= \mathbf{E}\left[\sum_{k=1}^{m} Y_k \Delta X_k \mid \mathscr{F}_n\right] \\
&= \sum_{k=1}^{n} \mathbf{E}\left[Y_k \Delta X_k \mid \mathscr{F}_n\right] + \sum_{k=n+1}^{m} \mathbf{E}\left[Y_k \Delta X_k \mid \mathscr{F}_n\right] \\
&= \sum_{k=1}^{n} Y_k \mathbf{E}\left[(X_k - X_{k-1}) \mid \mathscr{F}_n\right] + \sum_{k=n+1}^{m} \mathbf{E}\left[\mathbf{E}\left[Y_k \Delta X_k \mid \mathscr{F}_{k-1}\right] \mid \mathscr{F}_n\right] \\
&= \sum_{k=1}^{n} Y_k \left(X_k - X_{k-1}\right) + \sum_{k=n+1}^{m} \mathbf{E}\left[Y_k \mathbf{E}\left[\Delta X_k \mid \mathscr{F}_{k-1}\right] \mid \mathscr{F}_n\right] \\
&= \sum_{k=1}^{n} Y_k \Delta X_k = (Y \circ X)_n. \qquad \square
\end{aligned}
$$

Note that the key condition of a multiplier sequence $\{Y_n\}$ is that it is $\{\mathscr{F}_n\}$-predictable. The uniform bound condition on $\{Y_n\}$ is requested to ensure the integrability of the transform and we could ask for weaker conditions, for example, ask that $Y_n \Delta X_n$ to be integrable, for any n. Simple cases of multiplier sequences, but very important ones, are the constants sequences $\{Y_n : Y_n \in [-1, 1]\}$ or even simpler

$\{Y_n : Y_n \in \{-1, 1\}\}$. Additionally, the case $\{Y_n : Y_n \in \{0, 1\}\}$ have a long history and an interesting gambling interpretation see Halmos [93], Doob [69], and some of the earlier work referred to in [93].

Martingale transforms gives us a general method for building new martingales out of old ones and they have become a powerful tool in the interplay between probability and harmonic analysis. More recently, as we will see on Chap. 5, they are an important tool in geometry of Banach spaces. Additionally, martingale transforms are discrete versions of Itô's stochastic integrals, and for more details, see in Chap. 4, the section on discrete Itô's formula.

Remark 3.2 By Proposition 3.3 it is clear that $\mathcal{M}$ is closed under the martingale transform operation by multipliers sequences. We will denote by $\mathcal{M}(X)$ the set of all martingale transforms of a fixed $\{\mathcal{F}_n\}$-martingale $X = \{X_n\}$. Trivially, $\mathcal{M}(X) \subset \mathcal{M}$.

Example 3.2

(1) Let us consider the partial sums of a Walsh expansion , i.e.,

$$W_N = \sum_{m=1}^{N-1} c_m \psi_m(t),$$

where $\{c_m\}$ is a numerical sequence (real or complex) and $\{\psi_m\}$ are Walsh functions, Gundy [86] observed that for any numerical sequence $\{c_m\}$ $\{W_{2^N}\}$ is a martingale, since

$$W_{2^N} = W_{2^{N-1}} + \sum_{m=2^{N-1}}^{2^N-1} c_m \psi_m(t) = W_{2^{N-1}} + \left(\sum_{k=0}^{2^{N-1}-1} c_k \psi_k(t) \right) r_N(t).$$

Then,

$$\mathbf{E}\left[W_{2^N} \mid \sigma\left(\left\{ \psi_k : k \leqslant 2^{N-1} \right\} \right) \right] = \mathbf{E}\left[W_{2^N} \mid \sigma\left(\{ r_k : k \leqslant N-1 \} \right) \right] = W_{2^{N-1}},$$

by the independence of the Rademacher functions. Moreover, note that, from the discussion of Walsh functions in Chap. 1, we can write

$$W_{2^N}(f) = \sum_{n=1}^{N} a_n(t) r_n(t),$$

where $a_n(t) = f_n(r_1(t), r_2(t), \ldots, r_{n-1}(t)) \in \sigma\left(\{ r_k : k \leqslant n-1 \} \right)$, which although not necessarily uniformly bounded, they are such that $a_n(t) r_n(t)$ are integrable on $[0, 1)$ for all n. Therefore the Walsh expansions can be considered as martingale transform of the martingale $\left\{ \sum_{n=0}^{N} r_n(t) \right\}_N$, the partial sums of Rademacher series $\sum_{n=0}^{\infty} r_n(t)$. Then, the fact that they are martingales can be obtained as a immediate consequence of the Proposition 3.3.

(2) Analogously, given $H_N = c_0 + \sum_{n=1}^{N-1} \sum_{k=1}^{2^{n-1}} c_k^{(n)} \chi_k^{(n)}(t)$, where $\{c_k^{(n)}\}$ is a numerical sequence (real o complex) and $\{\chi_k^{(n)}\}$ are the Haar functions, it can be proved that they are martingales, either directly with a similar argument as above or by using Walsh's observation mentioned in Chap. 1 that W_{2^N} coincides with H_{2^N}.

The concept of transform can also be extended to $\{\mathcal{F}_n\}$-submartingales or $\{\mathcal{F}_n\}$-supermartingales. However, in these cases, for the transform to preserve its character it is necessary to require that Y_n is positive for all n.

The martingale transform is a very elementary notion and, as we already said, it is just the discrete version of stochastic integrals for continuous-time martingales. Nevertheless, as we will see in the next chapter, it is a very powerful notion. On the other hand, the martingale transform has an interesting interpretation in game theory. A martingale is the model of a fair game in which one dollar is bet on each move, so the quantity $\sum_{k=1}^{n} \Delta X_k$ represents the net gain until time n. If instead of betting one dollar on each play, we bet Y_n dollars in the n play (where only the information of what happened up to the time $n-1$ is available to decide the net gain up to time $n-1$) then, the net gain with the "strategy" $Y = \{Y_n\}$ is then equal to

$$(Y \circ X)_n = \sum_{k=1}^{n} Y_k \Delta X_k.$$

Thus, from what we have seen above, we have those strategies that cannot change the nature of the game (whether fair or unfair).[1]

The concept of transform has been known in game theory for a long time. However, Donald Burkholder, starting with his 1966 seminal paper [29], systematically studied this concept, opening up a whole new field of research in martingale theory, which will be discussed in the next chapter.

3.2   The Optional Stopping Time Theorem and Its Consequences

Definition 3.5 A stopping times for the filtration $\{\mathcal{F}_n\}$ is a measurable function

$$\tau : \Omega \to \overline{\mathbb{N}} = \mathbb{N} \cup \{\infty\}$$

such that

$$\{\tau = n\} \in \mathcal{F}_n \tag{3.11}$$

---

[1] A similar interpretation can be done for *investment strategies* see for instance, Steele's book [150]

for any n. Clearly, this condition is equivalent to asking that

$$\{\tau \leqslant n\} \in \mathscr{F}_n \tag{3.12}$$

for any n.

Proposition 3.4 *If τ, σ are $\{\mathscr{F}_n\}$-stopping times then $\tau \wedge \sigma, \tau \vee \sigma, \tau + \sigma$ are also stopping times too.*

Proof

$$\{\tau \wedge \sigma \leqslant n\} = \{\tau \leqslant n\} \cup \{\sigma \leqslant n\} \in \mathscr{F}_n$$
$$\{\tau \vee \sigma \leqslant n\} = \{\tau \leqslant n\} \cap \{\sigma \leqslant n\} \in \mathscr{F}_n$$
$$\{\tau + \sigma = n\} = \bigcup_{k=0}^{n}\{\tau = n - k, \sigma = k\} \in \mathscr{F}_n. \qquad \square$$

Note that, in general, $\tau - \sigma$ is not a stopping time since

$$\{\tau - \sigma = n\} = \bigcup_{k=0}^{n}\{\tau = n + k, \sigma = k\} \notin \mathscr{F}_n.$$

Proposition 3.4 has an immediate generalization.

Proposition 3.5 *If $\{\tau_n\}$ is a sequence of $\{\mathscr{F}_n\}$-stopping times then $\sup_n \tau_n, \inf_n \tau_n, \limsup_n \tau_n, \liminf_n \tau_n$ are also $\{\mathscr{F}_n\}$-stopping times.*

Proof Exercise. $\qquad \square$

Definition 3.6 Let τ be a $\{\mathscr{F}_n\}$-stopping time, now we define

$$\mathscr{F}_\tau = \{A \in \mathscr{F} : A \cap \{\tau = n\} \in \mathscr{F}_n\} = \{A \in \mathscr{F} : A \cap \{\tau \leqslant n\} \in \mathscr{F}_n\} \tag{3.13}$$

and and it is called the σ-algebra associated to τ.

We then have the following result:

Proposition 3.6

(i) $\mathscr{F}_\tau$ *is a σ-algebra.*
(ii) *If $\sigma \leqslant \tau$ then $\mathscr{F}_\sigma$ is contained in $\mathscr{F}_\tau$.*
(iii) *τ is $\mathscr{F}_\tau$-measurable.*

Proof

(i) - If $\{A_k\}_k \in \mathscr{F}_\tau$ then

$$\left(\bigcup_{k=1}^{\infty} A_k\right) \cap \{\tau \leqslant n\} = \bigcup_{k=1}^{\infty} (A_k \cap \{\tau \leqslant n\}) \in \mathscr{F}_n.$$

- If $A \in \mathscr{F}_\tau$ then $A \cap \{\tau \leqslant n\} \in \mathscr{F}_n$ and therefore

$$A^c \cap \{\tau \leqslant n\} = \{\tau \leqslant n\} \cap (A \cap \{\tau \leqslant n\})^c \in \mathscr{F}_n.$$

Thus, $\mathscr{F}_\tau$ is a σ-algebra.

(ii) If $\sigma \leqslant \tau$ then $\{\sigma \leqslant n\} \supseteq \{\tau \leqslant n\}$. If $A \in \mathscr{F}_\sigma$, we have, by definition $A \cap \{\sigma \leqslant n\} \in \mathscr{F}_n$, and hence

$$A \cap \{\tau \leqslant n\} = (A \cap \{\sigma \leqslant n\}) \cap \{\tau \leqslant n\} \in \mathscr{F}_n,$$

then, $A \in \mathscr{F}_\tau$.

(iii) τ is $\mathscr{F}_\tau$ measurable

$$\{\tau \leqslant k\} \cap \{\tau \leqslant n\} = \begin{cases} \{\tau \leqslant k\} \in \mathscr{F}_k, & k < n \\ \{\tau \leqslant n\} \in \mathscr{F}_n, & k \geqslant n, \end{cases}$$

therefore $\{\tau \leqslant k\} \cap \{\tau \leqslant n\} \in \mathscr{F}_n$. □

There is an important relationship between stopping times, and the notion of martingale transform , which is clear from the following definition.

Definition 3.7 Let ν, μ be $\{\mathscr{F}_n\}$-stopping times and let $X = \{X_n\}$ be a $\{\mathscr{F}_n\}$-martingale, ($\{\mathscr{F}_n\}$-supermartingale,$\{\mathscr{F}_n\}$-submartingale) then:

- The martingale (supermartingale, submartingale) stopped at ν and denoted by $X^\nu = \{X_n^\nu\}$, is defined, for $n \geqslant 1$, by:

$$X_n^\nu = \sum_{k=1}^{n} \chi_{\{\nu \geqslant k\}} \Delta X_k. \tag{3.14}$$

- The martingale (supermartingale, submartingale) started at μ and denoted by $^\mu X = \{^\mu X_n\}$, is defined, for $n \geqslant 1$, as:

$$^\mu X_n = \sum_{k=1}^{n} \chi_{\{\mu < k\}} \Delta X_k. \tag{3.15}$$

- The martingale (supermartingale, submartingale) started at en μ and stopped at ν and denoted by ${}^{\mu}X^{\nu} = \{{}^{\mu}X_n^{\nu}\}$, is defined, for $n \geq 1$, by

$$
{}^{\mu}X_n^{\nu} = \sum_{k=1}^{n} \chi_{\{\mu < k \leq \nu\}} \Delta X_k. \tag{3.16}
$$

Note that $\{\chi_{\{\nu \geq k\}}\}_k$, $\{\chi_{\{\mu < k\}}\}_k$ and $\{\chi_{\{\mu < k \leq \nu\}}\}_k$ are positive multiplier sequences, hence if $X = \{X_n\}$ is a $\{\mathscr{F}_n\}$-martingale, by Proposition 3.3, $X^{\nu} = \{X_n^{\nu}\}$, ${}^{\mu}X = \{{}^{\mu}X_n\}$, and ${}^{\mu}X^{\nu} = \{{}^{\mu}X_n^{\nu}\}$, are actually martingales. Therefore, the notion of martingale transforms includes those of stopped martingale, started martingale, and started and stopped martingale.

In addition, note that $X_n^{\nu} = X_{\nu \wedge n}$, because if $\omega \in \{\nu < n\}$,

$$
X_n^{\nu}(\omega) = \sum_{k=1}^{\nu(\omega)} \Delta X_k(\omega) = X_{\nu(\omega)}(\omega)
$$

and if $\omega \in \{\nu \geq n\}$,

$$
X_n^{\nu}(\omega) = \sum_{k=1}^{n} \Delta X_k(\omega) = X_n(\omega).
$$

Analogously, we have ${}^{\mu}X = X_{\mu \vee n}$ and ${}^{\mu}X^{\nu} = X_{(\mu \vee n) \wedge \nu}$.

Proposition 3.7 *Let μ and ν be two $\{\mathscr{F}_n\}$-stopping times and let $X = \{X_n\}_n$ be a $\{\mathscr{F}_n\}$-martingale, then the following relationships are verified*

(i) $X = X^{\nu} + {}^{\nu}X$.
(ii) $(X^{\mu})^{\nu} = X^{\mu \wedge \nu}$.
(iii) ${}^{\mu}({}^{\nu}X) = {}^{\mu \wedge \nu}X$.
(iv) ${}^{\mu}X^{\nu} = {}^{\mu}X - {}^{\nu}X = X^{\nu} - X^{\mu}$.
(v) $X^{\mu} - X^{\nu} = {}^{\nu}X^{\mu} - {}^{\mu}X^{\nu}$.

Proof

(i) In order to prove that $X_n = X_n^{\nu} + {}^{\nu}X_n$ note that

$$
X_n = \sum_{k=1}^{n} \Delta X_k = \sum_{k=1}^{n} \left[\chi_{\{\nu \geq k\}} + \chi_{\{\nu < k\}} \right] \Delta X_k
$$

$$
= \sum_{k=1}^{n} \chi_{\{\nu \geq k\}} \Delta X_k + \sum_{k=1}^{n} \chi_{\{\nu < k\}} \Delta X_k = X_n^{\nu} + {}^{\nu}X_n.
$$

(ii) Let us now prove that $\left(X_n^\mu\right)^\nu = X_n^{\mu\wedge\nu}$ we have $X_n^\mu = \sum_{k=1}^n \chi_{\{\mu\geqslant k\}}\Delta X_k$, that is $(Y\circ X)_n = X_n^\mu$ where $Y_j = \chi_{\{\mu\geqslant j\}}$.

$$(Y\circ X)_n^\nu = \sum_{k=1}^n \chi_{\{\nu\geqslant k\}}\Delta(Y\circ X)_k, \quad \text{where } \Delta\,(Y\circ X)_k = (Y\circ X)_k^\mu - (Y\circ X)_{k-1}^\mu$$

$$= \sum_{k=1}^n \chi_{\{\nu\geqslant k\}}\chi_{\{\mu\geqslant k\}}\Delta X_k = \sum_{k=1}^n \chi_{\{\mu\wedge\nu\geqslant k\}}\Delta X_k = X_n^{\mu\wedge\nu},$$

since $(Y\circ X)_n^\nu = \left(X_n^\mu\right)^\nu$ we have $(X^\mu)^\nu = X^{\mu\wedge\nu}$.

(iii) This is proved in the same way as (ii).

(iv) By definition,

$$^\nu X^\mu = \sum_{k=1}^n \chi_{\{\nu<k\leqslant\mu\}}\Delta X_k = \sum_{k=1}^n \chi_{\{\nu<k\}}\Delta X_k - \sum_{k=1}^n \chi_{\{\mu<k\}}\Delta X_k = {}^\nu X - {}^\mu X.$$

Similarly,

$$^\nu X^\mu = \sum_{k=1}^n \chi_{\{\nu<k\leqslant\mu\}}\Delta X_k = \sum_{k=1}^n \chi_{\{k\leqslant\mu\}}\Delta X_k - \sum_{k=1}^n \chi_{\{k\leqslant\nu\}}\Delta X_k = X^\mu - X^\nu.$$

(v) By definition, we have that

$$^\nu X_n^\mu - {}^\mu X_n^\nu = \sum_{k=1}^n \chi_{\{\nu<k\leqslant\mu\}}\Delta X_k - \sum_{k=1}^n \chi_{\{\mu<k\leqslant\nu\}}\Delta X_k.$$

If $\mu > \nu$ then so $\chi_{\{\mu<k\leqslant\nu\}} = 0$ and in this case

$$^\nu X_n^\mu - {}^\mu X_n^\nu = \sum_{k=1}^n \chi_{\{\nu<k\leqslant\mu\}}\Delta X_k = \sum_{k=1}^n \left[\chi_{\{\mu\geqslant k\}} - \chi_{\{\nu\geqslant k\}}\right]\Delta X_k = X_n^\mu - X_n^\nu.$$

If $\nu > \mu$, then $\chi_{\{\nu<k\leqslant\mu\}} = 0$ and we obtain

$$^\nu X_n^\mu - {}^\mu X_n^\nu = -\sum_{k=1}^n \chi_{\{\mu<k\leqslant\nu\}}\Delta X_k = \sum_{k=1}^n \left[\chi_{\{\mu\geqslant k\}} - \chi_{\{\nu\geqslant k\}}\right]\Delta X_k = X_n^\mu - X_n^\nu. \qquad \Box$$

Remark 3.3 From the previous discussion, it is clear that the notion of martingale transform generalizes the notion of stopped martingale. Additionally, by Remark 3.2 $\mathcal{M}$ and $\mathcal{M}(X)$ are closed under the operations of stopping, starting, and starting and stopping martingales.

There are several possibilities to generalize the martingale concept. A very useful generalization is the following:

Definition 3.8 A stochastic processes $X = \{X_n\}_n$ is a $\{\mathscr{F}_n\}$ local martingale ($\{\mathscr{F}_n\}$ local supermartingale, $\{\mathscr{F}_n\}$ local submartingale) if there is a sequence $\{\tau_n\}$ of $\{\mathscr{F}_n\}$ stopping times such that $\tau_n \leqslant \tau_{n+1}$, a.e., $\tau_n \uparrow \infty$ as $n \to \infty$ and at each stopped sequence $\left\{ X_{\tau_k \wedge n} \chi_{\{\tau_k > 0\}} \right\}_n$ is a $\{\mathscr{F}_n\}$-martingale ($\{\mathscr{F}_n\}$-supermartingale, $\{\mathscr{F}_n\}$-submartingale).

It can be proved (see Shiryayev [148], Chap. VII) that the class of local martingales coincides with the class of generalized martingales defined there. Moreover, each local martingale can be obtained as the transform of a martingale by a process that is $\{\mathscr{F}_n\}$-predictable.

Let us now discuss an important result regarding stopping times and martingales.

Theorem 3.1 (Doob's optional stopping time) *Let $X = \{X_n\}_n$ be a $\{\mathscr{F}_n\}$-martingale ($\{\mathscr{F}_n\}$-supermartingale, $\{\mathscr{F}_n\}$-submartingale) and let σ and τ be a.e. bounded $\{\mathscr{F}_n\}$-stopping times (that is to say there is $M \in \mathbb{R}$ such that $\sigma(\omega)$ and $\tau(\omega) \leqslant M$ a.e. $\omega \in \Omega$) and such that $\sigma(\omega) \leqslant \tau(\omega)$ for a.e. $\omega \in \Omega$. Then*

$$\mathbf{E}[X_\tau \mid \mathscr{F}_\sigma] = X_\sigma (\leqslant, \geqslant \text{ respectively) a.e.} \tag{3.17}$$

In particular,

$$\mathbf{E}[X_\tau] = \mathbf{E}[X_\sigma] \quad (\leqslant, \geqslant \text{ respectively).} \tag{3.18}$$

Proof First, we will prove (3.18). Let

$$Y_n = \chi_{\{\sigma < n \leqslant \tau\}} = \chi_{\{n \leqslant \tau\}} - \chi_{\{n \leqslant \sigma\}}, \quad n = 1, 2, \cdots$$

Then $Y = \{Y_n\}$ is a multiplier sequence and we have already seen that the transform $(Y \circ X)_n$ is equal to $X_{\{\tau \wedge n\}} - X_{\{\sigma \wedge n\}}$. In particular,

$$(Y \circ X)^\tau_M = X_{\tau \wedge M} - X_{\sigma \wedge M} = X_\tau - X_\sigma$$

so, since $(Y \cdot X)^\tau_n$ is a $\{\mathscr{F}_n\}$-martingale ($\{\mathscr{F}_n\}$-supermartingale, $\{\mathscr{F}_n\}$-submartingale), using the fact that

$$\mathbf{E}[(Y \circ X)_1] = \mathbf{E}[X_{\tau \wedge 1}] - \mathbf{E}[X_{\sigma \wedge 1}] = \mathbf{E}[X_1] - \mathbf{E}[X_1] = 0.$$

we have

$$\mathbf{E}[X_\tau] - \mathbf{E}[X_\sigma] = \mathbf{E}[(Y \circ X)] = \mathbf{E}[(Y \circ X)_1] = 0 \quad (\leqslant, \geqslant \text{ respectively).}$$

Therefore,

$$\mathbf{E}\left[X_\tau\right] = \mathbf{E}\left[X_\sigma\right] \quad (\leqslant, \geq \text{ respectively}).$$

In order to verify (3.17) it suffices to prove that if $A \in \mathscr{F}_\sigma$ then

$$\int_A X_\tau d\mathbf{P} = \int_A X_\sigma d\mathbf{P}. \quad (\leqslant, \geqslant \text{ respectively}).$$

Let

$$\sigma_A(\omega) = \begin{cases} \sigma(\omega) & \omega \in A \\ M & \omega \in A^c \end{cases} \quad \text{and} \quad \tau_A(\omega) = \begin{cases} \tau(\omega) & \omega \in A \\ M & \omega \in A^c \end{cases}.$$

Then σ_A and τ_A are $\{\mathscr{F}_n\}$-bounded stopping times, let us see that σ_A is a stopping time, the case of τ_A is analogous.

$$\{\sigma_A \leqslant n\} = \begin{cases} \Omega & \text{if } n \geqslant M \\ \{\sigma \leqslant n\} \cap A \in \mathscr{F}_n & \text{if } n < M \end{cases}$$

and then $\{\sigma_A \leqslant n\} \in \mathscr{F}_n$. Analogously for τ_A. Therefore, from what has already been proven

$$\mathbf{E}\left[X_{\sigma_A}\right] = \mathbf{E}\left[X_{\tau_A}\right] \quad (\leqslant, \geqslant \text{ respectively}),$$

but this means that

$$\int_A X_\tau d\mathbf{P} + \int_{A^c} X_M d\mathbf{P} = \int_A X_\sigma d\mathbf{P} + \int_{A^c} X_M d\mathbf{P} (\leqslant, \geqslant \text{ respectively})$$

and thus

$$\int_A X_\tau d\mathbf{P} = \int_A X_\sigma d\mathbf{P}. \qquad \qquad \square$$

The conditions on stopping times can be weakened by requiring that they be satisfied a.e. (see Corollary 3.3). Moreover, this theorem is a weak version of the following theorem. In the more general version, instead of requiring that the stopping times be bounded, a decay condition of the process with respect to the stopping times is required:

Theorem 3.2 (Doob) *Let $X = \{X_n\}$ be a $\{\mathscr{F}_n\}$-martingale ($\{\mathscr{F}_n\}$-supermartingale, $\{\mathscr{F}_n\}$-submartingale) and τ_1 and τ_2 be two $\{\mathscr{F}_n\}$-stopping times for which*

$$\mathbf{E}\left[|X_{\tau_k}|\right] < \infty, \ k = 1, 2, \tag{3.19}$$

$$\varliminf_{n \to \infty} \int_{\{\tau_m \geqslant n\}} |X_n| \, d\mathbf{P} = 0 \ \ m = 1, 2. \tag{3.20}$$

Then,

$$\mathbf{E}\left[X_{\tau_2} \mid \mathscr{F}_{\tau_1}\right] = X_{\tau_1} \text{ in } \{\tau_2 \geqslant \tau_1\} \text{ a.e. } (\leqslant, \geqslant \text{ respectively}). \tag{3.21}$$

If in addition, $\mathbf{P}\{\tau_1 \geqslant \tau_2\} = 1$, *then*

$$\mathbf{E}\left[X_{\tau_2}\right] = \mathbf{E}\left[X_{\tau_1}\right] \quad (\leqslant, \geqslant \text{ respectively}). \tag{3.22}$$

Proof To prove (3.21) it suffices to show that for each $A \in \mathscr{F}_{\tau_1}$

$$\int_{A \cap \{\tau_2 \geqslant \tau_1\}} X_{\tau_2} d\mathbf{P} = \int_{A \cap \{\tau_2 \geqslant \tau_1\}} X_{\tau_1} d\mathbf{P} \quad (\leqslant, \geqslant \text{ repectively})$$

and for that purpose, it is sufficient to prove that, for every $n \geq 0$

$$\int_{A \cap \{\tau_2 \geqslant \tau_1\} \cap \{\tau_1 = n\}} X_{\tau_2} d\mathbf{P} = \int_{A \cap \{\tau_2 \geqslant \tau_1\} \cap \{\tau_1 = n\}} X_{\tau_1} d\mathbf{P} \quad (\leqslant, \geqslant \text{ repectively}),$$

i.e., to prove that

$$\int_{B \cap \{\tau_2 \geqslant n\}} X_{\tau_2} d\mathbf{P} = \int_{B \cap \{\tau_2 \geqslant n\}} X_n d\mathbf{P} \quad (\leqslant, \geqslant \text{ repectively}),$$

where $B = A \cap \{\tau_1 = n\} \in \mathscr{F}_n$. We have,

$$\int_{B \cap \{\tau_2 \geqslant n\}} X_n d\mathbf{P} = \int_{B \cap \{\tau_2 = n\}} X_n d\mathbf{P} + \int_{B \cap \{\tau_2 > n\}} X_n d\mathbf{P}$$

$$= \int_{B \cap \{\tau_2 = n\}} X_n d\mathbf{P} + \int_{B \cap \{\tau_2 > n\}} E\left[X_{n+1} \mid F_n\right] d\mathbf{P}$$

$$= \int_{B \cap \{\tau_2 = n\}} X_n d\mathbf{P} + \int_{B \cap \{\tau_2 > n\}} X_{n+1} d\mathbf{P} \quad (\geqslant, \leqslant \text{ respectively}).$$

Writing $\{\tau_2 > n\} = \{\tau_2 = n + 1\} \cup \{\tau_2 > n + 1\}$ and observing that $\{\tau_2 > n + 1\} \in \mathscr{F}_{n+1}$, we obtain

$$= \int_{B \cap \{n \leqslant \tau_2 \leqslant n+1\}} X_{\tau_2} d\mathbf{P} + \int_{B \cap \{\tau_2 > n+1\}} X_{n+2} d\mathbf{P} \quad (\geqslant, \leqslant \text{ respectively}).$$

Iterating this argument $m - n$ times, we have that:

$$= \int_{B \cap \{n \leqslant \tau_2 \leqslant m\}} X_{\tau_2} d\mathbf{P} + \int_{B \cap \{\tau_2 > m\}} X_m d\mathbf{P} \quad (\geqslant, \leqslant \text{ respectively}),$$

then

$$= \int_{B \cap \{n \leqslant \tau_2 \leqslant m\}} X_{\tau_2} d\mathbf{P} = \int_{B \cap \{\tau_2 \geqslant n\}} X_n d\mathbf{P} - \int_{B \cap \{m > \tau_2\}} X_m d\mathbf{P} \quad (\geqslant, \leqslant \text{ respectively}).$$

Then since $X_m = 2X_m^+ - |X_m|$ we have, by (3.20)

$$\int_{B \cap \{\tau_2 \geqslant n\}} X_{\tau_2} d\mathbf{P} = \overline{\lim}_{m \to \infty} \left[\int_{B \cap \{\tau_2 \geqslant n\}} X_n d\mathbf{P} - \int_{B \cap \{m > \tau_2\}} X_m d\mathbf{P} \right]$$

$$= \int_{B \cap \{\tau_2 \geqslant n\}} X_n d\mathbf{P} - \lim_{m \to \infty} \int_{B \cap \{m > \tau_2\}} X_m d\mathbf{P}$$

$$= \int_{B \cap \{\tau_2 \geqslant n\}} X_n d\mathbf{P} \quad (\leqslant, \geqslant \text{ repectively}).$$

Then (3.21) has been proved and then (3.22) is immediate. $\qquad\qquad\square$

The version with weaker conditions of Theorem 3.1 can now be obtained as a corollary of the previous result:

Corollary 3.3 *Let* $X = \{X_n\}$ *be a* $\{\mathscr{F}_n\}$-*martingale* ($\{\mathscr{F}_n\}$-*supermartingale,* $\{\mathscr{F}_n\}$-*submartingale) and* τ_1 *and* τ_2 *be two* $\{\mathscr{F}_n\}$-*stopping times for which there is a constant* $M > 0$ *such that* $\mathbf{P}\{\tau_1 \leqslant M\} = \mathbf{P}\{\tau_2 \leqslant M\} = 1$, *then (3.19) y (3.20) are satisfied and therefore the conclusions of Theorem 3.2 are verified.*

Theorem 3.2 can be extended to stopping times sequences, as the following result, due to Doob states.

Corollary 3.4 (Optional stopping times) *Let* $X = \{X_n\}$ *be a* $\{\mathscr{F}_n\}$-*martingale* ($\{\mathscr{F}_n\}$-*supermartingale,* $\{\mathscr{F}_n\}$-*submartingale) and let* $\{\tau_m\}$ *be an increasing sequence of* $\{\mathscr{F}_n\}$-*stopping times that verifies (3.19) and (3.20) then the process* $\{X_{\tau_n}\}_n$ *is a* $\{\mathscr{F}_{\tau_n}\}$-*martingale* ($\{\mathscr{F}_{\tau_n}\}$-*supermartingale,* $\{\mathscr{F}_{\tau_n}\}$-*submartingale, repectively).*

Observe that, in particular by the previous corollary, any $X = \{X_n\}$ be a $\{\mathscr{F}_n\}$-martingale ($\{\mathscr{F}_n\}$- supermartingale, $\{\mathscr{F}_n\}$-submartingale) that verifies (3.19) and (3.20) is then a $\{\mathscr{F}_n\}$ local martingale ($\{\mathscr{F}_n\}$ local supermartingale, $\{\mathscr{F}_n\}$ local submartingale).

An interesting result, suggested by the proof of Theorem 3.1, is the following characterization of the $\{\mathscr{F}_n\}$-martingale ($\{\mathscr{F}_n\}$-supermartingale, $\{\mathscr{F}_n\}$-submartingale):

Corollary 3.5 *Let* $X = \{X_n\}$ *be a processes satisfying:*

(i) X_n *in integrable for each n.*
(ii) X_n *is* $\{\mathscr{F}_n\}$-*adapted.*

Then, X *is a* $\{\mathscr{F}_n\}$-*martingale* ($\{\mathscr{F}_n\}$-*supermartingale,* $\{\mathscr{F}_n\}$-*submartingale) if and only if for each couple of bounded stopping times* τ, σ *such that* $\sigma \leqslant \tau$

$$\mathbf{E}[X_\tau] = \mathbf{E}[X_\sigma] \quad (\leqslant, \geqslant \text{ repectively}).$$

Proof The direct implication is immediate by Theorem 3.1. Now, for the converse, let $m > n$. It is sufficient to define for $A \in \mathscr{F}_n$

$$\sigma_A(\omega) = \begin{cases} n & \text{si } \omega \in A \\ m & \text{si } \omega \in A^c \end{cases} \quad \text{and} \quad \tau_A(\omega) = m$$

for all $\omega \in \Omega$. Clearly τ_A is a $\{\mathscr{F}_n\}$-stopping time since it is constant,, and σ_A also a stopping time because

$$\{\sigma_A \leqslant k\} = \begin{cases} \Omega, & \text{if } k \geqslant m \\ A, & \text{if } n \leqslant k \leqslant m \\ \emptyset, & \text{if } k < n \end{cases},$$

and this implies that $\{\sigma_A \leqslant k\} \in \mathscr{F}_k$. Then, by hypothesis, given that τ_A and σ_A are bounded and $\sigma_A \leqslant \tau_A$, we have

$$\int_A X_{\tau_A} d\mathbf{P} + \int_{A^C} X_{\tau_A} d\mathbf{P} = \int_A X_{\sigma_A} d\mathbf{P} + \int_{A^C} X_{\sigma_A} d\mathbf{P} \quad (\leqslant, \geqslant \text{ respectively})$$

and by the definition of the times τ_A and σ_A

$$\int_A X_m d\mathbf{P} + \int_{A^C} X_m d\mathbf{P} = \int_A X_n dP + \int_{A^C} X_m d\mathbf{P} \quad (\leqslant, \geqslant \text{ respectively})$$

but this implies

$$\int_A X_m d\mathbf{P} = \int_A X_n d\mathbf{P}, \quad (\leqslant, \geqslant \text{ respectively})$$

that is to say

$$\mathbf{E}[X_m \mid \mathscr{F}_n] = X_n \text{ a.e.} \quad (\leqslant, \geqslant \text{ repectively}). \qquad \square$$

Recall that, by the classical theory of integration, given a random variable X, if it is integrable, i.e., $\mathbf{E}[|X|] < \infty$, then by the convergence dominated theorem

$$\lim_{M \to \infty} \int_{\{|X|>M\}} |X| d\mathbf{P} = 0,$$

or equivalently, for any $\varepsilon > 0$ there exists $\delta > 0$ such that if $\mathbf{P}(A) < \delta$ then

$$\int_A |X| d\mathbf{P} < \varepsilon. \tag{3.23}$$

This property is known as the *absolute continuity* property of the expected value. We will generalize this notion to a set of random variables:

Definition 3.9 A collection of random variables $\{X_n\}$ is said to be uniformly integrable if

$$\lim_{M \to \infty} \left[\sup_n \left(\int_{\{|X_n| > M\}} |X_n| \, d\mathbf{P} \right) \right] = 0.$$

We have an equivalent characterization of the uniform integrability.

Proposition 3.8 $\{X_n\}_n$ *is uniformly integrable if and only if*

(i) $\sup_n \mathbf{E}\left[|X_n|\right] < \infty$.
(ii) For any $\varepsilon > 0$ there exists $\delta > 0$ such that if $\mathbf{P}(A) < \delta$ then $\int_A |X_n| \, dP < \varepsilon$ for all n.

Proof Let us prove that uniform integrability implies conditions (i) and (ii). By hypothesis, given $\varepsilon > 0$ there exits $M > 0$ such that

$$\sup_n \left(\int_{\{|X_n| > M\}} |X_n| \, d\mathbf{P} \right) < \varepsilon.$$

Therefore

$$\mathbf{E}\left[|X_n|\right] = \int_{\{|X_n| \leqslant M\}} |X_n| \, d\mathbf{P} + \int_{\{|X_n| > M\}} |X_n| \, d\mathbf{P}$$
$$\leqslant M\mathbf{P}\{|X_n| \leqslant M\} + \varepsilon \leqslant M + \varepsilon,$$

for all n and therefore we have (i).

On the other hand, if $k \geqslant M$

$$\int_A |X_n| \, d\mathbf{P} = \int_{A \cap \{|X_n| \leqslant k\}} |X_n| \, d\mathbf{P} + \int_{A \cap \{|X_n| > k\}} |X_n| \, d\mathbf{P}$$
$$\leqslant k\mathbf{P}(A) + \int_{\{|X_n| > k\}} |X_n| \, d\mathbf{P} \leqslant k\mathbf{P}(A) + \varepsilon \leqslant 2\varepsilon$$

if $\mathbf{P}(A) < \delta$ and $\delta \leqslant k/\varepsilon$.

Now, we prove that conditions (i) and (ii) imply the uniform integrability of $\{X_n\}$.

$\mathbf{E}\left[|X_n|\right] \leqslant \sup_n \mathbf{E}\left[|X_n|\right] < \infty$, thus $X_n < \infty$ a.e. and therefore $\mathbf{P}\{|X_n| > M\} \to 0$ as $M \to \infty$. Then, by (ii) we can conclude , for any n

$$\int_{\{|X_n| > M\}} |X_n| \, \mathbf{dP} < \varepsilon,$$

if M is big enough. $\qquad\qquad\square$

Proposition 3.9 *Let $X \in L^1(\Omega)$ if we define $X_n = \mathbf{E}[X \mid \mathscr{F}_n]$, then the $\{\mathscr{F}_n\}$-martingale $\{X_n\}$ is uniformly integrable.*

Proof By the absolute continuity of the expected value (3.23), given $\varepsilon > 0$ consider the corresponding $\delta > 0$ and take M big enough such that $\frac{\mathbf{E}\|X\|}{M} \leqslant \delta$. Then, by Jensen's inequality and the definition of conditional expectation, we have

$$\int_{\{|\mathbf{E}[X|\mathscr{F}_n]|>M\}} |X_n|\, d\mathbf{P} = \int_{\{|\mathbf{E}[X|\mathscr{F}_n]|>M\}} |\mathbf{E}[X \mid \mathscr{F}_n]|\, d\mathbf{P} \leqslant \int_{\{|\mathbf{E}[X|\mathscr{F}_n]|>M\}} \mathbf{E}[|X| \mid \mathscr{F}_n]\, d\mathbf{P}$$

$$= \int_{\{|\mathbf{E}[X|\mathscr{F}_n]|>M\}} |X|\, d\mathbf{P},$$

and since, by Tchebychev's inequality, we have:

$$\mathbf{P}\{|\mathbf{E}[X \mid \mathscr{F}_n]| > M\} \leqslant \frac{\mathbf{E}[|\mathbf{E}[X \mid \mathscr{F}_n]]|}{M} \leqslant \frac{\mathbf{E}[\mathbf{E}[|X| \mid \mathscr{F}_n]]}{M} = \frac{\mathbf{E}[|X|]}{M} < \delta.$$

Therefore, for this M, it turns out then

$$\sup_n \left(\int_{\{|\mathbf{E}[X|\mathscr{F}_n]|>M\}} |X_n|\, d\mathbf{P} \right) < \varepsilon. \qquad \square$$

Definition 3.10 Let $\{X_n\}$ be a $\{\mathscr{F}_n\}$-martingale. $\{X_n\}$ is called a last-term $\{\mathscr{F}_n\}$-martingale, if there exists a random variable $X \in L^1(\Omega)$, such that

$$X_n = \mathbf{E}[X \mid \mathscr{F}_n].$$

Observe that by Proposition 3.9 a $\{X_n\}_n$ last-term $\{\mathscr{F}_n\}$-martingale is uniformly integrable.

A result more general than the one given in the previous proposition is the following:

Lemma 3.1 (First Lemma of Uniform Integrability) *Let $\{X_n\}_{n=1}^N$ be a non-negative and finite $\{\mathscr{F}_n\}$-submartingale then it is uniformly integrable.*

Proof Note that, by definition of conditional expectation, for every $A \in \mathscr{F}_n$

$$\int_A X_N\, d\mathbf{P} = \int_A \mathbf{E}[X_N \mid \mathscr{F}_n]\, d\mathbf{P} \geqslant \int_A X_n\, d\mathbf{P}$$

then, taking $A = \{X_n > M\} \in \mathscr{F}_n$,

$$\int_{\{X_n>M\}} X_N\, d\mathbf{P} \geqslant \int_{\{X_n>M\}} X_n\, d\mathbf{P}$$

and, by Tchebychev's inequality, we have

$$\mathbf{P}\{X_n > M\} \leqslant \frac{1}{M} \int_\Omega X_n d\mathbf{P} \leqslant \frac{1}{M} \int_\Omega X_N d\mathbf{P} \to 0 \text{ if } M \to \infty$$

i.e., $\mathbf{P}\{X_n > M\} \to 0$ uniformly in n when $M \to \infty$ and then

$$\lim_{M \to \infty} \left(\sup_n \int_{\{|X_n| > M\}} |X_n| \, d\mathbf{P} \right) = 0. \qquad \square$$

A finite martingale $\{X_n\}_{n=1}^N$ has a last term and is, in short, a particular case of martingales generated by an integrable random variable, since for all $n \leqslant N$, $X_n = \mathbf{E}[X_N \mid \mathscr{F}_n]$ and therefore, Lemma 3.1 is indeed an immediate consequence of Proposition 3.9.

Corollary 3.6 *Let $\{X_n\}_n$ be a $\{\mathscr{F}_n\}$-martingale ($\{\mathscr{F}_n\}$-supermartingale, $\{\mathscr{F}_n\}$-submartingale), if it is also uniformly integrable, then (3.19) and (3.20) are satisfied and therefore the conclusions of Theorem 3.2 are verified for any pair of $\{\mathscr{F}_n\}$-stopping times τ_1 y τ_2.*

Proof This is left as an exercise. $\qquad \square$

3.3   Fundamental Inequalities

We are going to study now, the main inequalities for martingales, for this we need to define the following operators on the set of martingales $\mathscr{M}$.

Definition 3.11 Let $\{X_n\}$ be a $\{\mathscr{F}_n\}$-martingale, let us define X_n^* as the maximal function of order n

$$X_n^* = \max_{0 \leqslant k \leqslant n} |X_k|$$

and define X^* as

$$X^* = \sup_k |X_k| \tag{3.24}$$

which is called the maximal function of de X.

Let us consider the L^p-norm of these operators. First we need the following definition.

Definition 3.12 For any $0 < p < \infty$ and for any $\{X_n\}$ a $\{\mathscr{F}_n\}$-martingale, we define

$$\|X_n\|_p = \left(\mathbf{E}\left[|X_n|^p\right]\right)^{1/p} \tag{3.25}$$

for all n and then

$$\|X\|_p = \sup_n \|X_n\|_p. \tag{3.26}$$

We say that $X = \{X_n\}_n$ is L^p bounded if $\|X\|_p < \infty$.

Theorem 3.7 (Doob's maximal inequality)[2]

(i) Let $X = \{X_n\}_n$ be a non-negative $\{\mathscr{F}_n\}$-submartingale. Then, for any $\lambda > 0$ and for all n

$$\lambda \mathbf{P}\left\{X_n^* \geqslant \lambda\right\} \leqslant \int_{\{X_n \geqslant \lambda\}} X_n d\mathbf{P} \leqslant \mathbf{E}[X_n] = \|X_n\|_1. \tag{3.27}$$

(ii) Let $X = \{X_n\}_n$ be a non-negative $\{\mathscr{F}_n\}$-supermartingale. Then, for any $\lambda > 0$ and for all n

$$\lambda \mathbf{P}\left\{X_n^* \geqslant \lambda\right\} \leqslant \int_{\{X_n \geqslant \lambda\}} X_0 d\mathbf{P} \leqslant \mathbf{E}[X_0] = \|X_0\|_1. \tag{3.28}$$

Proof It is enough to prove the first inequality (3.27). Let us define

$$\sigma = \begin{cases} \min\{k \leqslant n, X_k \geqslant \lambda\} & \text{if } \{k \leqslant n, X_k \geqslant \lambda\} \neq \emptyset \\ n & \text{if } \{k \leqslant n, X_k \geqslant \lambda\} = \emptyset. \end{cases}$$

Clearly, σ is a $\{\mathscr{F}_n\}$-stopping times, since

$$\{\sigma = k\} = \{X_1 < \lambda, \cdots, X_{k-1} < \lambda, X_k \geqslant \lambda\} \in \mathscr{F}_k.$$

Furthermore, then by definition $\sigma \leq n$ i.e., $\{\sigma \leqslant n\} = \left\{X_n^* \geqslant n\right\}$, and since $\{X_n\}_n$ is a $\{\mathscr{F}_n\}$-submartingale

$$\mathbf{E}[X_n] \geqslant \mathbf{E}[X_\sigma] = \int_{\{X_n^* \geqslant \lambda\}} X_\sigma d\mathbf{P} + \int_{\{X_n^* < \lambda\}} X_\sigma d\mathbf{P}$$

$$\geqslant \lambda \mathbf{P}\left\{X_n^* \geqslant \lambda\right\} + \int_{\{X_n^* < \lambda\}} X_n d\mathbf{P}$$

since $X_\sigma = n$ on $\left\{X_n^* < \lambda\right\} \subset \{X_n < \lambda\}$, and then

$$\lambda \mathbf{P}\left\{X_n^* \geqslant \lambda\right\} \leqslant \mathbf{E}[X_n] - \int_{\{X_n^* < \lambda\}} X_n d\mathbf{P}$$

$$\leqslant \mathbf{E}[X_n] - \int_{\{X_n < \lambda\}} X_n d\mathbf{P} = \int_{\{X_n \geqslant \lambda\}} X_n d\mathbf{P}.$$

---

[2] For other versions of Doob's maximal inequalities see Shiryayev [148, Chapter 7 §3 Theorem 1.]

The proof of (3.28) is similar but using the following stopping time,

$$\tau = \begin{cases} \min\{k \leqslant n, X_k \geqslant \lambda\} & \text{if } \{k \leqslant n, X_k \geqslant \lambda\} \neq \emptyset \\ 0 & \text{if } \{k \leqslant n, X_k \geqslant \lambda\} = \emptyset. \end{cases}$$

Additionally, we can prove directly the second inequality of (3.27). By Theorem 3.1 since $X = \{X_n\}_n$ is a $\{\mathscr{F}_n\}$-submartingale then $X = \{X_{n \wedge \sigma}\}_n$ is also a $\{\mathscr{F}_n\}$-submartingale. In particular, $\mathbf{E}[X_{n \wedge \sigma}] \leqslant \mathbf{E}[X_n]$.

On the set $\{\sigma \leqslant n\}$ we have $X_\sigma \geqslant \lambda$ and therefore, $X_{n \wedge \sigma} \geqslant \lambda \chi_{\{\sigma \leqslant n\}}$ then, taking expected value and using the previous inequality, we get

$$\mathbf{E}[X_n] \geqslant \mathbf{E}[X_{n \wedge \sigma}] \geqslant \lambda \mathbf{P}\{\sigma \leqslant n\} = \lambda \mathbf{P}\{X_n^* \geqslant n\}.$$

The proof of the second inequality of (3.28) is entirely similar to the first part of the proof, using the same stopping time σ, the only change needed is that in that case we have $\mathbf{E}[X_{n \wedge \sigma}] \leqslant \mathbf{E}[X_0]$. $\qquad\square$

Note that if X is L^1-bounded then by the maximal inequality we have

$$P\{X^* \geqslant \lambda\} \leqslant \frac{1}{\lambda} \|X\|_1. \tag{3.29}$$

Theorem 3.8 (Minimal Inequality) *Let $X = \{X_n\}$ be a $\{\mathscr{F}_n\}$-submartingale. Then, for each N and for $\lambda > 0$*

$$\lambda \mathbf{P}\left\{\left(\min_{1 \leqslant n \leqslant N} X_n\right) \leqslant -\lambda\right\} \leqslant -\mathbf{E}[X_1] + \int_{\{(\min_{1 \leqslant n \leqslant N} X_n) > -\lambda\}} X_n d\mathbf{P}.$$

Proof This inequality is obtained by considering the stopping times

$$\tau = \begin{cases} \min\{k \leqslant n, X_k \leqslant -\lambda\} & \text{if } \{k \leqslant n, X_k \leqslant -\lambda\} \neq \emptyset \\ n & \text{if } \{k \leqslant n, X_k \leqslant -\lambda\} = \emptyset. \end{cases}$$

Using analogous argument as in Theorem 3.7, since $\mathbf{E}[X_1] \leqslant \mathbf{E}[X_\tau]$, we obtain

$$\mathbf{E}[X_1] \leqslant \mathbf{E}[X_\tau] = \int_{\{(\min_{1 \leqslant n \leqslant N} X_n) \leqslant -\lambda\}} X_\tau d\mathbf{P} + \int_{\{(\min_{1 \leqslant n \leqslant N} X_n) > -\lambda\}} X_\tau d\mathbf{P}$$

$$\leqslant -\lambda \mathbf{P}\left\{\left(\min_{1 \leqslant n \leqslant N} X_n\right) \leqslant -\lambda\right\} + \int_{\{(\min_{1 \leqslant n \leqslant N} X_n) > -\lambda\}} X_n d\mathbf{P}.$$

since $X_\tau = n$ on $\{(\min_{1 \leqslant n \leqslant N} X_n) > -\lambda\}$. Thus,

$$\lambda \mathbf{P}\left\{\left(\min_{1 \leqslant n \leqslant N} X_n\right) \leqslant -\lambda\right\} \leqslant -\mathbf{E}[X_1] + \int_{\{(\min_{1 \leqslant n \leqslant N} X_n) > -\lambda\}} X_n d\mathbf{P}. \qquad\square$$

An important application of the Theorem 3.7 is the following result, that we will call *Doob's L^p inequality*.

Corollary 3.9 (Doob's L^p inequality) *Let $X = \{X_n\}_n$ be a non-negative $\{\mathscr{F}_n\}$-submartingale which is L^p bounded for some $p \geqslant 1$. Then, for all n*

$$\mathbf{P}\left\{X_n^* \geqslant \lambda\right\} \leqslant \frac{1}{\lambda^p}\mathbf{E}\left[|X_n|^p\right] = \frac{1}{\lambda^p}\|X_n\|_p^p. \tag{3.30}$$

In addition, if $p > 1$

$$\|X_n\|_p \leqslant \|X_n^*\|_p \leqslant \frac{p}{p-1}\|X_n\|_p, \tag{3.31}$$

but if $p = 1$

$$\|X_n\|_1 \leqslant \|X_n^*\|_1 \leqslant \frac{e}{e-1}\left(1 + \|X_n \log^+ X_n\|_1\right). \tag{3.32}$$

Proof In order to prove (3.30) recall that by Proposition 3.2, $\left\{|X_n|^p\right\}_n$ is a $\{\mathscr{F}_n\}$-submartingale and therefore the maximal inequality (3.27) holds for this submartingale, thus

$$\mathbf{P}\left\{X_n^* \geqslant \lambda\right\} = \mathbf{P}\left\{\max_{0\leqslant k\leqslant n}|X_k|^p \geqslant \lambda^p\right\} \leqslant \frac{1}{\lambda^p}\mathbf{E}\left[|X_n|^p\right].$$

To prove (3.31), the first inequality is immediate and for the second one, we will use the classical Marcinkiewicz argument. First, from the maximal inequality (3.27) we have

$$\lambda\mathbf{P}\left\{X_n^* \geqslant \lambda\right\} \leqslant \int_{\{X_n \geqslant \lambda\}} X_n d\mathbf{P}.$$

Then, using Fubini's Theorem:

$$\mathbf{E}\left[(X_n^*)^p\right] = p\int_\Omega d\mathbf{P}\int_0^{X_n^*}\lambda^{p-1}d\lambda = p\int_\Omega d\mathbf{P}\int_0^\infty \chi_{\{X_n^* \geqslant \lambda\}}\lambda^{p-1}d\lambda$$

$$= p\int_0^\infty \lambda^{p-1}\mathbf{P}\left\{X_n^* \geqslant \lambda\right\}d\lambda \leqslant p\int_0^\infty \lambda^{p-2}\left(\int_\Omega \chi_{\{X_n^* \geqslant \lambda\}}|X_n|d\mathbf{P}\right)d\lambda$$

$$= \frac{p}{p-1}\int_\Omega (X_n^*)^{p-1}|X_n|d\mathbf{P}.$$

Finally, using Hölder's inequality, we have

$$\mathbf{E}\left[(X_n^*)^p\right] \leqslant \frac{p}{p-1}\left(\mathbf{E}\left[|X_n|^p\right]\right)^{1/p}\left(\mathbf{E}\left[(X_n^*)^p\right]\right)^{p-1/p},$$

and if $\left\| X_n^* \right\|_p < \infty$, then

$$\left\| X_n^* \right\|_p = \left(\mathbf{E}\left[(X_n^*)^p \right] \right)^{1/p} \leqslant \frac{p}{p-1} \, \mathbf{E}\left[|X_n|^p \right].$$

If $\left\| X_n^* \right\|_p = \infty$ consider $\left(X_n^* \wedge M \right)$, with $M > 0$ a constant and by the same argument as above we obtain

$$\mathbf{E}\left[(X_n^* \wedge M)^p \right] \leqslant \frac{p}{p-1} \left(\mathbf{E}\left[|X_n|^p \right] \right)^{1/p} \left(\mathbf{E}\left[(X_n^* \wedge M)^p \right] \right)^{p-1/p}$$

and as $\mathbf{E}\left[(X_n^* \wedge M)^p \right] < \infty$ then

$$\mathbf{E}\left[(X_n^* \wedge M)^p \right]^{1/p} \leqslant \frac{p}{p-1} \, \left\| X_n \right\|_p,$$

and by the monotone convergence theorem, we obtain

$$\left(\mathbf{E}\left[|X_n^*|^p \right] \right)^{1/p} = \lim_{M \to \infty} \left(\mathbf{E}\left[(X_n^* \wedge M)^p \right] \right)^{1/p} \leqslant \frac{p}{p-1} \, \left\| X_n \right\|_p.$$

To prove (3.32), again the first inequality is obvious, and for the second inequality we have, by the previous argument

$$\mathbf{E}\left[X_n^* - 1 \right] \leqslant \mathbf{E}\left[(X_n^* - 1)^+ \right] = \int_0^\infty \mathbf{P}\left\{ X_n^* - 1 \geqslant \lambda \right\} d\lambda$$

$$\leqslant \int_0^\infty \frac{1}{\lambda+1} \int_{\{X_n^* > \lambda+1\}} X_n d\mathbf{P} = E\left[X_n \int_0^{X_n^*-1} \frac{1}{\lambda+1} d\lambda \right] = \mathbf{E}\left[X_n \log X_n^* \right].$$

As, for real numbers $a, b > 0$, $a \log b \leqslant a \log^+ a + b e^{-1}$, by using the concavity of $\log x$ at $x = e$. Then, we have

$$\mathbf{E}\left[X_n^* - 1 \right] \leqslant \mathbf{E}\left[X_n \log^+ X_n \right] + e^{-1} \mathbf{E}\left[X_n^* \right].$$

If $\mathbf{E}\left[X_n^* \right] < \infty$ we obtain the inequality and if $\mathbf{E}\left[X_n^* \right] = \infty$ we consider $\left(X_n^* \wedge M \right)$ and proceed as before. $\qquad\square$

Observe that (3.30) immediately implies,

$$\mathbf{P}\left\{ X^* \geqslant \lambda \right\} \leqslant \frac{1}{\lambda^p} \| X^* \|_p^p, \tag{3.33}$$

and (3.31) implies

$$\| X \|_p \leqslant \| X^* \|_p \leqslant \frac{p}{p-1} \, \| X \|_p. \tag{3.34}$$

Burkholder [30] has a more general version of Corollary 3.9 which establishes a clear link with the interpolation theory in harmonic analysis.

Proposition 3.10 (Burkholder) *Let X and Y be two non-negative random variables which satisfy, for some constants $\alpha > 0$, $\beta \geqslant 1$,*

$$\lambda \mathbf{P}\{Y > \beta\lambda\} \leqslant \alpha \int_{\{Y>\lambda\}} X d\mathbf{P}, \tag{3.35}$$

for all $\lambda > 0$ then

$$\|Y\|_p \leqslant \alpha q \beta^p \|X\|_p \tag{3.36}$$

with $\frac{1}{p} + \frac{1}{q} = 1$ and $1 < p < \infty$.

Proof Note that if Y satisfies the above inequality, then so does $(Y \wedge n)$ then, without loss of generality, we can assume that Y is finite. Thus, using the Fubini's theorem and Hölder's inequality, we have

$$\|Y\|_p^p = \int_\Omega Y^p d\mathbf{P} = p \int_\Omega \left(\int_0^Y \lambda^{p-1} d\lambda\right) d\mathbf{P} = p \int_\Omega \left(\int_0^\infty \chi_{\{Y>\lambda\}} \lambda^{p-1} d\lambda\right) d\mathbf{P}$$

$$= p \int_0^\infty \lambda^{p-1} \mathbf{P}\{Y > \lambda\} d\lambda \leqslant \alpha p \int_0^\infty \lambda^{p-2} \left(\int_{\{Y>\lambda/\beta\}} X d\mathbf{P}\right) d\lambda$$

$$= \alpha p \int_0^\infty \lambda^{p-2} \int_\Omega X \chi_{\{Y>\lambda/\beta\}} d\mathbf{P} d\lambda = \alpha p \int_\Omega X \left(\int_0^{\beta Y} \lambda^{p-2} d\lambda\right) d\mathbf{P}$$

$$= \alpha p \int_\Omega X \frac{Y^{p-1}}{p-1} \beta^{p-1} d\mathbf{P} = \alpha \frac{p}{p-1} \beta^{p-1} \int_\Omega X Y^{p-1} d\mathbf{P}$$

$$\leqslant \alpha q \left(\int_\Omega X^p d\mathbf{P}\right)^{1/p} \left(\int_\Omega \beta^p Y^p d\mathbf{P}\right)^{1/q},$$

and therefore

$$\|Y\|_p^p \leqslant \alpha q \beta^{p-1} \|X\|_p \|Y\|_p^{p/q},$$

that is to say,

$$\|Y\|_p \leqslant \alpha q \beta^{p-1} \|X\|_p \leqslant \alpha q \beta^p \|X\|_p. \qquad \square$$

The argument above is the basis of the classical argument of Marcinkiewicz's interpolation theorem widely used in harmonic analysis. The inequality (3.30) of Corollary 3.9 is now a consequence of the maximal inequality (3.27) and Proposition 3.10 for the case $\alpha = \beta = 1$.

The following is a generalization of Doob's L^p inequality (Corollary 3.9) due to Garsia [81]:

Theorem 3.10 *Let $X = \{X_n\}$ be a non-negative $\{\mathscr{F}_n\}$-submartingale and let Φ be a nondecreasing function in $[0, \infty)$, $\Phi(0) = 0$. Then for each n and $\lambda > 0$*

$$\mathbf{E}\left[\int_{\{X_n \geq \lambda\}} \eta \, d\Phi(\eta)\right] \leq \mathbf{E}\left[X_n \Phi\left(X_n^*\right)\right]. \tag{3.37}$$

Proof

$$\mathbf{E}\left[\int_{\{X_n^* \geq \lambda\}} \eta \, d\Phi(\eta)\right] = \sum_{k=1}^{n} \mathbf{E}\left[\int_{X_{k-1}^*}^{X_k^*} \eta \, d\Phi(\eta)\right] \leq \sum_{k=1}^{n} \mathbf{E}\left[X_k^*\left(\Phi\left(X_k^*\right) - \Phi\left(X_{k-1}^*\right)\right)\right].$$

Now, as $X_n^* = X_n$ when $\Phi\left(X_n^*\right) > \Phi\left(X_{n-1}^*\right)$ we conclude that

$$\mathbf{E}\left[\int_{\{X_n \geq \lambda\}} \eta \, d\Phi(\eta)\right] \leq \sum_{k=1}^{n} \mathbf{E}\left[X_k\left(\Phi\left(X_k^*\right) - \Phi\left(X_{k-1}^*\right)\right)\right].$$

Additionally, since X is a non-negative submartingale, we have

$$\sum_{k=1}^{n} \mathbf{E}[X_k(\Phi\left(X_k^*\right) - \Phi(X_{k-1}^*))] \leq \sum_{k=1}^{n} \mathbf{E}\left[X_k \Phi\left(X_k^*\right) - X_{k-1}\Phi\left(X_{k-1}^*\right)\right] = \mathbf{E}\left[X_k \Phi\left(X_k^*\right)\right]$$

From this last inequality, we obtain (3.37). $\qquad \square$

The inequality (3.30) of Corollary 3.9 is obtained by taking $\Phi(\lambda) = \lambda^{p-1}$ and using Hölder's inequality. In particular, using (3.30) for the case $p = 2$, we have

Corollary 3.11 *Let $X = \{X_n\}_n$ be a square integrable $\{\mathscr{F}_n\}$-martingale, i.e., $\mathbf{E}\left[X_n^2\right] < \infty$. Then $\left\{X_n^2\right\}$ is a submartingale and*

$$P\left\{X_n^* \geq \lambda\right\} \leq \frac{\mathbf{E}\left[X_n^2\right]}{\lambda^2}. \tag{3.38}$$

In particular, if $X_n = \xi_0 + \xi_1 + \ldots + \xi_n$ where $\{\xi_k\}$ is a sequence of independent random variables with

$$\mathbf{E}\left[\zeta_j\right] = 0 \quad and \quad \mathbf{E}\left[\xi_j^2\right] < \infty,$$

then inequality (3.38) is the classical Kolmogorov's inequality, see Theorem 5.1.

From inequality (3.38) we obtain immediately

Corollary 3.12 *Let $X = \{X_n\}_n$ be a square integrable $\{\mathscr{F}_n\}$-martingale, then*

$$\mathbf{E}\left[(X_n^*)^2\right] \leqslant 4\mathbf{E}\left[X_n^2\right]. \tag{3.39}$$

To conclude this section, let us consider the famous *Doob's upcrossing inequality*. First, given $X = \{X_n\}$ a $\{\mathscr{F}_n\}$-adapted stochastic process and $[a, b]$ an interval on the real line, we define:

$$\tau_1(\omega) = \min\{n > 0 : X_n(\omega) \leqslant a\}$$
$$\tau_2(\omega) = \min\{n > \tau_1 : X_n(\omega) \geqslant b\}$$

$$\cdot$$

$$\cdot$$

$$\tau_{2k-1}(\omega) = \min\{n > \tau_{2k-2} : X_n(\omega) \leqslant a\}$$
$$\tau_{2k}(\omega) = \min\{n > \tau_{2k-1} : X_n(\omega) \geqslant b\}$$

and $\tau_k = \infty$ if the corresponding set is empty.

The sequence $\{\tau_n\}$ is a sequence of successive stopping times where $\{X_n\}$ falls below a and above b. It is clear that $\{\tau_n\}$ is an increasing sequence (exercise). We define now for each $N \geqslant 1$, the random variable $U_N^X(a, b)(\omega) = \max\{k : \tau_{2k}(\omega) \leqslant N\}$, clearly $U_N^X(a, b)$ is the number of up-crossings of the interval $[a, b]$ of $\{X_n\}$ up to time N. Thus, we have the famous *Doob's up-crossing inequality*.

Theorem 3.13 (Doob) *Let $X = \{X_n\}$ be a $\{\mathscr{F}_n\}$-submartingale. Then, for each N and every couple of real numbers a and b, such that $a < b$, we obtain*

$$\mathbf{E}\left[U_N^X(a, b)\right] \leqslant \frac{1}{b-a}\left(\mathbf{E}\left[(X_N - a)^+\right]\right). \tag{3.40}$$

Proof It is clear that the number of upward crossings of the interval $[a, b]$ of $\{X_n\}$ is equal to the number of upward crossings of the non-negative $\{\mathscr{F}_n\}$-submartingale $Y = \{Y_n\}$ defined as $Y_n = (X_n - a)^+$, $n \in \mathbb{N}$, of the interval $[0, b - a]$, this is

$$U_N^X(a, b) = U_N^Y(0, b - a).$$

Then it suffices to assume that X is a non-negative submartingale, $a = 0$ and to prove

$$\mathbf{E}\left[U_N^X(0, b)\right] \leqslant \frac{\mathbf{E}[X_N]}{b}.$$

Let us consider $X_0 = 0$, $\mathscr{F}_0 = \{\emptyset, \Omega\}$, and for $k = 1, 2, \ldots$ let

$$\psi_k = \begin{cases} 1 & \text{if } \tau_n < k \leqslant \tau_{n+1} \text{ for some } n \text{ odd} \\ 0 & \text{if } \tau_n < k \leqslant \tau_{n+1} \text{ for some } n \text{ even} \end{cases}.$$

Then it is clear that

$$bU_N^X(0, b) \leqslant \sum_{k=1}^{N} \psi_k (X_k - X_{k-1})$$

and

$$\{\psi_k = 1\} = \bigcup_{n \text{ odd}} \left[\{\tau_n < k\} \setminus \{\tau_{n+1} < k\} \right] \in \mathscr{F}_{k-1}.$$

Therefore,

$$\begin{aligned} b\mathbf{E}\left[U_N^X(0, b) \right] &\leqslant \mathbf{E}\left[\sum_{k=1}^{N} \psi_k (X_k - X_{k-1}) \right] = \sum_{k=1}^{N} \int_{\{\psi_k=1\}} (X_k - X_{k-1}) \, d\mathbf{P} \\ &= \sum_{k=1}^{N} \int_{\{\psi_k=1\}} \mathbf{E}\left[(X_k - X_{k-1}) \mid \mathscr{F}_{k-1} \right] d\mathbf{P} \\ &= \sum_{k=1}^{N} \int_{\{\psi_k=1\}} \left(\mathbf{E}\left[X_k \mid \mathscr{F}_{k-1} \right] - X_{k-1} \right) d\mathbf{P} \\ &\leqslant \sum_{k=1}^{N} \int_{\Omega} \left(\mathbf{E}\left[X_k \mid \mathscr{F}_{k-1} \right] - X_{k-1} \right) d\mathbf{P} = \mathbf{E}\left[X_N \right]. \qquad \square \end{aligned}$$

For a full discussion of crossover inequalities, see Panzone [136].

3.4 Convergence Theorems for Martingales

The following result, which is fundamental to all convergence problems, can be seen as an analogue to the fact that every monotone sequence of real numbers that is bounded has a finite limit:

Theorem 3.14 (Doob's convergence theorem) *Let* $X = \{X_n\}$ *be a* $\{\mathscr{F}_n\}$-*submartingale, such that*

$$\sup_n \mathbf{E}\left[X_n^+ \right] < \infty, \tag{3.41}$$

then X_n *converges to a limit* X_∞ *a.e. with*

$$\mathbf{E}\left[|X_\infty| \right] < \infty.$$

Observe that since

$$\mathbf{E}\left[|X_n|\right] = 2\mathbf{E}\left[X_n^+\right] - \mathbf{E}\left[X_n\right] \leqslant 2\mathbf{E}\left[X_n^+\right] - \mathbf{E}\left[X_1\right]$$

we have that $\sup_n \mathbf{E}\left[|X_n|\right] < \infty$, by hypothesis, and therefore the condition (3.41) is equivalent to asking that the submartingale X be L^1 bounded.

Proof Since X is L^1 bounded, if $X_\infty = \lim_{n\to\infty} X_n$ exists a.e., then X_∞ is necessarily integrable by Fatou's lemma. Then, to prove the a.e. existence, we need to consider the divergence set

$$\left\{\varliminf_{n\to\infty} X_n < \varlimsup_{n\to\infty} X_n\right\} = \bigcup_{a,b\in\mathbb{Q}} \left\{\varliminf_{n\to\infty} X_n < a < b < \varlimsup_{n\to\infty} X_n\right\}.$$

We need to prove that this set has zero probability and for this, it is sufficient to prove that for any pair of rational numbers $a < b$ the set

$$\left\{\varliminf_{n\to\infty} X_n < a < b < \varlimsup_{n\to\infty} X_n\right\}$$

has probability zero.

Let $U_\infty^X(a,b) = \lim_{N\to\infty} U_N^X(a,b)$, note that $U_N^X(a,b)$ is not decreasing on N and then by the crossing's inequality (Theorem 3.13), for all N

$$\mathbf{E}\left[U_N^X(a,b)\right] \leqslant \frac{1}{b-a}\left(\mathbf{E}\left[(X_n - a)^+\right]\right) \leqslant \frac{1}{b-a}\left(\mathbf{E}\left[|X_n| + |a|\right]\right) < \infty.$$

Then, using the monotone convergence theorem, we have

$$\mathbf{E}\left[U_\infty^X(a,b)\right] \leqslant \lim_{N\to\infty} \mathbf{E}\left[U_N^X(a,b)\right] < \infty,$$

and hence $U_\infty^X(a,b) < \infty$ a.e. for all real numbers a and b such that $a < b$. Therefore, if in the set $\left\{\varliminf_{n\to\infty} X_n < a < b < \varlimsup_{n\to\infty} X_n\right\}$ there are infinite crossings upwards of the interval [a,b] then it must have probability zero. Then $\varliminf_{n\to\infty} X_n = \varlimsup_{n\to\infty} X_n$ a.e., which means that $X_\infty = \lim_{n\to\infty} X_n$ exists almost everywhere. $\qquad\square$

As we will discuss later in detail, L. Baez-Duarte has an alternative proof of the convergence theorem using ultimately the maximal inequality instead of the crossing's inequality, see [4].

From the previous theorem, we obtain, immediately

Corollary 3.15 *Let $\{X_n\}$ be a $\{\mathscr{F}_n\}$-submartingale ($\{\mathscr{F}_n\}$-supermartingale) such that $X_n \leqslant 0 (\geqslant 0)$, then $X_n \to X_\infty$ a.e. as $n \to \infty$.*

Corollary 3.16 *If $\{X_n\}$ is a non-positive $\{\mathscr{F}_n\}$-submartingale then $\{X_n\}_n \cup \{X_\infty\}$ is a non-negative $\{\mathscr{F}_n\} \cup \{\mathscr{F}_\infty\}$-submartingale, with $X_\infty = \lim_{n\to\infty} X_n$ and $\mathscr{F}_\infty = \sigma\left(\bigcup_n \mathscr{F}_n\right)$.*

Proof For Fatou's Lemma,

$$\mathbf{E}\left[X_\infty\right] = \lim_{n\to\infty} \mathbf{E}\left[X_n\right] \geqslant \limsup_{n\to\infty} \mathbf{E}\left[X_n\right] \geqslant \mathbf{E}\left[X_1\right] > -\infty$$

and

$$\mathbf{E}\left[X_\infty \mid \mathscr{F}_m\right] = \mathbf{E}\left[\lim_{n\to\infty} X_n \mid \mathscr{F}_m\right] \geqslant \limsup_{n\to\infty} \mathbf{E}\left[X_n \mid \mathscr{F}_m\right] \geqslant X_m. \qquad \square$$

Corollary 3.17 *Let $\{X_n\}$ be a non-negative $\{\mathscr{F}_n\}$-martingale, then $X_n \to X_\infty$ a.e. as $n \to \infty$.*

Additionally, we have the following result L^p,

Corollary 3.18 *Let $\{X_n\}$ be a $\{\mathscr{F}_n\}$-martingale, L^p-bounded for some $1 < p < \infty$, then there exists a random variable X_∞ such that $X_n \to X_\infty$ in L^p-norm as $n \to \infty$.*

Proof As $\{|X_n|^p\}$ is a L^1-bounded submartingale then, from Theorem 3.14 we have that $|X_n|^p \to |X_\infty|^p$ a.e. as $n \to \infty$ and from Doob's L^p inequality (Corollary 3.9) we have that $X^* = \sup_n |X_n| \in L^p$ and then by the dominated convergence theorem, $\mathbf{E}|X_n - X_\infty|^p \to 0$ as $n \to \infty$. $\qquad \square$

Observe that the converse of this result is trivial.

Example 3.3 Given a Rademacher series $\sum_{k=0}^{\infty} c_k r_k(t)$, we have already know, from Example 3.1 (1), that its partial sums are a martingale, and then, by Khintchine's inequality, Theorem 1.3, for $p = 1$ we get that it is L^1 bounded by $\sum_{n=0}^{\infty} c_n^2 < \infty$ and therefore, by Theorem 3.14 it converges a.e. This provides another proof of the first part of Theorem 1.1.

From Corollary 3.18 remains the question, what happens for the case $p = 1$? The convergence theorem guarantees a.e. convergence. As we will see in the following example, in general, there can be a.e. convergence without convergence in L^1 norm.

Example 3.4 Let $\{\zeta_n\}$ be a sequence of independent random variables identically distributed, with Bernoulli distribution $\left(\mathbf{P}\{\xi_n = 1\} = \mathbf{P}\{\xi_n = -1\} = \frac{1}{2}\right)$.

We know that, the sequence of partial sums $\{X_n = \sum_{k=1}^{n} \xi_k\}$ is a martingale, and consider

$$Y_n = \begin{cases} \frac{1}{2^{n-1}} & \text{if } X_{n-1} = n - 1 \\ 0 & \text{otherwise} \end{cases},$$

for each n, then $\{Y_n\}$ is a multiplier sequence, and considering the martingale $Z_n = 1 + (Y \circ X)_n \, n \geqslant 1$, then it can be proved that $\{Z_n\}$ is non-negative (check!) and therefore

$$\mathbf{E}\left[|Z_n|\right] = \mathbf{E}\left[Z_n\right] = 1.$$

On the other hand, $\mathbf{P}\{Z_n > 0\} = \frac{1}{2^n} \to 0$ as $n \to \infty$ and therefore there exists a subsequence that converges to zero a.e. but not in L^1. $\qquad\square$

To study the behavior of the limit in L^1 norm of a martingale in general, let us first study a special case.

Observe that by Proposition 3.9 a last-term martingale $\{X_n\}_n$ is uniformly integrable. Then, we have the following result.

Proposition 3.11 (P. Lévy) *Let $\{X_n\}_n$ be a last-term $\{\mathscr{F}_n\}$-martingale, i.e., there exists $X \in L^1(\Omega)$ such that $X_n = \mathbf{E}[X \mid \mathscr{F}_n]$, then $X_n \to X_\infty$ a.e. and in L^1, as $n \to \infty$. In addition, $X_\infty = \mathbf{E}[X \mid \mathscr{F}_\infty]$ a.e. where $\mathscr{F}_\infty = \sigma\left(\bigcup_n \mathscr{F}_n\right)$. Thus, if $\mathscr{F}_n \uparrow \mathscr{F}_\infty$ then $\mathbf{E}[X \mid \mathscr{F}_n] \to \mathbf{E}[X \mid \mathscr{F}_\infty]$ a.e. and in L^1.*

Proof Since

$$\sup_n \mathbf{E}\left[X_n^+\right] \leqslant \sup_n \mathbf{E}\left[|X_n|\right] \leqslant \mathbf{E}[|X|] < \infty,$$

then, by Theorem 3.14 $X_\infty = \lim_{n \to \infty} X_n$ a.e. and by the dominated convergence theorem in L^1-norm. The problem is to identify the limit with $\mathbf{E}[X \mid \mathscr{F}_\infty]$. By the uniform integrability, we have for any set $A \in \mathscr{F}$, $X_n \chi_A \to X_\infty \chi_A$ in L^1 as $n \to \infty$ and hence

$$\int_A X_\infty d\mathbf{P} = \lim_{n \to \infty} \int_A X_n d\mathbf{P}.$$

Also, if $A \in \mathscr{F}_k$ and $n \geqslant k$. Also,

$$\int_A X_n d\mathbf{P} = \int_A \mathbf{E}[X \mid \mathscr{F}_n] d\mathbf{P} = \int_A X d\mathbf{P}.$$

Thus,

$$\int_A X_n d\mathbf{P} = \int_A X d\mathbf{P}$$

for all $A \in \bigcup_n \mathscr{F}_n$ and as $X_\infty \in \mathscr{F}_\infty$ and

$$\int_A X_\infty d\mathbf{P} = \int_A X d\mathbf{P},$$

then we have that $X_\infty = \mathbf{E}[X \mid \mathscr{F}_\infty]$ a.e. $\square$

Alternatively, to prove the L^1-convergence, one may use that for any $X \in \bigcup_{n=1}^\infty L^1(\mathscr{F}_n)$, $\mathbf{E}[X \mid \mathscr{F}_n]$ is constant after some n. But it can be proved that $\bigcup_{n=1}^\infty L^1(\mathscr{F}_n)$ is dense in $L^1(\mathscr{F})$ and the operators $\mathbf{E}[\cdot \mid \mathscr{F}_n]$ are contractions, see [4].

Therefore, last-term martingales generalize the finite martingales (Lemma 3.1) having as last term $X_\infty = \mathbf{E}[X \mid \mathscr{F}_\infty]$ a.e. (because $\mathbf{E}[X_\infty \mid \mathscr{F}_n] = X_n$ a.e.). Note that if $\mathscr{F}_\infty = \mathscr{F}$ then $X_\infty = X$ a.e.

This was the first result obtained on martingale convergence.

Example 3.5 We know, from Example 3.1 (3) that the partial sums of Haar expansions form a martingale since $\mathscr{F}_\infty = \sigma\left(\bigcup_N \mathscr{F}_N\right) = \mathscr{B}([0, 1))$ then, by the above result, for a function $f \in L^1([0, 1), m)$ we have $\sum_{k=1}^N d_K \chi_k(t) \to f(x)$, as $N \to \infty$, a.e. and also in L^1.

It is clear that in general, we need an additional condition to guarantee convergence in L^1 norm. That condition will be precisely the condition of uniform integrability. To prove that we need first the following technical result.

Lemma 3.2 (Second Lemma of Uniform Integrability) *Given a stochastic process $\{X_n\}$, if $X_n \to X$ a.e. for some random variable X, then the following statements are equivalent*

(i) $\{X_n\}_n$ is uniformly integrable.
(ii) $X_n \to X$ in L^1 as $n \to \infty$.
(iii) $\mathbf{E}[|X_n|] \to \mathbf{E}[|X|] < \infty$ as $n \to \infty$.

Proof

- (i) $\Rightarrow$ (ii)

$$\phi_M(x) = \begin{cases} M & \text{if } M \leqslant x \\ x & \text{if } -M \leqslant x \leqslant M \\ -M & \text{if } x \leqslant -M. \end{cases}$$

It can be proved (checking all possible cases) that

$$\mathbf{E}[|X_n - X|] \leqslant |\phi_M(X_n) - \phi_M(X)| + \int_{\{|X_n|>M\}} |X_n| d\mathbf{P} + \int_{\{|X|>M\}} |X| d\mathbf{P},$$

as $n \to \infty$, the first term tends to 0 by the dominated convergence theorem, if $\varepsilon > 0$ and M is sufficiently large, by uniform integrability, the second term is less than ε.

Finally, to bound the third term, let us note that uniform integrability implies that $\sup_n \mathbf{E}\left[|X_n|\right] < \infty$ so Fatou's Lemma implies that $\mathbf{E}[|X|] < \infty$ and taking M sufficiently large we have that the third term is less than ε. Then we have that

$$\limsup_n \mathbf{E}\left[|X_n - X|\right] \leqslant 2\varepsilon.$$

- (ii) $\Rightarrow$ (iii) Simply, by the triangle inequality,

$$\left|\mathbf{E}\left[|X_n|\right] - \mathbf{E}[|X|]\right| \leqslant \mathbf{E}\left[\left||X_n| - |X|\right|\right] \leqslant \mathbf{E}\left[|X_n - X|\right].$$

- (iii) $\Rightarrow$ (i) Let

$$\psi_M(x) = \begin{cases} x & \text{in } [0, M-1] \\ 0 & \text{in } [M, \infty] \\ \text{lineal} & \text{in } [M-1, M] \end{cases}.$$

If M is sufficiently large $\mathbf{E}[|X|] - \mathbf{E}\left[\psi_M(|X|)\right] < \varepsilon/2$, then the dominated convergence theorem implies that

$$\mathbf{E}\left[\psi_M\left(|X_n|\right)\right] \to \mathbf{E}\left[\psi_M(|X|)\right]$$

as $n \to \infty$, so, if $n \geqslant n_0$

$$\int_{\{|X_n|>M\}} |X_n|\, d\mathbf{P} \leqslant \mathbf{E}\left[|X_n|\right] - \mathbf{E}\left(\psi_M\left(|X_n|\right)\right) < \varepsilon$$

Taking M sufficiently large we can make $\int_{\{|X_n|>M\}} |X_n|\, d\mathbf{P} < \varepsilon$ for $0 \leqslant n < n_0$, then $\{X_n\}$ is uniformly integrable. $\qquad\qquad\qquad\qquad\qquad\square$

We want to see that the general case is analogous to the last term martingales. Let's consider first the case of submartingales. Using the Lemma 3.2 we obtain the following result.

Corollary 3.19 *If $\{X_n\}$ is a $\{\mathscr{F}_n\}$-submartingale, the following statements are equivalent:*

(i) $\{X_n\}$ is uniformly integrable.
(ii) $\{X_n\}$ converges in L^1 as $n \to \infty$.

Proof

- (i) $\Rightarrow$ (ii) The uniform integrability implies that $\sup_n \mathbf{E}[|X_n|] < \infty$, therefore, by the dominated convergence theorem $X_n \to X$ a.e. and then Lemma 3.2 implies $X_n \to X$ in L^1 as $n \to \infty$.
- (ii) $\Rightarrow$ (i) is an immediate consequence of Lemma 3.2. $\qquad\square$

Now, let us consider the case of L^1-bounded martingales.

Proposition 3.12 *If $\{X_n\}$ is a $\{\mathscr{F}_n\}$-martingale the following statements are equivalents*

(i) $\{X_n\}$ *is uniformly integrable.*
(ii) $\{X_n\}$ *converges in L^1 as $n \to \infty$.*
(iii) $\{X_n\}$ *has a last term, i.e. there exists a random variable X, such that*

$$X_n = \mathbf{E}[X \mid \mathscr{F}_n] \ a.e.$$

Proof

- (i) $\Rightarrow$ (ii) It is a consequence of Corollary 3.19.
- (ii) $\Rightarrow$ (iii) Let $X = \lim_n X_n$ be the limit in L^1, then by the definition of martingale, if $m > n$ and $A \in \mathscr{F}_n$

$$\int_A X_m d\mathbf{P} = \int_A X_n d\mathbf{P}.$$

  However, $X_n \chi_A \to X \chi_A$ in L^1 as $n \to \infty$, and therefore $\int_A X_n d\mathbf{P} = \int_A X d\mathbf{P}$, for all $A \in \mathscr{F}_n$. It follows that $X_n = \mathbf{E}[X \mid \mathscr{F}_n]$ a.e.
- (iii) $\Rightarrow$ (i) It is immediate from Proposition 3.9. $\qquad\square$

Therefore, by the previous proposition, in the case $p = 1$, uniformly integrable martingales always have a last term, which is precisely the martingale's L^1-limit. Similarly, in the case of $p > 1$, by Corollary 3.18 L^p-bounded martingales are L^p-convergent and therefore, by Hölder's inequality, also converges in L^1 thus, they are also last-term martingales.

On the other hand, if a martingale is L^1-convergent, then it is L^1-bounded and then by Theorem 3.14, it converges a.e. This can also be proved directly using Doob's maximal inequality (3.27), see [4].

Proposition 3.13 *Let $\{X_n\}$ be a $\{\mathscr{F}_n\}$-martingale L^1-convergent, then it converges $\mathbf{P}$-almost everywhere.*

Proof Let $X \in L^1(\Omega)$ such that $\|X - X_n\|_1 \to 0$. Choose integers $\{m_k\}$ so that $m_k \uparrow \infty$ and such that

$$\sum_{k=1}^{\infty} k \|X - X_{m_k}\|_1 < \infty.$$

For each k, $\{X_n - X_{m_k} : n \geqslant m_k\}$ is a martingale, so Doob's maximal inequality (3.27) implies

$$\mathbf{P}\left\{ \sup_{n \geqslant m_k} |X_n - X_{m_k}| > \frac{1}{k} \right\} \leqslant k \|X - X_{m_k}\|_1,$$

then, adding for $k = 1, 2, \ldots$ we get

$$\sum_{k=1}^{\infty} \mathbf{P}\left\{ \sup_{n \geqslant m_k} |X_n - X_{m_k}| > \frac{1}{k} \right\} < \infty,$$

this implies

$$\lim_{k \to \infty} \mathbf{P}\left\{ \sup_{n \geqslant m_k} |X_n - X_{m_k}| > \frac{1}{k} \right\} = 0$$

and therefore

$$\lim_{k \to \infty} \left(\sup_{n > m_k} |X_n - X_{m_k}| \right) = 0, \quad \mathbf{P}\text{-a.e.}$$

In other words, $\{X_n(\omega)\}$ is a Cauchy sequence $\mathbf{P}$-almost everywhere. $\qquad\square$

Additionally, in [4] L. Báez-Duarte has another example of a L^1 bounded martingale that is not uniformly integrable and therefore it has not a last term. Actually, a large class of such martingales arises in the theory of abstract differentiation as follows. Assume that the σ-algebras $\{\mathscr{F}_n\}$ are generated by a sequence $\{\pi_n\}$ of finite refining partitions with no atoms of probability zero. Furthermore let μ be a finite signed measure on $\mathscr{F}$, then

$$X_n = \sum_{A \in \pi_n} \frac{\mu(A)}{\mathbf{P}(A)} \chi_A, \quad n \geqslant 1,$$

defines an L^1-bounded martingale. $\{X_n\}$ turns out to have a last term if and only if μ is absolutely continuous with respect to $\mathbf{P}$. This example suggests the following definition.

Definition 3.13 A $\{\mathscr{F}_n\}$- martingale $\{X_n\}$ is said to be measure-dominated if there is a finite signed measure μ on $\mathscr{F}$ whose restriction to each $\mathscr{F}_n$ coincides with the measure μ_n defined as

$$\mu_n(A) = \int_A X_n d\mathbf{P}, \quad A \in \mathscr{F}_n. \tag{3.42}$$

In this case, μ is said to dominate the martingale. If a martingale is dominated by a singular measure with respect to $\mathbf{P}$, then we say it is a singular $\{\mathscr{F}_n\}$-martingale.

For measure-dominated martingales, Báez-Duarte obtained a pointwise convergence result using Doob's maximal inequality (3.27) as we are going to discuss in detail now. First, observe that we have the following decomposition.

Theorem 3.20 *Let $\{X_n\}$ be a measure-dominated $\{\mathscr{F}_n\}$-martingale, then $\{X_n\}$ is the sum of a last-term $\{\mathscr{F}_n\}$-martingale plus a singular $\{\mathscr{F}_n\}$-martingale, i.e.*

$$X_n = Y_n + Z_n$$

where $\{Y_n\}$ is a last-term $\{\mathscr{F}_n\}$-martingale and Z_n is a singular $\{\mathscr{F}_n\}$-martingale.

Proof Let $\{X_n\}$ be a $\{\mathscr{F}_n\}$-martingale dominated by μ. By Lebesgue's decomposition theorem, there exists an integrable function X and singular measure v with respect to $\mathbf{P}$, such that

$$d\mu = Xd\mathbf{P} + dv.$$

Now, set $Y_n = \mathbf{E}[X \mid \mathscr{F}_n]$, and $Z_n = X_n - Y_n$. Clearly, $\{Z_n\}$ is a martingale, being the difference of two martingales, and it remains to prove that it is singular. A trivial computation gives, for $A \in \mathscr{F}_n$,

$$\int_A Z_n d\mathbf{P} = \int_A X_n d\mathbf{P} - \int_A Y_n d\mathbf{P} = \int_A X_n d\mathbf{P} - \int_A X d\mathbf{P} = \mu(A) - \int_A X d\mathbf{P} = \int_A dv = v(A).\;\square$$

From Proposition 3.11 we know that any last-term martingale converges a.e. and in L^1. For singular martingales, we have the following convergent result.

Theorem 3.21 (L. Báez-Duarte) *Let $\{X_n\}$ be a singular $\{\mathscr{F}_n\}$-martingale, then $X_n \to 0$ $\mathbf{P}$-almost everywhere.*

Proof Let v be a singular measure, with respect to $\mathbf{P}$, which dominates $\{X_n\}$; since there exists $A \in \mathscr{F}$ such that $\mathbf{P}(A) = |v|(A^c) = 0$, it is easy to show (check!) that given any $\delta_1 > 0$ and $\delta_2 > 0$, there is a set $E \in \bigcup_{n=1}^{\infty} \mathscr{F}_n$ such that $\mathbf{P}(E) < \delta_1$ and $|v|(E^c) < \delta_2$. Therefore for fixed arbitrary $\varepsilon > 0$ one can find a sequence $\{E_k\} \subset \bigcup_{n=1}^{\infty} \mathscr{F}_n$ such that

$$\sum_{k=1}^{\infty} \mathbf{P}(E_k) < \varepsilon, \tag{3.43}$$

and

$$\sum_{k=1}^{\infty} k|\nu|\left(E_k^c\right) < \infty. \tag{3.44}$$

For each k there is an integer m_k such that $E_k \in \mathscr{F}_{m_k}$, and without loss of generality we may assume $m_k \uparrow \infty$. Now note that

$$\left\{ X_n \chi_{E_k^c} : n \geqslant m_k \right\}$$

is a martingale for each k. By Doob's maximal inequality (3.27) and the fact that $\{X_n\}$ is dominated by ν,

$$\mathbf{P}\left\{ \sup_{m_k \leqslant n \leqslant p} |X_n|\,\chi_{E_k^c} > \frac{1}{k} \right\} \leqslant 2k \int_{E_k^c} |X_p|\,d\mathbf{P} \leqslant 2k|\nu|\left(E_k^c\right), \quad k = 1, 2, \dots.$$

Let $p \to \infty$ in each of these inequalities, and add on $k = 1, 2, \dots$ to obtain, taking into account (3.44),

$$\sum_{k=1}^{\infty} \mathbf{P}\left\{ \sup_{n \geqslant m_k} |X_n|\,\chi_{E_k^c} > \frac{1}{k} \right\} < \infty.$$

Therefore

$$\lim_{k \to \infty} \mathbf{P}\left\{ \sup_{n \geqslant m_k} |X_n|\,\chi_{E_k^c} > 1/k \right\} = 0,$$

and this shows that

$$\lim_{k \to \infty} \left(\sup_{n \geqslant m_k} |X_n|\,\chi_{E_k^c} \right) = 0, \quad \mathbf{P} - \text{ a.e.}$$

Hence, $X_n \to 0$ $\mathbf{P}$-a.e. in the complement of the set $E(\varepsilon) = \bigcup_{k=1}^{\infty} E_k$, with $\mathbf{P}(E(\varepsilon)) < \varepsilon$ by virtue of (3.43). But then $X_n \to 0$ in the complement of $\bigcap_{j=1}^{\infty} E(1/j)$ whose probability is zero. $\qquad \square$

Corollary 3.22 *Let $\{X_n\}$ be a measure-dominated $\{\mathscr{F}_n\}$-martingale, then $\{X_n\}$ converges $\mathbf{P}$-a.e. to the Radon–Nikodým density of the $\mathbf{P}$- absolutely continuous part of the dominating measure.*

Unfortunately, not all L^1-bounded martingales are measure-dominated. Nevertheless, L. Báez-Duarte shows that any L^1-bounded martingale is measure-dominated if Ω is a product of compact spaces and the $\{\mathscr{F}_n\}$ are chosen appropriately, Theorem 3.2 of [4]; and that any L^1-bounded martingale can be equivalently realized on such a space, Theorem 3.3 of [4]. From this, he gets another proof of Doob's martingale convergence theorem, Theorem 3.14.

3.5   Reverse Martingales

Let (Ω, F, P) be a probability space, let $\{\mathscr{F}_n\}_{n\in\mathbb{Z}^-}$ be a family of sub$-\sigma$-algebras of $\mathscr{F}$ such that $\mathscr{F}_n \supseteq \mathscr{F}_{n+1}$ for all $n \in \mathbb{Z}^- = \{-1, -2, -3, \ldots\}$.

Definition 3.14 A stochastic process $X = \{X_n\}_{n\in\mathbb{Z}^-}$ is called a reverse martingale with respect to $\{\mathscr{F}_n\}_{n\in\mathbb{Z}^-}$(supermartingale, submartingale), if

 (i) For each $n \in \mathbb{Z}^-$, $X_n \in L^1(\Omega)$.
 (ii) $\{X_n\}$ is $\{\mathscr{F}_n\}_{n\in\mathbb{Z}^-}$ adapted.
(iii) For all $m, n \in \mathbb{Z}^-$ if $m > n$, then

$$\mathbf{E}\left[X_m \mid \mathscr{F}_n\right] = X_n \text{ a.e. } \quad (\leqslant, \geqslant \text{ repectively}). \tag{3.45}$$

Note that for each $n \in \mathbb{Z}^-$, $\{X_{-n}, \ldots, X_{-1}\}$ is a finite martingale (supermartingale, submartingale, respectively) with respect to the family $\{\mathscr{F}_{-n}, \ldots, \mathscr{F}_{-1}\}$, with last term X_{-1}.

For reverse martingales, we have the following convergence theorem.

Theorem 3.23 *Let $X = \{X_n\}_{n\in\mathbb{Z}^-}$ be a $\{\mathscr{F}_n\}_{n\in\mathbb{Z}^-}$-reverse martingale, then* $\lim_{n\to-\infty} X_n = X$ *exists a.e. and is integrable* $\mathbf{E}[X] = \mathbf{E}[X_n]$ *for all* $n \in \mathbb{Z}^-$.

Proof The proof of this result is analogous to the proof of Theorem 3.14. Let us consider the set

$$\left\{ \varliminf_{n\to-\infty} X_n < \varlimsup_{n\to-\infty} X_n \right\} = \bigcup_{a,b\in\mathbb{Q}} \left\{ \varliminf_{n\to-\infty} X_n < a < b < \varlimsup_{n\to-\infty} X_n \right\}$$

again by the up-crossing inequality (Theorem 3.13), if U_n is the number of up-crossings of $[a, b]$ of the martingale $\{X_{-n}, \ldots, X_{-1}\}$ then

$$\mathbf{E}\left[U_n\right] \leqslant \frac{\mathbf{E}\left[|X_{-1}| + |a|\right]}{b - a}$$

therefore $\mathbf{E}\left[U_n\right]$ is bounded and so $\sup_n U_n$ is finite a.e. the set under consideration has zero probability and therefore $\lim_n \ X_{-n} = X_{-\infty}$ exists a.e. but as $X_{-n} = \mathbf{E}\left[X_{-1} \mid \mathscr{F}_n\right]$ a.e. for all $n \in \mathbb{Z}^-$, then $\{X_n\}_{n\in\mathbb{Z}^-}$ is uniformly integrable, and hence $X_{-\infty}$ is integrable and $\mathbf{E}\left[X_{-\infty}\right]$ is the limit of $\mathbf{E}\left[X_n\right]$ (which are constants). $\square$

Theorem 3.24 *Let $X = \{X_n\}_{n\in\mathbb{Z}^-}$ be a $\{\mathscr{F}_n\}_{n\in\mathbb{Z}^-}$-reverse submartingale such that*

$$\inf_{n\in\mathbb{Z}^-} \mathbf{E}\left[X_n\right] > -\infty \tag{3.46}$$

then X is uniformly integrable and $\lim_{n\to\infty} X_n = X_{-\infty}$ exists a.e. and in L^1.

Proof By the above theorem, it suffices to prove uniform integrability. As $\mathbf{E}\,[X_n]$ is decreasing, (3.46) implies that $\lim_{n\to\infty} X_n$ exists and is finite. Let $M > 0$, and let us consider

$$\int_{\{|X_n|>M\}} |X_n|\,d\mathbf{P} = \int_{\{X_n>M\}} |X_n|\,d\mathbf{P} + \int_{\{-X_n>M\}} |X_n|\,d\mathbf{P}$$

$$= \int_{\{X_n>M\}} |X_n|\,d\mathbf{P} + \int_{\{X_n\geqslant -M\}} |X_n|\,d\mathbf{P} - \int_{\Omega} |X_n|\,d\mathbf{P}.$$

Let $\varepsilon > 0$, and let k be such that

$$\mathbf{E}\,[X_k] - \lim_{n\to\infty} \mathbf{E}\,[X_n] < \varepsilon.$$

Then, by the above inequality, we have for $n < k$

$$\int_{\{|X_n|>M\}} |X_n|\,d\mathbf{P} \leqslant \int_{\{X_n>M\}} |X_n|\,d\mathbf{P} + \int_{\{X_n\geqslant -M\}} |X_k|\,d\mathbf{P} - \int_{\Omega} |X_k|\,d\mathbf{P} + \varepsilon$$

$$\leqslant 2\int_{\{X_n\geqslant -M\}} |X_k|\,d\mathbf{P} + \varepsilon$$

then, by Tchebychev's inequality

$$\mathbf{P}\,\{X_k \geqslant M\} \leqslant P\,\{|X_n| \geqslant M\} \leqslant \frac{1}{M}\mathbf{E}\,[X_n] = \frac{1}{M}\left\{\mathbf{E}\left[X_n^+\right] - \mathbf{E}\,[X_n]\right\}$$

$$\leqslant \frac{1}{M}\left\{\mathbf{E}\left[X_0^+\right] - \mathbf{E}\,[X_n]\right\} \to 0,$$

as $M \to \infty$ uniformly in n. $\qquad\square$

Finally, we have a result analogous to the Proposition 3.11. Let $\{\mathscr{F}_n\}$ be a family of σ-algebras satisfying $\mathscr{F}_n \supseteq \mathscr{F}_{n+1}$ for all n, then the intersection $\bigcap_{n=1}^{\infty} \mathscr{F}_n = \mathscr{F}_0$ is σ-algebra too and we say that $\mathscr{F}_n \downarrow \mathscr{F}_0$.

Proposition 3.14 *If* $\mathscr{F}_n \downarrow \mathscr{F}_0$, $X \in L^1(\Omega)$ $\mathbf{E}\,[X \mid \mathscr{F}_n] \to \mathbf{E}\,[X \mid \mathscr{F}_0]$ *a.e.*

Proof If $X_{-n} = \mathbf{E}\,[X \mid \mathscr{F}_n]$, then $\{X_{-n}\}_n$ is a uniformly integrable martingale in reverse. By the convergence theorem for reverse martingales $X_{-n} \to X_{-\infty}$, as $n \to \infty$ and then $X_{-\infty}$ is $\mathscr{F}_n-$ measurable for all n therefore $X_{-\infty}$ is $\mathscr{F}_0$ -measurable, thus the uniform integrability implies that if $A \in \mathscr{F}_0$

$$\int_A X_{-\infty}\,d\mathbf{P} = \lim_n \int_A \mathbf{E}\,[X \mid \mathscr{F}_n]\,d\mathbf{P} = \lim_n \int_A \mathbf{E}\,[\mathbf{E}\,[X \mid \mathscr{F}_n] \mid \mathscr{F}_0]\,d\mathbf{P} = \int_A \mathbf{E}\,[X \mid \mathscr{F}_0]\,d\mathbf{P}$$

and so $X_{-\infty} = \mathbf{E}\,[X \mid \mathscr{F}_0]$ a.e. $\qquad\square$

3.6   Exercises

(1) Let $\xi_1, \xi_2 \ldots$ be independent random variables with mean value 0. Let $X_1 = \xi_1$, and

$$X_{n+1} = X_n + \xi_{n+1} f_n (\xi_1, \ldots, \xi_n)$$

assuming that f_n is integrable. Prove that $\{X_n\}$ is a martingale. The martingales that model games are of this form, see for instance [50, Chap. 2].

(2) Let $\xi_1, \xi_2 \ldots$ be independent random variables with mean value 0 and variance $\sigma^2 > 0$, and consider the sums $S_n = \xi_1 + \ldots + \xi_n$. We define $X_n = \left(\sum_{k=1}^n \xi_k\right)^2 - n\sigma^2$. Prove that $\{X_n\}$ is a martingale.

(3) *Product martingales.* Let $\xi_1, \xi_2 \ldots$ be positive independent random variables such that $\mathbf{E}[\xi_n] = K$. Let's define

$$X_n = \xi_1 \xi_2 \cdots \xi_n = \prod_{k=1}^n \xi_k.$$

 (i) Prove that $\{X_n\}$ is a martingale that converges with probability 1 to an integrable random variable X, if $K = 1$. Also prove that $\{X_n\}$ is a submartingale if $K \geqslant 1$ and it is supermartingale if $K \leqslant 1$.

(ii) Suppose specifically that ξ_n assumes values $1/2$ and $3/2$ with probability $1/2$ each. If $X = lim_n X_n$ prove that $X = 0$ with probability 1. This gives an example where

$$\mathbf{E}\left[\prod_{n=1}^{\infty} \xi_n\right] \neq \prod_{n=1}^{\infty} \mathbf{E}[\xi_n]$$

for independent, positive, and integrable random variables. Prove, however, that it is always true that

$$\mathbf{E}\left[\prod_{n=1}^{\infty} \xi_n\right] \leqslant \prod_{n=1}^{\infty} \mathbf{E}[\xi_n].$$

(4) Prove Proposition 3.5.

(5) Prove Corollary 3.6.

(6) Show that the τ_n defined in the Doob's upcrossing inequality, Theorem 3.13, are indeed stopping times.

(7) Suppose that $\{X_n\}$ is a martingale satisfying $\mathbf{E}[X_n] = 0$ and $\mathbf{E}[X_n^2] < \infty$. Prove that

$$\mathbf{E}\left[(X_{n+r} - X_n)^2\right] = \sum_{k=1}^{r} \mathbf{E}\left[(X_{n+k} - X_{n+k-1})^2\right]$$

i.e., the variance of the sum is the sum of the variances. Assume that

$$\sum_{k=1}^{r} \mathbf{E}\left[(X_{n+k} - X_{n+k-1})^2\right] < \infty.$$

Prove that $\{X_n\}$ converges with probability 1.

(8) If U_n is the number of up-crossings of $[r_1, r_2]$ by a $\{\mathscr{F}_k\}$-submartingale $\{X_k\}_{k=1}^{n}$ with $\mathscr{F}_k = \sigma\left\{X_j, 1 \leqslant j \leqslant k\right\}$, prove that

$$\mathbf{P}\{U_n \geqslant K\} \leqslant (r_2 - r_1)^{-1}\, \mathbf{E}\left[(X_n - r_1)^+ \chi_{\{U_n=k\}}\right], \ k \geqslant 1.$$

(9) If $\{X_n\}$ is a martingale that is bounded above or below, prove that

$$\sup_{n} \mathbf{E}\left[|X_n|\right] < \infty.$$

(10) Given $X = \{X_n\}$ a $\{\mathscr{F}_n\}$-martingale, $\{\Delta X_n\}$ are its martingale differences with $\mathbf{E}[\Delta X_n] = 0, \mathbf{E}\left[|\Delta X|^p\right] < \infty$ for some $p \geqslant 2$, prove that $\left\{\left|X_n/n^{1/2}\right|^p, n \geqslant 1\right\}$ are uniformly integrable.

[Hint: Consider $\mathbf{E}\left[\left|\sum_{k=1}^{n} \Delta X_n'\right|^{p+1}\right]$ and $\mathbf{E}\left[\left|\sum_{k=1}^{n} \Delta X_n''\right|^{p}\right]$ where

$$\Delta X_n' = \Delta X_n \chi_{[|\Delta X_n|\leqslant M]} - \mathbf{E}\left[\Delta X_n \chi_{[|\Delta X_n|\leqslant M]} \mid \Delta X_1 \cdots \Delta X_{n-1}\right]$$

and $\Delta X_n'' = \Delta X_n - \Delta X_n'$.]

(11) *Exponential martingales.* Let $\{\xi_n\}$ be a sequence of independent identically distributed random variables with normal standard $N(0, 1)$ distribution. Let $S_n = \xi_1 + \cdots + \xi_n$ and denote for any real t different from zero,

$$X_n(t) = \exp\left(t S_n - \frac{nt^2}{2}\right).$$

Prove that $\{X_n(t)\}$ is a martingale which converges almost surely to zero, and prove that nevertheless $\{X_n(t)\}$ does not converge in L^1.

(12) *Autoregressive martingales.* Let $\{X_n\}$ be an stochastic process defined, for all $n \geqslant 0$, by

$$X_{n+1} = \theta X_n + (1 - \theta)\varepsilon_{n+1}$$

where $X_0 = p$ with $0 < p < 1$ and the parameter $0 < \theta < 1$. Assume that the conditional distribution of ε_{n+1} given $\mathscr{F}_n$ is Bernoulli distributed with parameter X_n. Prove, by induction, that $0 < X_n < 1$. Moreover, prove that $\{X_n\}$ is a L^1-bounded martingale which converges almost surely to a random variable X. Also prove that this convergence also holds in L^1 and X has Bernoulli distribution with parameter p. This process is called *autoregresive process*.

(13) *Poisson martingales* A stochastic process $\{N_t\}_{t \geq 0}$ is said to be a *Poisson process* with mean λ, if for all $k \geq 1$ and all $t_0 < t_1 < \cdots < t_k$, $N_{t_j} - N_{t_{j-1}}, j \leqslant k$ are independent, random variables having Poisson distribution with parameters $\lambda \left(t_i - t_{j-1}\right)$ respectively. Let $X_n := N_n - n\lambda$, and $\mathscr{F}_n := \sigma\left(N_t : t \leqslant n\right)$, $n \geqslant 1$. Prove that $\{X_n\}$ is a $\{\mathscr{F}_n\}$-martingale.

(14) *Bienaymé–Galton–Watson branching process* Suppose we have independent identically distributed random variables $\left\{\xi_{n,k} : n \geq 1, k \geq 1\right\}$ which take values in $\{0, 1, 2, \ldots\}$, and have a finite mean μ. Let $X_0 := 1$ and for $n \geq 1$,

$$X_n := \begin{cases} \xi_{n,1} + \xi_{n,2} + \cdots + \xi_{n,Z_{n-1}} & \text{if } Z_{n-1} > 0 \\ 0 & \text{if } Z_{n-1} = 0 \end{cases}$$

The idea is that we start with a single parent X_0, called the root. The kth parent in the $(n-1)$th generation produces $\xi_{n,k}$ offspring. At the nth generation, the total number of progeny is X_n. Let $\mathscr{F}_0 := \{\emptyset, \Omega\}$ and $\mathscr{F}_n := \sigma\left(\xi_{j,k} : j \leq n, 1 \leqslant k < \infty\right)$. Prove that $\left\{\mu^{-n} X_n\right\}$ is a $\{\mathscr{F}_n\}$-martingale.

(15) Let $([0, 1), \mathscr{B}[0, 1), m)$, as usual, and $\{\mathscr{F}_n\}_n$ the σ-algebras generated by those $f \in C([0, 1); \mathbb{C})$ that are periodic with period 2^{-n} for each $x \in \mathbb{N}$, and

$$\{\mathbf{e}_k = \exp(2\pi k i x) : k \in \mathbb{Z}\},$$

as in Exercise 7 of Chap. 2. Next, take $S_m = \{(2k + 1)2^m : k \in \mathbb{Z}\}$ for each $m \in \mathbb{N}$, and, for $f \in L^2([0, 1), m; \mathbb{C})$, set

$$\Delta_m(f) = \sum_{\ell \in S_m} \langle f, \mathbf{e}_\ell \rangle_{L^2([0,1))} \mathbf{e}_\ell$$

where the convergence is in $L^2([0, 1), m; \mathbb{C})$. Then, by Exercise 7 of Chap. 2,

$$f - \mathbf{E}\left[f \mid \mathscr{F}_{n+1}\right] = \sum_{m=0}^{n} \Delta_m(f).$$

Since $\{\mathscr{F}_n\}_n$ is non-increasing, use the Theorem 3.23, on convergence for reversed martingales, to prove that the expansion

$$f = \langle f, 1 \rangle_{L^2([0,1))} + \sum_{m=0}^{\infty} \Delta_m(f)$$

converges both almost everywhere as well as in $L^2([0, 1), m; \mathbb{C})$.[3]

---

[3] Therefore, the Fourier series of $f \in L^2([0, 1), m; \mathbb{C})$ converges almost everywhere, as mentioned by Strook [155] "this result that was discovered originally by Kolmogorov. Ever since L. Carleson's definitive theorem on the almost every convergence of the Fourier series of an arbitrary square integrable function, the interest in this result of Kolmogorov is mostly historical."

Advanced Topics in Martingale Theory

4

In this chapter, we will extend the results of the classical martingale theory. The main references for this chapter are D. L. Burkholder's *Martingale Transform* [29], which opened a new era in the development of the martingale theory, *Distribution Function Inequalities* [30], and D. Burkholder and R. Gundy *Extrapolation and interpolation and quasi-linear operators on martingales* [44].

Recall from the previous chapter, see Definitions 3.1 and 3.4, that given a probability space $(\Omega, \mathscr{F}, \{\mathscr{F}_n\}, \mathbf{P})$ with $\{\mathscr{F}_n\}$ a filtration on Ω, i.e., an increasing family of sub-σ algebras of $\mathscr{F}$, and given $X = \{X_n\}$ a $\{\mathscr{F}_n\}$-martingale, the sequence of *martingale differences* $\{\Delta X_n\}$ is defined as $\Delta X_1 = X_1$, and for $n \geqslant 2$, $\Delta X_n = X_n - X_{n-1}$. Then, it is clear that

$$X_n = \sum_{k=1}^{n} \Delta X_k.$$

On the other hand, given $\{Y_n\}$ be a $\{\mathscr{F}_n\}$-predictable sequence, which means that for any n, Y_n is $\mathscr{F}_{n-1}$-measurable, we have that the *transform* of X by Y, denoted as $(Y \circ X) = \{(Y \circ X)_n\}_n$, is defined as

$$(Y \circ X)_n = \sum_{k=1}^{n} Y_k \Delta X_k.$$

The transform $\{(Y \circ X)_n\}$ is not necessarily a martingale. It is easy to see that $(Y \circ X)$ is a martingale if and only if

$$\mathbf{E}\left[\left|(Y \circ X)_n\right|\right] < \infty$$

© The Author(s), under exclusive license to Springer Nature Switzerland AG 2025
W. Urbina-Romero and R. Rios, *An Introduction to the Modern Martingale Theory and Applications*, Texts in Applied Mathematics 81,
https://doi.org/10.1007/978-3-031-88903-5_4

for each n and this condition is satisfied if for instance every Y_n is bounded, see Proposition 3.3. In what follows we assume the sequence $\{Y_n\}$ is uniformly bounded (most of the time by 1, for simplicity) and that is what we call a *multiplier sequence*.

Later, in [43] Burkholder extended the notion of martingale transform as follows

Definition 4.1 Let $X = \{X_n\}$ and $Z = \{Z_n\}$ be $\{\mathscr{F}_n\}$-adapted and integrable stochastic processes and consider their increments $\{\Delta X_n\}$ and $\{\Delta Z_n\}$. If

$$|\Delta Z_n| \leqslant |\Delta X_n| \text{ for } n \geqslant 0 \tag{4.1}$$

is satisfied, then Z is said to be differentially subordinate to X.

If

$$\left|\mathbf{E}\left[\Delta Z_n \mid \mathscr{F}_{n-1}\right]\right| \leqslant \left|\mathbf{E}\left[\Delta X_n \mid \mathscr{F}_{n-1}\right]\right| \text{ for } n \geqslant 1 \tag{4.2}$$

is satisfied, then Z is conditionally differentially subordinate to X.

If both of the conditions (4.1) and (4.2) are satisfied, then Z is strongly differentially subordinate to X or, more simply, Z is strongly subordinate to X.

Of course, if X and Z are $\{\mathscr{F}_n\}$-martingales, then both sides of (4.2) vanish and therefore (4.2) is trivially satisfied. If $\Delta Z_n = Y_n \Delta X_n$, where $\{Y_n\}$ is a multiplier sequence, then Z is strongly subordinate to X which coincides precisely with the notion of martingale transform, i.e., $Z = Y \circ X$. However, strong subordination is less restrictive and leads to a wider class of applications. Many of the results for martingale transforms carry over to this case. Nevertheless, we are not going to consider strong subordination in detail in these notes. For the martingale case we refer to the original papers by Burkholder [41,42], and more recent works by Bañuelos and Osękowski, [11,12].

4.1 Decomposition Theorems

The results on decomposition have proven to be very important both in analysis and in probability. We will study several martingale decomposition theorems that will be very useful in later sections. First, let us consider a classical operator defined on martingales.

Definition 4.2 On $\mathscr{M}$, the set of all $\{\mathscr{F}_n\}$-martingales defined on the probability space $(\Omega, \mathscr{F}, \mathbf{P})$, given $X \in \mathscr{M}$ for each $n \geqslant 1$, define the operator $S_n(X)$, known as the quadratic function of order n of X, as

$$S_n(X) = \left(\sum_{k=1}^{n} (\Delta X)_k^2\right)^{1/2}, \tag{4.3}$$

and the operator $S(X)$, known as the quadratic function of X,[1] as

$$S(X) = \left(\sum_{k=1}^{\infty} (\Delta X)_k^2 \right)^{1/2}, \tag{4.4}$$

if the series converges, otherwise it would be infinite.

As we have already mentioned in Chap. 1, for a discussion of the role of the quadratic function in both analysis and probability, we recommend E. Stein's article *The development of square functions in the work of Zygmund* [151]. As we shall see in the next section, the quadratic function plays a central role in one of the basic inequalities of contemporary martingale theory.

In addition, let us define the *quadratic variation* for a martingale.

Definition 4.3 Given $\{X_n\}$ a $\{\mathscr{F}_n\}$-martingale, the quadratic variation of X of order n is defined as

$$\langle X \rangle_n = \sum_{k=1}^{n} (\Delta X_k)^2, \tag{4.5}$$

and the quadratic variation of X as

$$\langle X \rangle = \sum_{k=1}^{\infty} (\Delta X_k)^2, \tag{4.6}$$

if the series converges, otherwise it would be infinite.

Even though the quadratic variation is simply the square of the quadratic function, it plays an important role in stochastic integration as well for the law of large numbers that will be discussed later.

We also define here the quadratic covariation which is going to be very important for the formulation of the *discrete Itô's formula,* see (4.35) in Sect. 4.3.

Definition 4.4 Let $X = \{X_n\}_{0 \leqslant n \leqslant N}$ and $Y = \{Y_n\}_{0 \leqslant n \leqslant N}$ be two sequences of random variables on the probability space $(\Omega, \mathscr{F}, \mathbf{P})$, $X_0 = Y_0 = 0$, the quadratic covariation of X and Y, $\langle X, Y \rangle = \{\langle X, Y \rangle_n\}_{0 \leqslant n \leqslant N}$, is defined for $n \geqslant 1$, as

$$\langle X, Y \rangle_n = \sum_{k=1}^{n} \Delta X_k \Delta Y_k. \tag{4.7}$$

---

[1] Also called *square function.*

The martingale differences $\{\Delta X_n\}_n$ of any L^2 bounded martingale are then orthogonal and therefore it follows trivially that $\|S(X)\|_2 = \|X\|_2$ which implies $S(X) < \infty$ a.e.

In 1966, Austin in [3], strengthened that by proving the following result.

Theorem 4.1 (Austin) *Let $X = \{X_n\}$ be a L^1-bounded $\{\mathscr{F}_n\}$-martingale, then $S(X)$ is finite a.e.*

Proof Let $E_n = \bigcap_{k<n} \{|X_k| < M\}$ and $E = \bigcap_n E_n$, where M is a positive number and define $\hat{X}_n = X_n \chi_{\{|X_n| < M\}}$ a truncation of X_n. Then, by the martingale property

$$\int_{E_{n-1}} X_n X_{n-1} d\mathbf{P} = \int_{E_{n-1}} X_{n-1}^2 d\mathbf{P} = \int_{E_{n-1}} \hat{X}_{n-1}^2 d\mathbf{P},$$

and this implies

$$\int_{E_{n-1}} \left(\hat{X}_n - \hat{X}_{n-1}\right)^2 d\mathbf{P} = \int_{E_{n-1}} \left(\hat{X}_n^2 - \hat{X}_{n-1}^2\right) d\mathbf{P} + 2 \int_{E_{n-1}} \left(X_n X_{n-1} - \hat{X}_n \hat{X}_{n-1}\right) d\mathbf{P}.$$

Let N be a positive integer, then

$$\sum_{n \leqslant N} \int_{E_{n-1}} \left(\hat{X}_n^2 - \hat{X}_{n-1}^2\right) d\mathbf{P} = \int_{E_{N-1}} \hat{X}_N^2 d\mathbf{P} + \sum_{n=1}^{N-1} \int_{E_{n-1} \setminus E_n} \hat{X}_n^2 d\mathbf{P}$$

$$= \int_{E_{N-1}} \hat{X}_N^2 d\mathbf{P} \leqslant M^2.$$

Using the fact that $X_{n-1} = \hat{X}_{n-1}$ in E_{n-1}, $\left|X_n - \hat{X}_n\right| \leqslant |X_n|$ and that $\{|X_n|\}$ is a submartingale, we have

$$\int_{E_{n-1}} \left(X_n X_{n-1} - \hat{X}_n \hat{X}_{n-1}\right) d\mathbf{P} \leqslant M \int_{E_{n-1}} \left|X_n - \hat{X}_n\right| d\mathbf{P} = M \int_{E_{n-1} \setminus E_n} \left|X_n - \hat{X}_n\right| d\mathbf{P}$$

$$\leqslant M \int_{E_{n-1} \setminus E_n} |X_n| d\mathbf{P} \leqslant M \int_{E_{n-1} \setminus E_n} |X_N| d\mathbf{P},$$

if $n \leqslant N$. Then,

$$\sum_{n \leqslant N} \int_{E_{n-1}} \left(X_n X_{n-1} - \hat{X}_n \hat{X}_{n-1}\right) d\mathbf{P} \leqslant M \int |X_N| d\mathbf{P} \leqslant M \|X\|_1,$$

and therefore by the monotone convergence theorem

$$\int_E \sum_n (X_n - X_{n-1})^2 d\mathbf{P} = \int_E \sum_n \left(\hat{X}_n - \hat{X}_{n-1}\right)^2 d\mathbf{P} = \sum_n \int_E \left(\hat{X}_n - \hat{X}_{n-1}\right)^2 d\mathbf{P}$$

$$\leqslant \sum_n \int_{E_{n-1}} \left(\hat{X}_n - \hat{X}_{n-1}\right)^2 d\mathbf{P} \leqslant M^2 + 2M \|X\|_1 < \infty.$$

Thus, $S(X)$ is finite a.e on of the event E. Now, by the martingale convergence theorem $\sup_n |X_n| < \infty$ a.e. and therefore, taking M big enough the probability of the event E tends to 1, hence $S(X)$ is finite a.e. $\qquad\square$

Note that, in particular, this result implies that the integral of $S^2(X)$ over the set $\{X^* \leqslant \lambda\}$ is finite, for any $\lambda > 0$. The following lemma can be considered as a refinement of it.

Lemma 4.1 *Assume that $X = \{X_n\}$ is either a $\{\mathscr{F}_n\}$-martingale or a non-negative, L^1-bounded $\{\mathscr{F}_n\}$-submartingale. Define μ, a stopping time as*

$$\mu = \inf\{n : |X_n| > \lambda\},$$

where $\lambda > 0$. Then

$$\left\| S_{\mu-1}(X) \right\|_2^2 + \left\| X_{\mu-1} \right\|_2^2 \leqslant 2\mathbf{E}\left[X_\mu X_{\mu-1} \right] \leqslant 2\lambda \|X\|_1 \tag{4.8}$$

and equality holds in the first inequality if X is a martingale.

Recall that $\inf(\emptyset) = \infty$. Now, since X is L^1-bounded then $X_\infty = \lim_{n\to\infty} X_n$ exists a.e. by the martingale convergence theorem. Therefore $S_{\mu-1}(x)$ and $X_{\mu-1}$ are well defined.

Proof By construction $\left| X_{\mu-1} \right| \leqslant \lambda$, then

$$\mathbf{E}\left[X_\mu X_{\mu-1} \right] \leqslant \lambda \left\| X_\mu \right\|_1 \leqslant \lambda \|X\|_1.$$

But by a simple calculation, we have that

$$S_{n-1}^2(X) + X_{n-1}^2 = 2X_n X_{n-1} - 2\sum_{k=1}^{n} X_{k-1}\Delta X_k.$$

Let $v = \mu \wedge n$, then, given that $\chi_{\{\mu>k\}}$ is $\mathscr{F}_{k-1}$-measurable for any k, and that $X_{k-1}\chi_{\{\mu>k\}}$ is bounded, we get by the orthogonality of the increments,

$$\mathbf{E}\left[\sum_{k=1}^{v} X_{k-1}\Delta X_k \right] = \sum_{k=1}^{v} \mathbf{E}\left[X_{k-1}\chi_{\{\mu\geqslant k\}}\mathbf{E}\left[\Delta X_k \mid \mathscr{F}_{k-1} \right] \right] \geqslant 0,$$

therefore

$$\mathbf{E}\left[S_{\mu-1}^2(X) \right] + \mathbf{E}\left[X_{\mu-1}^2 \right] \leqslant 2\mathbf{E}\left[X_\mu X_{\mu-1} \right],$$

and equality holds if X is a martingale.

Since $\sup_n X_{v-1}^2 \leqslant \lambda^2$ and $\sup_n |X_v X_{v-1}| \leqslant \lambda^2 + \lambda |X_\mu|$ are integrable let $n \to \infty$, obtaining the left-hand side of the inequality. $\qquad\square$

Definition 4.5 Let $\{Y_n\}_{n=0}^{\infty}$ be a sequence of random variables, adapted to the filtration $\{\mathscr{F}_n\}$, $\{Y_n\}$ is said to be increasing, if for almost every $\omega \in \Omega$, we have

$$Y_0(\omega) \leqslant Y_1(\omega) \leqslant Y_2(\omega) \leqslant \cdots \quad \text{and} \quad \mathbf{E}[Y_n] < \infty$$

for all $n \in \mathbb{N}$.

It is clear that both $\{S_n(X)\}$ and $\{\langle X \rangle_n\}$ are increasing sequences. The following result is key to stochastic integration theory.

Proposition 4.1 *Let $\{X_n\}$ be a $\{\mathscr{F}_n\}$-martingale, then $\{X_n^2\}$ is a $\{\mathscr{F}_n\}$-submartingale and $\{X_n^2 - \langle X \rangle_n\}$ is a $\{\mathscr{F}_n\}$-martingale. Therefore, X_n^2 can be written as the sum of a martingale and an increasing process*

$$X_n^2 = \left(X_n^2 - \langle X \rangle_n\right) + \langle X \rangle_n, \tag{4.9}$$

and therefore we obtain

$$\mathbf{E}\left[X_n^2\right] = \mathbf{E}\left[\langle X \rangle_n\right]. \tag{4.10}$$

Proof Let $m > n$, then,

$$\mathbf{E}\left[X_m^2 - \langle X \rangle_m \mid \mathscr{F}_n\right] = \mathbf{E}\left[X_m^2 - \sum_{k=1}^{m} (\Delta X_k)^2 \mid \mathscr{F}_n\right]$$

$$= \mathbf{E}\left[X_m^2 - \sum_{k=1}^{n} (X_k - X_{k-1})^2 - \sum_{k=n+1}^{m} (X_k - X_{k-1})^2 \mid \mathscr{F}_n\right]$$

$$= \mathbf{E}\left[X_m^2 \mid \mathscr{F}_n\right] - \sum_{k=1}^{n} (X_k - X_{k-1})^2$$

$$\qquad - \mathbf{E}\left[\sum_{k=n+1}^{m} \left(X_k^2 - 2X_k X_{k-1} + X_{k-1}^2\right) \mid \mathscr{F}_n\right]$$

$$= \mathbf{E}\left[X_m^2 \mid \mathscr{F}_n\right] - \sum_{k=1}^{n} (X_k - X_{k-1})^2 - \sum_{k=n+1}^{m} \mathbf{E}\left[X_k^2 \mid \mathscr{F}_n\right]$$

$$\qquad + 2\sum_{k=n+1}^{m} \mathbf{E}\left[X_k X_{k-1} \mid \mathscr{F}_n\right] - \sum_{k=n+1}^{m} \mathbf{E}\left[X_{k-1}^2 \mid \mathscr{F}_n\right]$$

$$= \mathbf{E}\left[X_m^2 \mid \mathscr{F}_n\right] - \sum_{k=1}^{n} (X_k - X_{k-1})^2 - \sum_{k=n+1}^{m} \mathbf{E}\left[X_k^2 \mid \mathscr{F}_n\right]$$

$$\qquad + \sum_{k=n+1}^{m} \mathbf{E}\left[X_{k-1}^2 \mid \mathscr{F}_n\right]$$

$$= X_n^2 - \sum_{k=1}^{n} (X_k - X_{k-1})^2 = X_n^2 - \langle X \rangle_n,$$

since, using the properties of conditional expectation,

$$\sum_{k=n+1}^{m} \mathbf{E}\left[X_k X_{k-1} \mid \mathscr{F}_n\right] = \sum_{k=n+1}^{m} \mathbf{E}\left[\mathbf{E}\left[X_k X_{k-1} \mid \mathscr{F}_{k-1}\right] \mid \mathscr{F}_n\right]$$

$$= \sum_{k=n+1}^{m} \mathbf{E}\left[X_{k-1}\mathbf{E}\left[X_k \mid \mathscr{F}_{k-1}\right] \mid \mathscr{F}_n\right]$$

$$= \sum_{k=n+1}^{m} \mathbf{E}\left[X_{k-1}^2 \mid \mathscr{F}_n\right].$$

Formula (4.10) is called *Itô's isometry formula* in this discrete setting. In particular, if we consider $\{Y_k\}$ a multiplier sequence and a $\{X_n\}$ a $\{\mathscr{F}_n\}$-martingale, we have for the martingale transform $\{(Y \circ X)_n\}$, Itô's isometry formula is then

$$\mathbf{E}\left[\left(\sum_{k=1}^{n} Y_k \Delta X_k\right)^2\right] = \mathbf{E}\left[(Y \circ X)_n^2\right] = \mathbf{E}\left[\langle (Y \circ X)\rangle_n\right] = \mathbf{E}\left[\sum_{k=1}^{n} Y_k^2 \left(\Delta X_k\right)^2\right].$$

$$(4.11)$$

The following result is another key result for the development of the theory of stochastic integration.

Theorem 4.2 (Dobb's decomposition) *Let $X = \{X_n\}$ be a $\{\mathscr{F}_n\}$-submartingale, then X admits a unique decomposition*

$$X_n = Z_n + Y_n \tag{4.12}$$

for each $n \geqslant 1$, where $\{Z_n\}$ is a $\{\mathscr{F}_n\}$-martingale and $\{Y_n\}$ is an increasing process.

Proof Let us define

$$Y_0 = 0, \text{ and for } n \geqslant 1$$
$$Y_n = Y_{n-1} - X_{n-1} + \mathbf{E}\left[X_n \mid \mathscr{F}_{n-1}\right] = Y_{n-1} + \mathbf{E}\left[X_n - X_{n-1} \mid \mathscr{F}_{n-1}\right]$$
$$= \sum_{k=1}^{n} \mathbf{E}\left[X_k - X_{k-1} \mid \mathscr{F}_{k-1}\right].$$

Clearly, $\{Y_n\}$ is predictable and, by the submartingale property, an increasing process. Moreover, $Z_n = X_n - Y_n$ is a $\{\mathscr{F}_n\}$-martingale, since

$$\mathbf{E}\left[Z_n - Z_{n-1} \mid \mathscr{F}_{n-1}\right] = \mathbf{E}\left[X_n - Y_n - X_{n-1} + Y_{n-1} \mid \mathscr{F}_{n-1}\right]$$
$$= \mathbf{E}\left[X_n \mid \mathscr{F}_{n-1}\right] - X_{n-1} - Y_n + Y_{n-1} = 0.$$

The decomposition is unique since if $X_n = Z'_n + Y'_n$ is another decomposition, taking $Z_n - Z'_n = Y'_n - Y_n \in \mathscr{F}_{n-1}$, then

$$Z_n - Z'_n = E\left[Z_n - Z'_n \mid \mathscr{F}_{n-1}\right] = Z_{n-1} - Z'_{n-1}$$

and since $Z_0 - Z'_0 = X_0 - X_0 = 0$ then $Z_n - Z'_n = 0$, for all n. □

The continuous parameter version of this result, which is the necessary result for stochastic integration and is known as the *Doob-Meyer decomposition*, is a technically very difficult result, see [113].

Definition 4.6 An increasing sequence $\{Y_n\}$ is natural if for any bounded $\{\mathscr{F}_n\}$-martingale $\{X_n\}$ we have

$$E\left[X_n Y_n\right] = E\left[\sum_{k=1}^{n} X_{k-1}\left(Y_k - Y_{k-1}\right)\right]$$

for all $n \geqslant 1$.

Note that this condition is equivalent to ask that any martingale transform $(Y \circ X)$ by Y for any bounded martingale X satisfies $E\left[(Y \circ X)_n\right] = 0$ for all $n > 0$. Since, $Y_0 = 0$

$$(Y \circ X)_n = \sum_{k=1}^{n} Y_k\left(X_k - X_{k-1}\right) = \sum_{k=1}^{n} Y_k X_k - \sum_{k=0}^{n-1} Y_{k+1} X_k$$

$$= Y_n X_n - \sum_{k=1}^{n-1} X_k\left(Y_{k+1} - Y_k\right) - Y_1 X_0$$

$$= Y_n X_n - \sum_{k=1}^{n} X_{k-1}\left(Y_k - Y_{k-1}\right),$$

then

$$E\left[(Y \circ X)_n\right] = E\left[Y_n X_n\right] - E\left[\sum_{k=1}^{n} X_{k-1}\left(Y_k - Y_{k-1}\right)\right] = 0,$$

and this is equivalent to say that $\{Y_n\}$ is natural, since $\left\{(Y \circ X)_n\right\}$ is a martingale.

Thus, every predictable increasing sequence is natural. The two concepts are equivalent as we will prove next.

Proposition 4.2 *An increasing sequence $Y = \{Y_n\}$ is predictable if and only if it is natural.*

Proof Note that if Y is natural and X is a bounded martingale, then considering the martingale transform of X by Y,

$$\mathbf{E}\left[Y_n\left(X_n - X_{n-1}\right)\right] = \mathbf{E}\left[(Y \circ X)_n\right] - \mathbf{E}\left[(Y \circ X)_{n-1}\right] = 0$$

for all $n \geqslant 1$ and then we have

$$\mathbf{E}\left[X_n\left(Y_n - \mathbf{E}\left[Y_n \mid \mathscr{F}_{n-1}\right]\right)\right] = \mathbf{E}\left[(X_n - X_{n-1})\,Y_n\right] + \mathbf{E}\left[X_{n-1}\left(Y_n - \mathbf{E}\left[Y_n \mid \mathscr{F}_{n-1}\right]\right)\right]$$
$$-\mathbf{E}\left[(X_n - X_{n-1})\,\mathbf{E}\left[Y_n \mid \mathscr{F}_{n-1}\right]\right] = 0.$$

Let $n \geqslant 1$ be an arbitrary integer, let us prove that $Y_n \in \mathscr{F}_{n-1}$. Consider the identity above, then for the integer n fixed consider the martingale X, given by

$$X_k = \begin{cases} \operatorname{sgn}\left[Y_n - \mathbf{E}\left[Y_n \mid \mathscr{F}_{n-1}\right]\right] & k = n \\ Y_n & k > n \\ E\left[Y_n \mid \mathscr{F}_n\right] & k = 0, 1, \ldots, n - 1 \end{cases}$$

then we have $\mathbf{E}\left[Y_n - \mathbf{E}\left[Y_n \mid \mathscr{F}_{n-1}\right]\right] = 0$, and the proof is done. $\qquad\square$

In the continuous case, the definition of an increasing process, an integrable process, and a natural process are analogous to each other however the equivalence between a predictable process and a natural process is not true.

Now, analogous to the fact that a function can be decomposed into its positive and negative parts, we have the following decomposition for martingales

Theorem 4.3 (Krickerberg's decomposition) *Let $X = \{X_n\}$ a $\{\mathscr{F}_n\}$-submartingale that satisfies $\sup_n \mathbf{E}\left[X_n^+\right] < \infty$, then*

$$X_n = Y_n - Z_n,$$

where $Y = \{Y_n\}_n$ is a positive $\{\mathscr{F}_n\}$-martingale and $Z = \{Z_n\}$ is a positive $\{\mathscr{F}_n\}$-supermartingale.

Proof First, $\left\{X_n^+\right\}$ is a positive $\{\mathscr{F}_n\}$-submartingale, L^1-bounded, since

$$\mathbf{E}\left[X_{n+1}^+ \mid \mathscr{F}_n\right] \geqslant \mathbf{E}\left[X_{n+1} \mid \mathscr{F}_n\right] \geqslant X_n$$

and therefore, taking the positive part, we get[2]

$$\mathbf{E}\left[X_{n+1}^+ \mid \mathscr{F}_n\right] \geqslant X_n^+.$$

---

[2] Alternatively, we can get this directly using Jensen's inequality,

$$\mathbf{E}\left[X_{n+1}^+ \mid \mathscr{F}_n\right] \geqslant \left(\mathbf{E}\left[X_{n+1} \mid \mathscr{F}_n\right]\right)^+ \geqslant X_n^+.$$

Fixing n let us consider the sequence $\left\{ \mathbf{E}\left[X_m^+ \mid \mathscr{F}_n \right] \right\}_{m \geq n}$. It is a monotone sequence of positive random variables . If $m \geq n$

$$\mathbf{E}\left[X_{m+1}^+ \mid \mathscr{F}_n \right] = \mathbf{E}\left[\mathbf{E}\left[X_{m+1}^+ \mid \mathscr{F}_n \right] \mid \mathscr{F}_m \right] \geq \mathbf{E}\left[X_m^+ \mid \mathscr{F}_n \right].$$

Let us define $Y_n = \lim_{m \to \infty} \mathbf{E}\left[X_m^+ \mid \mathscr{F}_n \right]$. We will prove that $Y = \{Y_n\}$ is a positive $\{\mathscr{F}_n\}$-martingale L^1-bounded.

$$\mathbf{E}[Y_n] = \lim_{m \to \infty} \mathbf{E}\left[\mathbf{E}\left[X_m^+ \mid \mathscr{F}_n \right] \right] = \lim_{m \to \infty} \mathbf{E}\left[X_m^+ \right] < \sup_n \mathbf{E}\left[X_n^+ \right] < \infty$$

this implies that $Y_n < \infty$ a.e. and that $\{Y_n\}$ is L^1-bounded, moreover

$$\mathbf{E}\left[Y_{n+1} \mid \mathscr{F}_n \right] = \lim_{m \to \infty} \mathbf{E}\left[\mathbf{E}\left[X_n^+ \mid \mathscr{F}_{n+1} \right] \mid \mathscr{F}_n \right] = \lim_{m \to \infty} \mathbf{E}\left[X_m^+ \mid \mathscr{F}_n \right] = Y_n.$$

Now, define $Z_n = Y_n - X_n$. We will prove that $Z = \{Z_n\}$ is a positive $\{\mathscr{F}_n\}$-supermartingale, L^1-bounded. This is immediate, since

$$\mathbf{E}\left[Z_{n+1} \mid \mathscr{F}_n \right] = \mathbf{E}\left[Y_{n+1} \mid \mathscr{F}_n \right] - \mathbf{E}\left[X_{n+1} \mid \mathscr{F}_n \right] \leq Y_n - X_n = Z_n,$$

$Z_n \geq X_n^+ - \left(X_n^+ - X_n^- \right) = X_n^- \geq 0$ and $\sup_n \mathbf{E}\left[Z_n \right] < 2 \sup_n \mathbf{E}\left[X_n^+ \right] < \infty.$ $\square$

As an immediate consequence, we obtain the following result.

Corollary 4.4 *Let $X = \{X_n\}$ be a $\{\mathscr{F}_n\}$-martingale L^1-bounded, then there exists two positive $\{\mathscr{F}_n\}$-martingales L^1-bounded $Y = \{Y_n\}$ and $Z = \{Z_n\}$ such that*

$$X_n = Y_n - Z_n.$$

Now we consider *Gundy's decomposition* [88], which is the analog in martingale theory to the famous *Calderón-Zygmund decomposition lemma* in harmonic analysis. In both the idea is similar, to decompose the object into a "good part" and a "bad part." For a more detailed discussion of this see [88, 89]. The following is a simplified version of it due to Burkholder [30].

Theorem 4.5 (Gundy's decomposition) *Let $X = \{X_n\}_n$ be a $\{\mathscr{F}_n\}$-martingale L^1 bounded and λ a non-negative real number. Then, there exist $\{\mathscr{F}_n\}$-martingales Y, Z, W, such that*

(i) $X = Y + Z + W$.
(ii) $\|Y\|_2^2 \leq 2\lambda \|X\|_1$.
(iii) $\left\| \sum_{k=1}^{\infty} |\Delta Z_n| \right\|_1 \leq 4 \|X\|_1$.
(iv) $\mathbf{P}\{W^* > 0\} \leq \frac{\|X\|_1}{\lambda}$.

Proof Instead of Gundy's original proof, we will follow Burkholder's proof, which uses a single stopping time. Let $\lambda > 0$ and consider the stopping time

$$\mu = \inf\left\{n : |X_n| > \lambda\right\},$$

and let

$$Y_1 = X_1 \chi_{\{\mu>1\}}, \; Z_1 = X_1 \chi_{\{\mu=1\}}, \; W_1 = 0$$

and for $k \geqslant 2$ let

$$\Delta Y_k = \Delta X_k \chi_{\{\mu>k\}} - \mathbf{E}\left[\Delta X_k \chi_{\{\mu>k\}} \mid \mathscr{F}_{k-1}\right]$$
$$\Delta Z_k = \Delta X_k \chi_{\{\mu=k\}} - \mathbf{E}\left[\Delta X_k \chi_{\{\mu=k\}} \mid \mathscr{F}_{k-1}\right]$$
$$\Delta W_k = \Delta X_k \chi_{\{\mu<k\}}.$$

It is clear, by the orthogonality of the increments, that Y, Z, and W are $\{\mathscr{F}_n\}$-martingales and we have

$$Y_n + Z_n + W_n = \sum_{k=1}^{n} \left[\Delta X_k \chi_{\{\mu>k\}} - \mathbf{E}\left[\Delta X_k \chi_{\{\mu>k\}} \mid \mathscr{F}_{k-1}\right]\right]$$

$$+ \sum_{k=1}^{n} \left[\Delta X_k \chi_{\{\mu=k\}} - \mathbf{E}\left[\Delta X_k X_{\{\mu=k\}} \mid \mathscr{F}_{k-1}\right]\right]$$

$$+ \sum_{k=1}^{n} \Delta X_k - \sum_{k=1}^{n} \Delta X_k \chi_{\{\mu \geqslant k\}},$$

and since $\{\mu \geqslant k\} \in \mathscr{F}_{k-1}$, we get

$$\mathbf{E}\left[\Delta X_k \chi_{\{\mu>k\}} \mid \mathscr{F}_{k-1}\right] = \mathbf{E}\left[\Delta X_k \chi_{\{\mu \geqslant k\}} - \Delta X_k \chi_{\{\mu=k\}} \mid \mathscr{F}_{k-1}\right]$$
$$= -\mathbf{E}\left[\Delta X_k \chi_{\{\mu=k\}} \mid \mathscr{F}_{k-1}\right].$$

Therefore, we have

$$Y_n + Z_n + W_n = \sum_{k=1}^{n} \Delta X_k = X_n,$$

so condition (i) holds. On the other hand, observe that

$$\|\Delta Y_k\|_2^2 = \|\Delta X_k \chi_{\{\mu>k\}} - \mathbf{E}\left[\Delta X_k \chi_{\{\mu>k\}} \mid \mathscr{F}_{k-1}\right]\|_2^2$$

$$= \mathbf{E}\left[\left(\Delta X_k \chi_{\{\mu>k\}}\right)^2\right] - 2\mathbf{E}\left[\Delta X_k \chi_{\{\mu>k\}} \mathbf{E}\left[\Delta X_k \chi_{\{\mu>k\}} \mid \mathscr{F}_{k-1}\right]\right]$$

$$+ \mathbf{E}\left[\mathbf{E}\left[\Delta X_k \chi_{\{\mu>k\}} \mid \mathscr{F}_{k-1}\right]^2\right]$$

$$= \mathbf{E}\left[\left(\Delta X_k \chi_{\{\mu>k\}}\right)^2\right] - 2\mathbf{E}\left[\mathbf{E}\left[\Delta X_k \chi_{\{\mu>k\}} \mathbf{E}\left[\Delta X_k \chi_{\{\mu>k\}} \mid \mathscr{F}_{k-1}\right]\right] \mid \mathscr{F}_{k-1}\right]$$

$$+ \mathbf{E}\left[\mathbf{E}\left[\Delta X_k \chi_{\{\mu>k\}} \mid \mathscr{F}_{k-1}\right]^2\right]$$

$$= \mathbf{E}\left[\left(\Delta X_k \chi_{\{\mu>k\}}\right)^2\right] - \mathbf{E}\left[\mathbf{E}\left[\Delta X_k \chi_{\{\mu>k\}} \mid \mathscr{F}_{k-1}\right]^2\right]$$

$$\leqslant \|\Delta X_k \chi_{\{\mu>k\}}\|_2^2 \leqslant (2\lambda)^2.$$

Thus,

$$\|\Delta Y_k\|_2 \leqslant 2\lambda.$$

Moreover, by Lemma 4.1, we get

$$\|Y_n\|_2^2 = \sum_{k=1}^{n} \|\Delta Y_n\|_2^2 \leqslant \sum_{k=1}^{n} \|\Delta X_k \chi_{\{\mu>k\}}\|_2^2 \leqslant \|S_{\mu-1}(X)\|_2^2 \leqslant 2\lambda \|X\|_1.$$

Thus, the condition (ii) holds. Now,

$$\left\|\sum_{k=1}^{\infty} |\Delta Z_k|\right\|_1 \leqslant 2\sum_{k=1}^{\infty} \|\Delta X_k \chi_{\{\mu=k\}}\|_1 = 2\|\Delta X_\mu \chi_{\{\mu<\infty\}}\|_1 \leqslant 4\|X_\mu \chi_{\{\mu<\infty\}}\|_1 \leqslant 4\|X\|_1,$$

so (iii) holds. Finally, given that

$$W_n = \sum_{k=1}^{n} \Delta W_k = \sum_{k=1}^{n} \Delta X_k \chi_{(\mu<k\}}$$

we get, using Corollary 3.9 for $p = 1$,

$$\mathbf{P}\{W^* > 0\} \leqslant \mathbf{P}\{\mu < \infty\} \leqslant \mathbf{P}\{X^* > \lambda\} \leqslant \frac{C}{\lambda}\|X\|_1,$$

so condition (iv) holds. $\qquad\square$

The following decomposition is a bit more technical but is key to an important result that we will discuss in the next section (see [64]).

Theorem 4.6 (Davis' decomposition) *Let $X = \{X_n\}_n$ be a $\{\mathscr{F}_n\}$-martingale, then*

$$X = Y + Z,$$

where Y and Z are $\{\mathscr{F}_n\}$-martingales, such that Y has increments $\mathscr{F}_{k-1}$-bounded, i.e.,

$$|\Delta Y_k| \leqslant 4\Delta X_{k-1}^* \in \mathscr{F}_{k-1},$$

where $\Delta X_k^* = \max_{0<j\leqslant k} |\Delta X_j|$, $\Delta X_0 = 0$. Explicitly, we have that Y and Z are defined as

$$Y_n = \sum_{k=1}^{n} \Delta Y_k = \sum_{k=1}^{n} \left[Y_k^{\#} - \mathbf{E}\left[Y_k^{\#} \mid \mathscr{F}_{k-1} \right] \right], \quad \text{where } Y_k^{\#} = \Delta X_k \chi_{\{|\Delta X_k|\leqslant 2\Delta X_{k-1}^*\}}$$

$$Z_n = \sum_{k=1}^{n} \Delta Z_k = \sum_{k=1}^{n} \left[Z_k^{\#} + \mathbf{E}\left[Y_k^{\#} \mid \mathscr{F}_{k-1} \right] \right], \quad \text{where } Z_k^{\#} = \Delta X_k \chi_{\{|\Delta X_k|>2\Delta X_{k-1}^*\}}.$$

Moreover, Z is such that

$$\sum_{k=1}^{\infty} \left| Z_k^{\#} \right| \leqslant 2\Delta X^*. \qquad (4.13)$$

Proof It is clear that $X_n = Y_n + Z_n$ and that Y and Z are martingale, since

$$Y_n + Z_n = \sum_{k=1}^{n} \left(Y_k^{\#} + Z_k^{\#} \right) = \sum_{k=1}^{n} \Delta X_k = X_n, \quad \text{and}$$

$$\mathbf{E}\left[Y_k^{\#} - \mathbf{E}\left[Y_k^{\#} \mid \mathscr{F}_{k-1} \right] \mid \mathscr{F}_{k-1} \right] = \mathbf{E}\left[Y_k^{\#} \mid \mathscr{F}_{k-1} \right] - \mathbf{E}\left[Y_k^{\#} \mid \mathscr{F}_{k-1} \right] = 0 \quad \text{and}$$

$$\mathbf{E}\left[Z_k^{\#} + \mathbf{E}\left[Y_k^{\#} \mid \mathscr{F}_{k-1} \right] \mid \mathscr{F}_{k-1} \right] = \mathbf{E}\left[Z_k^{\#} \mid \mathscr{F}_{k-1} \right] + \mathbf{E}\left[Y_k^{\#} \mid \mathscr{F}_{k-1} \right]$$

$$= \mathbf{E}\left[\Delta X_k \mid \mathscr{F}_{k-1} \right] = 0.$$

Observe that the last equality also implies that

$$\mathbf{E}\left[Y_k^{\#} \mid \mathscr{F}_{k-1} \right] = -\mathbf{E}\left[Z_k^{\#} \mid \mathscr{F}_{k-1} \right],$$

and therefore

$$\left| \mathbf{E}\left[Y_k^{\#} \mid \mathscr{F}_{k-1} \right] \right| = \left| \mathbf{E}\left[Z_k^{\#} \mid \mathscr{F}_{k-1} \right] \right| \leqslant \mathbf{E}\left[\left| Z_k^{\#} \right| \mid \mathscr{F}_{k-1} \right].$$

On the other hand, it is clear that $\left| Y_k^{\#} \right| \leqslant 2\Delta X_{k-1}^*$, by definition and therefore

$$\left| Y_k^{\#} - \mathbf{E}\left[Y_k^{\#} \mid \mathscr{F}_{k-1} \right] \right| \leqslant \left| Y_k^{\#} \right| + \mathbf{E}\left[\left| Y_k^{\#} \right| \mid \mathscr{F}_{k-1} \right] \leqslant 4\Delta X_{k-1}^* \in \mathscr{F}_{k-1}$$

In addition, based on the above observations, the martingale Z is controlled by

$$\sum_{k=1}^{\infty} \left| \Delta Z_k \right| \leqslant \sum_{k=1}^{\infty} \left| Z_k^{\#} \right| + \sum_{k=1}^{\infty} \left| \mathbf{E}\left[Y_k^{\#} \mid \mathscr{F}_{k-1} \right] \right|,$$

$$\leqslant \sum_{k=1}^{\infty} \left| Z_k^{\#} \right| + \sum_{k=1}^{\infty} \mathbf{E}\left[\left| Z_k^{\#} \right| \mid \mathscr{F}_{k-1} \right],$$

and then, we have that, on the set $\left\{ \left| \Delta X_k \right| > 2\Delta X_{k-1}^* \right\}$

$$\left| Z_k^{\#} \right| + 2\Delta X_{k-1}^* = \left| \Delta X_k \right| + 2\Delta X_{k-1}^* \leqslant 2\left| \Delta X_k \right| \leqslant 2\Delta X_k^*,$$

therefore, on the set $\left\{ \left| \Delta X_k \right| > 2\Delta X_{k-1}^* \right\}$, we have

$$\left| Z_k^{\#} \right| \leqslant 2\left(\Delta X_k^* - \Delta X_{k-1}^* \right)$$

and then

$$\sum_{k=1}^{\infty} \left| Z_k^{\#} \right| \leqslant 2\Delta X^*. \qquad \qquad \square$$

4.2   Convergence Theorems

In this section, we will consider a result regarding the convergence for martingale transforms as well as a version of the law of large numbers for martingale and also a version of the central limit theorems. Recall the convergence theorem for martingales, Theorem 3.14, states that if X is a L^1-bounded $\{\mathscr{F}_n\}$-martingale (i.e., $\|X\|_1 = \sup_n \mathbf{E}\,[|X_n|] < \infty$) then $X_n \to X_\infty$ a.e. and additionally by Fatou's lemma, $\mathbf{E}\,[|X_\infty|] < \infty$. However, even though we will prove later that a martingale transform of a L^1-bounded $\{\mathscr{F}_n\}$-martingale converges a.e., $(Y \circ X)_n \to (Y \circ X)_\infty$ as $n \to \infty$ a.e., $\mathbf{E}\,[|(Y \circ X)_\infty|]$ may be infinite, as the next example shows.

Example 4.1 Let $\Omega = \mathbb{N}$ and $\mathbf{P}(\{k\}) = \frac{1}{k} - \frac{1}{k+1}$ then, it is clear that $\mathbf{P}$ is a probability measure on Ω since

$$\mathbf{P}(\Omega) = \lim_{n \to \infty} \sum_{k=1}^{n} \mathbf{P}(\{k\}) = \lim_{k \to \infty} \sum_{k=1}^{n} \left(\frac{1}{k} - \frac{1}{k+1} \right)$$
$$= \lim_{n \to \infty} \left(1 - \frac{1}{n+1} \right) = 1.$$

Define the stochastic process $\{X_n\}$ as follows: for each n, $X_n : \Omega \to \mathbb{R}$ is defined as

$$X_n(k) = \begin{cases} n & \text{if } n < k \\ -1 & \text{if } n \geqslant k \end{cases} = (n+1)\chi_{[n,\infty)}(k) - 1.$$

Let us prove that $\{X_n\}$ is a martingale L^1-bounded with limit X_∞ equal to -1. In fact, the σ-algebra generated by $\{X_n\}$ is given by

$$\mathscr{F}_n = \sigma\left(\{X_1, \ldots, X_n\}\right) = \sigma\left(\left\{X_i^{-1}(\{k\}) : i = 1, 2, \ldots, n;\ k \in \mathbb{N}\right\}\right)$$
$$= \sigma\left(\{\emptyset, \{1, 2, \ldots, i\}, \{i + 1, i + 2, \ldots\},\ i = 1, 2, \ldots, n\}\right).$$

Therefore, in order to prove that $\mathbf{E}\left[X_n \mid \mathscr{F}_{n-1}\right] = X_{n-1}$ it is enough to prove that

$$\int_A X_n d\mathbf{P} = \int_A X_{n-1} d\mathbf{P}$$

for A any generator of $\mathscr{F}_{n-1}$:

- If $A = \emptyset$ is trivial;

- if $A = \{1, 2, \ldots, i\}, 1 \leqslant i \leqslant n - 1$, we have

$$\int_A X_n d\mathbf{P} = \sum_{k=1}^{i} X_n(k)\mathbf{P}(\{k\}) = (-1)\sum_{k=1}^{i}\left(\frac{1}{k} - \frac{1}{k+1}\right) = (-1)\left(1 - \frac{1}{i+1}\right)$$
$$= -\frac{i}{i+1},$$
$$\int_A X_{n-1} d\mathbf{P} = \sum_{k=1}^{i} X_{n-1}(k)\mathbf{P}(\{k\}) = (-1)\sum_{k=1}^{i}\left(\frac{1}{k} - \frac{1}{k+1}\right) = (-1)\left(1 - \frac{1}{i+1}\right)$$
$$= -\frac{i}{i+1}.$$

- Finally, if $A = \{i+1, i+2, \ldots\},\ i \leqslant n - 1$, then

$$\int_A X_n d\mathbf{P} = \sum_{k=i+1}^{\infty} X_n(k)\mathbf{P}(\{k\}) = (-1)\sum_{k=i+1}^{n}\left(\frac{1}{k} - \frac{1}{k+1}\right) + n\sum_{k=n+1}^{\infty}\left(\frac{1}{k} - \frac{1}{k+1}\right)$$
$$= (-1)\left(\frac{1}{i+1} - \frac{1}{n+1}\right) + \frac{n}{n+1} = 1 - \frac{1}{i+1},$$
$$\int_A X_{n-1} d\mathbf{P} = \sum_{k=i+1}^{\infty} X_{n-1}(k)\mathbf{P}(\{k\}) = (-1)\sum_{k=i+1}^{n-1}\left(\frac{1}{k} - \frac{1}{k+1}\right)$$
$$+ (n-1)\sum_{k=n}^{\infty}\left(\frac{1}{k} - \frac{1}{k+1}\right)$$
$$= (-1)\left(\frac{1}{i+1} - \frac{1}{n}\right) + \frac{n-1}{n} = 1 - \frac{1}{i+1}.$$

Moreover, let us observe that

$$\mathbf{E}[X_n] = \sum_{k=1}^{\infty} X_n(k)P(\{k\}) = \sum_{k=1}^{n}(-1)\left(\frac{1}{k} - \frac{1}{k+1}\right) + n\sum_{k=n+1}^{\infty}\left(\frac{1}{k} - \frac{1}{k+1}\right)$$
$$= (-1)\left(1 - \frac{1}{n+1}\right) + \frac{n}{n+1} = 0$$

for any n, and from an analogous computation

$$\mathbf{E}[|X_n|] = 1 + \frac{n-1}{n+1} \leqslant 2,$$

then $\{X_n\}$ is L^1-bounded and as $X_n \to X_\infty = -1$, a.e. then applying the dominated convergence theorem $X_\infty \in L^1$.

Consider now the multiplier sequence $Y = \{1, -1, 1, -1, \ldots\}$, i.e., $Y_n = (-1)^{n-1}$ then, given that

$$\Delta X_n(k) = X_n(k) - X_{n-1}(k) = \begin{cases} n - (n-1) = 1 & \text{if } n < k \\ -1 - (n-1) = -k & \text{if } n = k \\ -1 - (-1) = 0 & \text{if } n > k \end{cases}$$

then, the martingale transform of X by Y is given by

$$(Y \circ X)_n(k) = \sum_{i=1}^{n} (-1)^{i-1} (\Delta X)_j(k) = \begin{cases} \delta_n & \text{if } n < k \\ \delta_{k-1} + (-1)^k k & \text{if } n \geqslant k, \end{cases}$$

where δ_n is defined as $\delta_{2n} = 0$, $n \geqslant 0$ and $\delta_{2n-1} = 1$, $n \geqslant 1$.

Therefore, assuming the a.e. convergence of the martingale transform holds, we have that the limit is defined as

$$(Y \circ X)_\infty(k) = \delta_{k-1} + (-1)^k k$$

and therefore

$$\mathbf{P}\{(Y \circ X)_\infty > k\} \geqslant \mathbf{P}\{k+1, \ldots\} = \frac{1}{k+1}.$$

Then,

$$\mathbf{E}[|(Y \circ X)_\infty|] \geqslant \sum_{k=1}^{\infty} \mathbf{P}\{|(Y \circ X)_\infty| > k\} \geqslant \sum_{k=1}^{\infty} \frac{1}{k+1} = \infty.$$

Moreover,

$$\begin{aligned}
\mathbf{E}[|(Y \circ X)_n|] &= \sum_{k=1}^{\infty} |(Y \circ X)_n(k)| \mathbf{P}(\{k\}) \\
&= \sum_{k=1}^{n-1} \delta_n \left(\frac{1}{k} - \frac{1}{k+1} \right) + \sum_{k=n}^{\infty} |\delta_{k-1} + (-1)^k k| \left(\frac{1}{k} - \frac{1}{k+1} \right) \\
&\geqslant \sum_{k=n}^{\infty} (k-1) \left(\frac{1}{k} - \frac{1}{k+1} \right) = \sum_{k=n}^{\infty} (k-1) \frac{1}{k(k+1)}.
\end{aligned}$$

Thus, $(Y \circ X)$ is not L^1-bounded, and $(Y \circ X)_\infty$ is not integrable.

Theorem 4.7 (Martingale Transform Convergence Theorem) *Let $X = \{X_n\}_n$ be a $\{\mathscr{F}_n\}$-martingale and let $Y = \{Y_n\}_n$ be a multiplier sequence, if X is L^1-bounded then $(Y \circ X) = \{(Y \circ X)_n\}_n$ the martingale transform of X by Y converges a.e.*

We will discuss the original proof due to Burkholder [29]. For an alternative proof, see Neveu [130].

Proof The proof will be done in several steps.

(i) Let us assume that X is L^2 bounded (i.e., $\sup_n \|X_n\|_2 = \|X\|_2 < \infty$), then by the orthogonality of the increments and the boundedness of Y, we have

$$\mathbf{E}\left[(Y \circ X)_n^2\right] = \mathbf{E}\left(\sum_{k=1}^{n} Y_k \Delta X_k\right)^2 = \mathbf{E}\left(\sum_{k=1}^{n} Y_k^2 \Delta X_k^2\right) + 2\mathbf{E}\left[\sum_{1 \leqslant k < i \leqslant n} (Y_k \Delta X_k)(Y_i \Delta X_i)\right]$$

$$= \mathbf{E}\left[\sum_{k=1}^{n} Y_k^2 \Delta X_k^2\right] \leqslant \mathbf{E}\left[\sum_{k=1}^{n} \Delta X_k^2\right] = E\left(\sum_{k=1}^{n} \Delta X_k\right)^2 = \mathbf{E}\left[X_n^2\right] < \infty.$$

Then, $(Y \circ X)_n$ is also L^2-bounded and therefore L^1-bounded, then by the convergence theorem for martingales, Theorem 3.14, $(Y \circ X)$ converges a.e. to $(Y \circ X)_\infty$.

(ii) Let us assume now that X is a $\{\mathscr{F}_n\}$-submartingale uniformly bounded. Without loss of generality, we may assume that for any n, $X_n \geqslant 0$; otherwise we may add a constant which clearly will not change the increments ΔX_n. Then, as

$$\mathbf{E}\left[X_{n-1}\Delta X_n\right] = \mathbf{E}\left[\mathbf{E}\left[X_{n-1}\Delta X_n \mid \mathscr{F}_{n-1}\right]\right] = \mathbf{E}\left[X_{n-1}\mathbf{E}\left[\Delta X_n \mid \mathscr{F}_{n-1}\right]\right] \geqslant 0,$$

and therefore

$$\begin{aligned}\mathbf{E}\left[X_n^2\right] &= \mathbf{E}\left[(X_{n-1} + \Delta X_n)^2\right] = \mathbf{E}\left[X_{n-1}^2\right] + 2\mathbf{E}\left[X_{n-1}\Delta X_n\right] + \mathbf{E}\left[\Delta X_n^2\right]\\ &\geqslant \mathbf{E}\left[X_{n-1}^2\right] + \mathbf{E}\left[\Delta X_n^2\right].\end{aligned}$$

Then, iterating this argument we get

$$\mathbf{E}\left[X_n^2\right] \geqslant \mathbf{E}\left[\sum_{k=1}^{n} \Delta X_k^2\right].$$

Define a martingale $\hat{X} = \left\{\hat{X}_n\right\}_n$ with increments

$$\Delta \hat{X}_n = \Delta X_n - \mathbf{E}[\Delta X_n \mid \mathscr{F}_{n-1}],$$

and consider a martingale transform of $\hat{X}$ by a multiplier sequence Y

$$\left(Y \circ \hat{X}\right)_n = \sum_{k=1}^{n} Y_k \Delta \hat{X}_k.$$

$\hat{X}$ is the martingale associated to X in Doob's decomposition (Theorem 4.2). Moreover, as

$$\mathbf{E}\left[\Delta X_n \mathbf{E}\left[\Delta X_n \mid \mathscr{F}_{n-1}\right]\right] = \mathbf{E}\left[\mathbf{E}\left[\Delta X_n \mathbf{E}\left[\Delta X_n \mid \mathscr{F}_{n-1}\right] \mid \mathscr{F}_{n-1}\right]\right]$$
$$= \mathbf{E}\left[(\mathbf{E}\left[\Delta X_n \mid \mathscr{F}_{n-1}\right])^2\right],$$

then

$$\mathbf{E}\left[(\Delta \hat{X}_n)^2\right] = \mathbf{E}\left[(\Delta X_n - \mathbf{E}\left[\Delta X_n \mid \mathscr{F}_{n-1}\right])^2\right]$$
$$= \mathbf{E}\left[\Delta X_n^2\right] - 2\mathbf{E}\left[\Delta X_n \mathbf{E}\left[\Delta X_n \mid \mathscr{F}_{n-1}\right]\right] + \mathbf{E}\left[\mathbf{E}\left[\Delta X_n \mid \mathscr{F}_{n-1}\right]^2\right]$$
$$= \mathbf{E}\left[\Delta X_n^2\right] - \mathbf{E}\left[\mathbf{E}\left[\Delta X_n \mid \mathscr{F}_{n-1}\right]^2\right] \leqslant \mathbf{E}\left[\Delta X_n^2\right].$$

Therefore,

$$\mathbf{E}\left[\hat{X}_n\right]^2 = \mathbf{E}\left[\sum_{k=1}^{n} \Delta \hat{X}_k\right]^2 = \mathbf{E}\left[\sum_{k=1}^{n} \Delta \hat{X}_k^2\right]$$
$$\leqslant \mathbf{E}\left[\sum_{k=1}^{n} \Delta X_k^2\right] \leqslant \mathbf{E}\left[\sum_{k=1}^{n} \Delta X_k\right]^2 = \mathbf{E}\left[X_n\right]^2 \leqslant \mathbf{E}\left[\sup_n |X_n|\right]^2 < \infty.$$

Then $\hat{X}$ is L^2-bounded and then, by (i) $\{\left(Y \circ \hat{X}\right)_n\}$ converges a.e. to $\left(Y \circ \hat{X}\right)_\infty$. Now, by the martingale convergence theorem (Theorem 3.14), we know that $\{X_n\}$ converges a.e. to X_∞ and that $\{\hat{X}_n\}$ converges a.e. to $\hat{X}_\infty$, thus

$$\sum_{k=1}^{n} E\left[\Delta X_n \mid \mathscr{F}_{n-1}\right] = \sum_{k=1}^{n}\left(\Delta X_n - \Delta \hat{X}_n\right) = X_n - \hat{X}_n \to X_\infty - \hat{X}_\infty \text{ a.e.}$$

as $n \to \infty$, i.e., $\sum_{k=1}^{n} \mathbf{E}\left[\Delta X_n \mid \mathscr{F}_{n-1}\right]$ converges a.e. and as each term of the sum is positive a.e. then also $\sum_{k=1}^{n} Y_k \mathbf{E}\left[\Delta X_n \mid \mathscr{F}_{n-1}\right]$ converges a.e. Therefore as

$$\sum_{k=1}^{n} Y_k \Delta X_k = \sum_{k=1}^{n} Y_k \Delta \hat{X}_k + \sum_{k=1}^{n} Y_k \mathbf{E}\left[\Delta X_k \mid \mathscr{F}_{n-1}\right],$$

then $(Y \circ X)_n = \sum_{k=1}^{n} Y_k \Delta X_k$ converges a.e.

(iii) Finally, let us assume X is L^1-bounded. By Krickerberg's decomposition (Theorem 4.3) we may assume that for any n, $X_n \geqslant 0$. Let $C > 0$ a constant and for each n let

$$\hat{X}_n = -\min\left(X_n, C\right).$$

Then, $\left\{\hat{X}_n\right\}$ is a uniformly bounded submartingale since $-C \leqslant \hat{X}_n \leqslant 0$ and then

$$\mathbf{E}\left[\hat{X}_{n+1} \mid \mathscr{F}_n\right] = -\mathbf{E}\left[\min\left(X_{n+1}, C\right) \mid \mathscr{F}_n\right] \geqslant -\min\left(X_n, C\right) = \hat{X}_n.$$

Thus, by (ii) $(Y \circ \hat{X})$ converges a.e. and $\left(Y \circ \hat{X}\right) = (Y \circ X)$ if $X^* < C$, that is to say $(Y \circ X)$ converges a.e on the set $\{X^* < C\}$. Now, again by the martingale convergence theorem (Theorem 3.14) $\{X_n\}$ converges a.e. and therefore $\mathbf{P}\{X^* < \infty\} = 1$ then it is enough to take $C \to \infty$ in order to obtain the result. $\qquad\square$

An immediate consequence of the previous theorem is that if X is a martingale is L^1-bounded, then given $\{\varepsilon_k\}$ a sequence of numbers in $\{-1, 1\}$, i.e., any random choice of signs,

$$\sum_{k=1}^{\infty} \varepsilon_k \Delta X_k$$

converges a.e.

Moreover, Theorem 4.7 allows to get an alternative proof to Austin's theorem (Theorem 4.1).

Corollary 4.8 *Let $X = \{X_n\}_n$ be a $\{\mathscr{F}_n\}$-martingale such that $\mathbf{E}[S(X)] < \infty$. Then, X converges a.e.*

Proof Let $\{r_k\}$ be the Rademacher functions on $[0, 1]$ and let $t \in [0, 1]$ fix. Consider the martingale transform $\left\{\sum_{k=1}^n r_k(t)\Delta X_k\right\}_n$. Since $\left\{\left|\sum_{k=1}^n r_k(t)\Delta X_k\right|\right\}_n$ is a submartingale, then $\left\{\mathbf{E}\left|\sum_{k=1}^n r_k(t)\Delta X_k\right|\right\}_n$ is a non-decreasing sequence. Thus, by Tonelli's theorem and Cauchy-Schwarz's inequality, we get

$$\int_0^1 \mathbf{E}\left|\sum_{k=1}^n r_k(t)\Delta X_k\right| dt = \mathbf{E}\int_0^1 \left|\sum_{k=1}^n r_k(t)\Delta X_k\right| dt \leqslant \mathbf{E}\left(\int_0^1 \left(\sum_{k=1}^n r_k(t)\Delta X_k\right)^2 dt\right)^{1/2}$$

$$= \mathbf{E}\left(\int_0^1 \left(\sum_{k=1}^n (\Delta X_k)^2\right) dt\right)^{1/2} = \mathbf{E}\left[S_n(X)\right],$$

by the orthogonality of the Rademacher functions on $[0, 1]$. Then, by the monotone convergence theorem we have

$$\int_0^1 \sup_n \mathbf{E}\left|\sum_{k=1}^n r_k(t)\Delta X_k\right| dt \leqslant \mathbf{E}[S(X)],$$

and this implies that for some t,

$$\sup_n \mathbf{E}\left|\sum_{k=1}^{n} r_k(t)\Delta X_k\right| \leqslant \mathbf{E}[S(X)] < \infty.$$

For this t then $\left\{\sum_{k=1}^{n} r_k(t)\Delta X_k\right\}_n$ is a martingale L^1-bounded and X is its martingale transform under the multiplier sequence $\{r_k(t)\}$. Hence, by Theorem 4.7, X converges a.e. $\qquad\square$

Now, let us discuss a version of the strong law of large numbers for square-integrable martingales, see Shiryayev [148, Chap. VII]. First, we need the following theorems.

Let $a > 0$, and define

$$\tau_a = \begin{cases} \min\{n \geqslant 1 : X_n > a\}, & \textit{if } \{n \geqslant 1 : X_n > a\} \neq \varnothing \\ \infty, & \textit{if } \{n \geqslant 1 : X_n > a\} = \varnothing. \end{cases}$$

Theorem 4.9 *Let* $X = \{X_n\}$ *be a* $\{\mathscr{F}_n\}$-*submartingale and*

$$X_n = Z_n + Y_n$$

be its Doob decomposition.

(i) If X is a non-negative submartingale, then a.e.

$$\{Y_\infty < \infty\} \subseteq \{\lim X_n \textit{ exists and is finite }\} \subseteq \{\sup X_n < \infty\}. \qquad (4.14)$$

(ii) If X satisfies

$$\mathbf{E}\left(\Delta X_{\tau_a}\right)^+ \chi_{\{\tau_a < \infty\}} < \infty$$

for every $a > 0$, then a.e.

$$\{\lim X_n \textit{ exists and is finite }\} = \{\sup X_n < \infty\} \subseteq \{Y_\infty < \infty\}. \qquad (4.15)$$

(iii) If X is a non-negative submartingale and satisfies

$$\mathbf{E}\left(\Delta X_{\tau_a}\right)^+ \chi\{\tau_a < \infty\} < \infty$$

for every $a > 0$, then a.e.

$$\{\lim X_n \textit{ exists and is finite }\} = \{\sup X_n < \infty\} = \{Y_\infty < \infty\}. \qquad (4.16)$$

Proof

(i) The second inclusion in (4.14) is obvious. To establish the first inclusion we introduce the stopping times

$$\sigma_a = \inf\{n \geqslant 1 : Y_{n+1} > a\}, \quad a > 0,$$

taking $\sigma_a = +\infty$ if $\{n \geqslant 1 : Y_{n+1} > a\} = \varnothing$. Then, $Y_{\sigma_a} \leqslant a$ and by Theorem 3.1, we have

$$\mathbf{E}[X_{n \wedge \sigma_a}] = \mathbf{E}[Y_{n \wedge \sigma_a}] \leqslant a.$$

Let $W_n^a = X_{n \wedge \sigma_a}$. Then $W^a = \{W_n^a\}$ is a $\{\mathscr{F}_n\}$- submartingale with $\sup \mathbf{E}[W_n^a] \leqslant a < \infty$. Since the martingale is non-negative, it follows from Theorem 3.2 that a.e.

$$\{Y_\infty \leqslant a\} = \{\sigma_a = \infty\} \subseteq \{\lim X_n \text{ exists and is finite }\}$$

Therefore a.e.

$$\{Y_\infty < \infty\} = \bigcup_{a>0} \{Z_\infty \leqslant a\} \subseteq \{\lim X_n \text{ exists and is finite }\}.$$

(ii) The first equality of (4.15) follows easily. To prove the second, we notice that

$$\mathbf{E}[Y_{\tau_a \wedge n}] = \mathbf{E}[X_{\tau_a \wedge n}] \leqslant \mathbf{E}[X_{\tau_\alpha \wedge n}^+] \leqslant 2a + \mathbf{E}\left[\left(\Delta X_{\tau_a}\right)^+ \chi_{\{\tau_a < \infty\}}\right]$$

and therefore

$$\mathbf{E}[Y_{\tau_\alpha}] = \mathbf{E}[\lim_n Y_{\tau_a \wedge n}] < \infty.$$

Hence $\{\tau_a = \infty\} \subseteq \{Y_\infty < \infty\}$ and we obtain the required conclusion since

$$\bigcup_{a>0} \{\tau_a = \infty\} = \{\sup X_n < \infty\}.$$

(iii) (4.16) is an immediate consequence of (i) and (ii). $\qquad\qquad\square$

Theorem 4.10 *Let $X = \{X_n\}$ be a square-integrable $\{\mathscr{F}_n\}$- martingale. Then*

$$\{\langle X \rangle_\infty < \infty\} \subseteq \{\lim X_n \text{ exists and is finite }\} \quad a.e. \tag{4.17}$$

If also

$$\mathbf{E}[\sup_n(|\Delta X_n|^2)] < \infty,$$

then,

$$\{\langle X \rangle_\infty < \infty\} = \{X_n \text{ exists and is finite }\} \quad a.e. \tag{4.18}$$

where

$$\langle X \rangle_\infty = \sum_{n=1}^{\infty} \mathbf{E}\left[(\Delta X_n)^2 \mid \mathscr{F}_{n-1} \right], \tag{4.19}$$

with $X_0 = 0$, $\mathscr{F}_0 = \{\emptyset, \Omega\}$.

Proof Consider the two submartingales $X^2 = \{X_n^2\}$ and $(X+1)^2 = \{(X+1)_n^2\}$. Let their Doob decompositions be

$$X_n^2 = Z_n' + Y_n', \quad (X_n + 1)^2 = Z_n'' + Y_n''.$$

Then Y_n' and Y_n'' are the same, since

$$Y_n' = \sum_{k=1}^{n} \mathbf{E}\left[\Delta X_k^2 \mid \mathscr{F}_{k-1} \right] = \sum_{k=1}^{n} \mathbf{E}\left[(\Delta X_k)^2 \mid \mathscr{F}_{k-1} \right]$$

and

$$Y_n'' = \sum_{k=1}^{n} \mathbf{E}\left[\Delta (X_k + 1)^2 \mid \mathscr{F}_{k-1} \right] = \sum_{k=1}^{n} \mathbf{E}\left[\Delta X_k^2 \mid \mathscr{F}_{k-1} \right]$$

$$= \sum_{k=1}^{n} \mathbf{E}\left[(\Delta X_k)^2 \mid \mathscr{F}_{k-1} \right].$$

Hence (4.14) implies that a.e.

$$\{\langle X \rangle_\infty < \infty\} = \{Y_\infty' < \infty\} \subseteq \left\{ \lim X_n^2 \text{ exists and is finite} \right\} \cap \left\{ \lim (X_n + 1)^2 \text{ exists and is finite} \right\}$$
$$= \{\lim X_n \text{ exists and is finite} \}.$$

By (4.16), Eq. (4.18) will be established if we show that the condition

$$\mathbf{E}[\sup |\Delta X_n|^2] < \infty$$

guarantees that X^2 satisfies

$$\mathbf{E}\left(\Delta X_{\tau_a}^2 \right)^+ \chi_{\{\tau_a < \infty\}} < \infty$$

for every $a > 0$, where $\tau_a = \inf \left\{ n \geq 1 : X_n^2 > a \right\}$. Then, on the set $\{\tau_a < \infty\}$,

$$\left| \Delta X_{\tau_a}^2 \right| = \left| X_{\tau_a}^2 - X_{\tau_a - 1}^2 \right| \leq \left| X_{\tau_a} - X_{\tau_a - 1} \right|^2$$
$$+ 2 \left| X_{\tau_a - 1} \right| \cdot \left| X_{\tau_a} - X_{\tau_a - 1} \right| \leq \left(\Delta X_{\tau_a} \right)^2 + 2 a^{1/2} \left| \Delta X_{\tau_a} \right|$$

whence

$$\mathbf{E}\left[\left|\Delta X_{\tau_a}^2\right| \chi_{\{\tau_a<\infty\}}\right] \leqslant \mathbf{E}\left[\left(\Delta X_{\tau_a}\right)^2 \chi_{\{\tau_a<\infty\}}\right] + 2a^{1/2}\sqrt{\mathbf{E}\left[\left(\Delta X_{\tau_a}\right)^2 \chi_{\{\tau_a<\infty\}}\right]}$$

$$\leqslant \mathbf{E}\left[\sup|\Delta X_n|^2\right] + 2a^{1/2}\sqrt{\mathbf{E}\left[\sup|\Delta X_n|^2\right]} < \infty, \qquad \square$$

At the end of this chapter, we will also consider local properties of martingales transforms.

Now, we have the following version of the strong law of large numbers for martingales, see [148].

Theorem 4.11 (Strong Law of large numbers for martingales) *Let $X = \{X_n\}$ be a square-integrable $\{\mathscr{F}_n\}$-martingale and let $Y = \{Y_n\}$ be a predictable increasing sequence with $Y_1 \geqslant 1$, $Y_\infty = \infty$ a.e. If*

$$\sum_{n=1}^\infty \frac{\mathbf{E}\left[(\Delta X_n)^2 \mid \mathscr{F}_{n-1}\right]}{Y_n^2} < \infty. \quad a.e. \tag{4.20}$$

then

$$\frac{X_n}{Y_n} \to 0, \ as \ n \to \infty \quad a.e. \tag{4.21}$$

In particular, if $Y = \{\langle X \rangle_n\}$ is the sequence of quadratic variations of X and $\langle X \rangle_\infty = \infty$ a.e. then,

$$\frac{X_n}{\langle X \rangle_n} \to 0, \quad n \to \infty. \quad a.e. \tag{4.22}$$

Proof Consider $Z = \{Z_n\}$ a square-integrable $\{\mathscr{F}_n\}$-martingale with

$$Z_n = \sum_{k=1}^n \frac{\Delta X_k}{Y_k}.$$

Then

$$\langle Z \rangle_n = \sum_{k=1}^n \frac{\mathbf{E}\left[(\Delta X_k)^2 \mid \mathscr{F}_{k-1}\right]}{Y_k^2},$$

since

$$\frac{X_n}{Y_n} = \frac{\sum_{k=1}^n Y_k \Delta Z_k}{Y_n}$$

we have, by Kronecker's lemma [148, Chap. IV, Sect. 3, Lemma 2], $X_n/Y_n \to 0$ a.e. if the limit $\lim_n Z_n$ exists a.e. (i.e., it is finite). By (4.17),

$$\{\langle Z \rangle_\infty < \infty\} \subseteq \{\lim Z_n \text{ exists and is finite}\}.$$

Therefore it follows that (4.20) is a sufficient condition for (4.21).

If now $Y_n = \langle X \rangle_n$, then (4.20) is automatically satisfied and consequently we have

$$\frac{X_n}{\langle X \rangle_n} \to 0 \quad \text{a.e.} \qquad \square$$

Additionally, in the version of the law of large numbers given by M. Duflo [72, Theorem 1.3.15], the following rate of convergence is given,

$$X_n / \langle X \rangle_n = \mathrm{o}\left(\left((\ln \langle X \rangle_n)^{1+\gamma} / \langle X \rangle_n \right)^{1/2} \right)$$

for any $\gamma > 0$.

Chow's condition allows to avoid the assumption that the martingale is square integrable in Theorem 4.11, see [53].

Theorem 4.12 (Chow's convergence theorem) *Let $\{X_n\}$ be a martingale such that for some $1 \leqslant p \leqslant 2$ and for all $n \geqslant 1$, $\mathbf{E}\left[|X_n|^p \right] < \infty$. On the set*

$$\Gamma = \left\{ \sum_{n=0}^{\infty} \mathbf{E}\left[|\Delta X_n|^p \mid \mathscr{F}_{n-1} \right] < \infty \right\},$$

$\{X_n\}$ converges almost everywhere.

Finally, let us consider a central limit theorem for martingales. The distributional limits of martingales need not be Gaussian, see for instance Example 7.2.1 of [21]. There are extremely general martingale CLT where the limits are quite complicated. See for example Hall and Heyde [95]. We focus only on a relatively simple situation where the limit is Gaussian. This result is very similar in spirit to the Lindeberg-Lévy CLT, see Theorem B.5 for triangle arrays of random variables.

Theorem 4.13 (Central limit theorem for martingales) *Suppose that $\left\{ X_{n,k} \right\}_{k \geq 0}$ be a $\left\{ \mathscr{F}_{n,k} \right\}_{k \geq 0}$ martingale for each $n \in \mathbb{N}$, with $X_{n,0} = 0$ and let $\Delta X_{nk} = X_{n,k} - X_{n,k-1}, k \geq 1$. Assume that*

$$\sum_{k=1}^{\infty} \mathbf{E}\left[\Delta X_{n,k}^2 \right] < \infty, \tag{4.23}$$

and

$$\lim_{n \to \infty} \sum_{k=1}^{\infty} \mathbf{E}\left[\Delta X_{n,k}^2 \, \chi_{\{|\Delta X_{n,k}| \geq \varepsilon\}} \right] = 0 \tag{4.24}$$

for all $\varepsilon > 0$ and for some constant $\sigma > 0$

$$\sum_{k=1}^{\infty} \mathbf{E}\left[\Delta X_{n,k}^2 \mid \mathscr{F}_{n,k-1}\right] \xrightarrow{P} \sigma^2, \ as \ n \to \infty. \tag{4.25}$$

Then, $X_n = \sum_{k=1}^{\infty} \Delta X_{n,k}$ converges in distribution to a normal distribution $N\left(0, \sigma^2\right)$ as $n \to \infty$.

Proof First note that for any fixed n, (4.23) implies that $\{X_{n,k} = \sum_{j=1}^{k} \Delta X_{n,j}\}$, as $k \to \infty$, converges to some X_n almost everywhere by an easy application of Corollary 3.9 with $p = 2$. That is, $X_n = \sum_{k=1}^{\infty} \Delta X_{n,k}$ is well defined.

Let the conditional variances and their partial sums be denoted by

$$\sigma_{n,k}^2 := \mathbf{E}\left[\Delta X_{n,k}^2 \mid \mathscr{F}_{n,k-1}\right], \ n, k \in \mathbb{N}$$

$$\Sigma_{n,k} := \sum_{j=1}^{k} \sigma_{n,j}^2, \ n \geq 1, 1 \leq k \leq \infty.$$

Note that condition (4.23) ensures that for every n,

$$\sum_{k=1}^{\infty} \Delta X_{n,k}^2 < \infty \ \text{and} \ \sum_{k=1}^{\infty} \sigma_{n,k}^2 < \infty \ \text{a.e.} \tag{4.26}$$

For the time being, let us assume an additional restriction (which shall be removed later). Assume that for some constant C,

$$\Sigma_{n,\infty} = \sum_{k=1}^{\infty} \sigma_{n,k}^2 \leq C < \infty, \ \text{for all } n \geq 1. \tag{4.27}$$

In order to prove the convergence in distribution, we use Lévy continuity theorem, see for instance [73] or [78], it suffices to show that

$$\lim_{n \to \infty} \mathbf{E}\left[e^{itX_n}\right] = e^{-t^2\sigma^2/2}, \ \text{for all } t \in \mathbb{R}. \tag{4.28}$$

Fix t and observe that

$$\left|\mathbf{E}\left[e^{itX_n}\right] - e^{-t^2\sigma^2/2}\right|$$

$$= \left|\mathbf{E}\left[e^{itX_n} - e^{itX_n+(\Sigma_{n,\infty}-\sigma^2)t^2/2}\right] + \mathbf{E}\left[e^{itX_n+(\Sigma_{n,\infty}-\sigma^2)t^2/2} - e^{-t^2\sigma^2/2}\right]\right|$$

$$\leq \mathbf{E}\left[\left|e^{itX_n}\left(1 - e^{(\Sigma_{n,\infty}-\sigma^2)t^2/2}\right)\right|\right] + e^{-t^2\sigma^2/2}\left|\mathbf{E}\left[e^{itX_n+\Sigma_{n,\infty}t^2/2}\right] - 1\right|$$

$$\leq \mathbf{E}\left[\left|1 - e^{(\Sigma_{n,\infty}-\sigma^2)t^2/2}\right|\right] + \left|\mathbf{E}\left[e^{itX_n+\Sigma_{n,\infty}t^2/2}\right] - 1\right|.$$

By Conditions (4.25) and (4.27), $\{\Sigma_{n,\infty} - \sigma^2\}$ remains uniformly bounded as $n \to \infty$, and converges to 0 in probability. Hence, by the dominated convergence theorem in probability, the first term on the right side above converges to 0. Thus (4.28) will follow if we show that

$$\lim_{n\to\infty} \mathbf{E}\left[e^{itX_n + \Sigma_{n,\infty}t^2/2}\right] = 1. \tag{4.29}$$

Fix n, again using (4.27), note that

$$\lim_{k\to\infty} \mathbf{E}\left[e^{itX_{n,k} + \Sigma_{n,k}t^2/2}\right] = \mathbf{E}\left[e^{itX_n + \Sigma_{n,\infty}t^2/2}\right].$$

This implies

$$\mathbf{E}\left[e^{itX_n + \Sigma_{n,\infty}t^2/2} - 1\right] = \sum_{k=1}^{\infty} \mathbf{E}\left[e^{itX_{n,k} + \Sigma_{n,k}t^2/2} - e^{itX_{n,k-1} + \Sigma_{nk-1}t^2/2}\right]$$

$$= \sum_{k=1}^{\infty} \mathbf{E}\left[e^{itX_{n,k-1}}e^{\Sigma_{n,k}t^2/2}\mathbf{E}\left[e^{it\Delta X_{n,k}} - e^{-\sigma_{n,k}^2 t^2/2} \mid \mathscr{F}_{n,k-1}\right]\right].$$

Hence, again by (4.27)

$$\left|\mathbf{E}\left[e^{itX_n + \Sigma_{n,\infty}t^2/2} - 1\right]\right| \leq e^{Ct^2} \sum_{k=1}^{\infty} \mathbf{E}\left[\left|\mathbf{E}\left[e^{it\Delta X_{n,k}} - e^{-\sigma_{n,k}^2 t^2/2} \mid \mathscr{F}_{n,k-1}\right]\right|\right]. \tag{4.30}$$

Fix $k \in \mathbb{N}$, using Taylor's expansion implies that

$$e^{it\Delta X_{n,k}} = 1 + it\Delta X_{n,k} - \frac{1}{2}t^2\Delta_{n,k}^2 + \theta,$$

where θ is a measurable function and

$$|\theta| \leq \left|t\Delta X_{n,k}\right|^3 \wedge \left(t\Delta X_{n,k}\right)^2.$$

Fix $\varepsilon > 0$, there exists a constant K_t, depending only on t, such that

$$|\theta| \leq K_t\left(\left|\Delta X_{n,k}\right|^3 \wedge \Delta X_{n,k}^2\right)$$

$$\leq K_t\left(\Delta X_{n,k}^2 \chi_{\{|\Delta X_{n,k}|\geq\varepsilon\}} + \left|\Delta X_{n,k}\right|^3\left(1 - \chi_{\{|\Delta X_{n,k}|\geq\varepsilon\}}\right)\right)$$

$$\leq K_t\left(\Delta X_{n,k}^2 \chi_{\{|\Delta X_{n,k}|\geq\varepsilon\}} + \varepsilon\Delta X_{n,k}^2\right). \tag{4.31}$$

Taylor's expansion also implies that

$$e^{-\sigma_{n,k}^2 t^2/2} = 1 - \frac{1}{2}t^2\sigma_{n,k}^2 + \theta',$$

where θ' is measurable, and for some constant $\bar{K}_t \geqslant K_t$,

$$|\theta'| \leqslant \bar{K}_t \sigma_{n,k}^4. \tag{4.32}$$

To estimate the individual terms in (4.30), recall that

$$\mathbf{E}\left[\Delta X_{n,k} \mid \mathscr{F}_{n,k-1}\right] = 0 \quad \text{and} \quad \mathbf{E}\left[\Delta X_{n,k}^2 \mid \mathscr{F}_{n,k-1}\right] = \sigma_{n,k}^2.$$

Hence, using (4.31) and (4.32),

$$\left|\mathbf{E}\left[e^{it\Delta X_{n,k}} - e^{-\sigma_{n,k}^2 t^2/2} \mid \mathscr{F}_{n,k-1}\right]\right| =$$

$$= \left|\mathbf{E}\left[e^{it\Delta X_{n,k}} - 1 - it\Delta X_{n,k} + \frac{1}{2}t^2 \Delta X_{n,k}^2 \mid \mathscr{F}_{n,k-1}\right]\right.$$

$$\left. - \mathbf{E}\left[e^{-\sigma_{n,k}^2 t^2/2} - 1 + \frac{1}{2}t^2 \sigma_{n,k} \mid \mathscr{F}_{n,k-1}\right]\right|$$

$$\leqslant \mathbf{E}\left[|\theta| + |\theta'| \mid \mathscr{F}_{n,k-1}\right]$$

$$\leqslant \bar{K}_t \left(\mathbf{E}\left[\Delta X_{n,k}^2 \chi_{\{|\Delta X_{n,k}|\geqslant\varepsilon\}} \mid \mathscr{F}_{n,k-1}\right] + \varepsilon\sigma_{n,k}^2 + \sigma_{n,k}^4\right).$$

Note that

$$\sigma_{nk}^2 = \mathbf{E}\left[\Delta X_{n,k}^2 \mid \mathscr{F}_{n,k-1}\right] \leqslant \varepsilon^2 + \mathbf{E}\left[\Delta X_{n,k}^2 \chi_{\{|\Delta X_{n,k}|\geqslant\varepsilon\}} \mid \mathscr{F}_{n,k-1}\right]$$

$$\leqslant \varepsilon^2 + \sum_{j=1}^{\infty} \mathbf{E}\left[\Delta X_{n,j}^2 \chi_{\{|\Delta X_{n,k}|\geqslant\varepsilon\}} \mid \mathscr{F}_{n,j-1}\right]$$

and hence

$$\mathbf{E}\left[\sup_{k\geq 1} \sigma_{nk}^2\right] \leqslant \varepsilon^2 + \sum_{k=1}^{\infty} \mathbf{E}\left[\Delta X_{n,k}^2 \chi_{\{|\Delta X_{n,k}|\geqslant\varepsilon\}}\right]. \tag{4.33}$$

Therefore,

$$\left|\mathbf{E}\left[e^{itX_n + \Sigma_{n\infty} t^2/2} - 1\right]\right|$$

$$\leqslant \bar{K}_t e^{Ct^2} \sum_{k=1}^{\infty} \mathbf{E}\left(\Delta X_{n,k}^2 \chi_{\{|\Delta X_{n,k}|\geq\varepsilon\}} + \varepsilon\sigma_{nk}^2 + \sigma_{nk}^4\right)$$

$$\leqslant \bar{K}_t e^{Ct^2} \mathbf{E}\left[\sum_{k=1}^{\infty} \Delta X_{n,k}^2 \chi_{\{|\Delta X_{n,k}|\geq\varepsilon\}} + \left(\varepsilon + \sup_{k\geqslant 1}\sigma_{n,k}^2\right)\Sigma_{n,\infty}\right]$$

$$\leqslant \bar{K}_t e^{Ct^2} \left(\sum_{k=1}^{\infty} \mathbf{E}\left[\Delta X_{n,k}^2 \chi_{\{|\Delta X_{n,k}|\geq\varepsilon\}}\right] + C\left(\varepsilon + \mathbf{E}\left[\sup_{k\geq 1}\sigma_{n,k}^2\right]\right)\right)$$

$$\leqslant \bar{K}_t e^{Ct^2}\left[(1+C)\sum_{k=1}^{\infty} \mathbf{E}\left(\Delta X_{n,k}^2 \chi_{\{|\Delta X_{n,k}|\geq\varepsilon\}}\right) + C\left(\varepsilon + \varepsilon^2\right)\right], \tag{4.34}$$

where the last inequality follows from (4.33). Letting $n \to \infty$ in (4.34) and using (4.24), we have

$$\limsup_{n \to \infty} \left| \mathbf{E}\left[e^{itX_n + \Sigma, \infty t^2/2} - 1 \right] \right| \leqslant \bar{K}_t e^{Ct^2} \left(\varepsilon + \varepsilon^2 \right) C.$$

Since ε was arbitrary, left side of (4.30) goes to zero as $n \to \infty$, that is, (4.29) holds, and this implies (4.28). Thus X_n converges in distribution to a normal distribution $N\left(0, \sigma^2\right)$ as $n \to \infty$, under the additional restriction (4.27).

If (4.27) does not hold, fix $C > \sigma^2$, and let

$$A_{n,k} := \left\{ \omega : \sum_{j=1}^{k} \sigma_{nj}^2(\omega) \leqslant C \right\}, \quad Z_{n,k} := \Delta X_{n,k} \chi_{A_{nk}}, \, 1 \leqslant n < \infty, 1 \leqslant k \leqslant \infty.$$

Since $\chi_{A_{n,k}} \in \mathscr{F}_{n,k-1}$, it follows that

$$\mathbf{E}\left[Z_{n,k} \mid \mathscr{F}_{n,k-1} \right] = 0, \ \text{for } k, n \in \mathbb{N}.$$

Let

$$\theta_{n,k}^2 = \mathbf{E}\left[Z_{n,k}^2 \mid \mathscr{F}_{n,k-1} \right] = \sigma_{n,k}^2 \chi_{A_{nk}}.$$

Observe that for n fixed, $A_{n,k} \downarrow A_{n,\infty}$ as $k \to \infty$, and hence

$$\Omega = A_{n,\infty} \bigcup \left(\bigcup_{j=1}^{\infty} \left(A_{n,j-1} \backslash A_{n,j} \right) \right),$$

where $A_{n,0} = \Omega$. So, for any $\omega \in \Omega$, either $\omega \in A_{n,\infty}$ or $\omega \in A_{n,j-1} \backslash A_{n,j}$ for some $j \in \mathbb{N}$. Moreover,

$$\sum_{j=1}^{\infty} \theta_{n,j}^2(\omega) = \begin{cases} \sum_{j=1}^{k} \sigma_{n,j}^2(\omega) & \text{if } \omega \in A_{n,k} \backslash A_{n,k+1} \text{ for some } k \geq 0 \\ \sum_{j=1}^{\infty} \sigma_{n,j}^2(\omega) & \text{if } \omega \in A_{n\infty} \end{cases}.$$

In any case, $\sum_{j=1}^{\infty} \theta_{n,j}^2(\omega) \leqslant C$ holds. Besides,

$$\chi_{A_{n,\infty}} \sum_{j=1}^{\infty} \theta_{n,j}^2 = \chi_{A_{n,\infty}} \sum_{j=1}^{\infty} \sigma_{n,j}^2.$$

Therefore,

$$\lim_{n \to \infty} \sum_{j=1}^{\infty} \theta_{n,j}^2 \xrightarrow{P} \sigma^2.$$

By previous case that has already been proved, $\sum_{k=1}^{\infty} Z_{n,k}$ converges in distribution to a normal distribution $N(0, \sigma^2)$. Now note that

$$X_n = X_n \chi_{A_{n\infty}} + X_n \chi_{A_{n,\infty}^c} = \chi_{A_{n,\infty}} \sum_{k=1}^{\infty} Z_{n,k} + X_n \chi_{A_{n,\infty}^c}$$

and from here easily follows, that X_n converges in distribution to a normal distribution $N(0, \sigma^2)$ and the proof is complete. $\qquad\square$

4.3   Discrete Version of Itô's Formula

In the stochastic analysis of Brownian motion and other related processes *Itô's change-of-variables formula* plays a key role. In this section, we present a discrete version of this formula and show briefly how Itô's formula for Brownian motion could be derived from it using a limiting procedure.

Consider F an absolutely continuous function, i.e.,

$$F(x) = F(0) + \int_0^x f(y)dy,$$

where f is a real value Borel function locally L^1.

The change-of-variables formula in which we are interested concerns the possibility of representing the sequence

$$F(X) = \{F(X_n)\}_{0 \leqslant n \leqslant N},$$

in terms of "natural" functionals of the sequence $X = \{X_n\}_{0 \leqslant n \leqslant N}$.

Consider the quadratic covariation $\langle X, f(X) \rangle$ of the sequences X and $f(X)$ of $\{X_n\}_{0 \leqslant n \leqslant N}$.

$$\langle X, f(X) \rangle_n = \sum_{k=1}^{n} \Delta f(X_k)\, \Delta X_k$$
$$= \sum_{k=1}^{n} (f(X_k) - f(X_{k-1}))\, (X_k - X_{k-1}).$$

We introduce two "discrete integrals":

$$I_n(X, f(X)) = \sum_{k=1}^{n} f(X_{k-1})\, \Delta X_k, \quad 1 \leqslant n \leqslant N,$$
$$\tilde{I}_n(X, f(X)) = \sum_{k=1}^{n} f(X_k)\, \Delta X_k, \quad 1 \leqslant n \leqslant N.$$

$I_n(X, f(X))$ is called "forward integral" and $\tilde{I}_n(X, f(X))$ is called "backward integral."

Then, for $n \geqslant 1$

$$\langle X, f(X) \rangle_n = \tilde{I}_n(X, f(X)) - I_n(X, f(X)),$$

and for $n = 0$, set $I_0 = \tilde{I}_0 = 0$.

Since for any function g we trivially have

$$g(X_{k-1}) + \frac{1}{2}\left[g(X_k) - g(X_{k-1})\right] - \frac{1}{2}\left[g(X_k) + g(X_{k-1})\right] = 0,$$

then, it is clear that, for F and f as above, we get

$$
\begin{aligned}
F(X_n) = \ & F(X_0) + \sum_{k=1}^{n} f(X_{k-1}) \Delta X_k + \frac{1}{2}\langle X, f(X)\rangle_n \\
& + \sum_{k=1}^{n}\left\{ (F(X_k) - F(X_{k-1})) - \frac{f(X_{k-1}) + f(X_k)}{2}\Delta X_k \right\} \\
= \ & F(X_0) + I_n(X, f(X)) + \frac{1}{2}\langle X, f(X)\rangle_n + R_n(X, f(X)), \quad (4.35)
\end{aligned}
$$

where

$$R_n(X, f(X)) = \sum_{k=1}^{n} \int_{X_{k-1}}^{X_k} \left[f(x) - \frac{f(X_{k-1}) + f(X_k)}{2} \right] dx. \qquad (4.36)$$

From analysis, it is well known that if the function $f''(x)$ is continuous, then the following formula (trapezoidal rule) holds:

$$
\begin{aligned}
\int_a^b \left[f(x) - \frac{f(a) + f(b)}{2} \right] dx &= \int_a^b (x - a)(x - b)\frac{f''(\xi(x))}{2!}dx \\
&= \frac{(b-a)^3}{2} \int_0^1 x(x-1) f''(\xi(a + x(b-a)))dx \\
&= \frac{(b-a)^3}{2} f''(\xi(a + \bar{x}(b-a))) \int_0^1 x(x-1)dx \\
&= -\frac{(b-a)^3}{12} f''(\eta),
\end{aligned}
$$

where $\xi(x)$, $\bar{x}$, and η are intermediate points in the interval $[a, b]$. Thus, in (4.36),

$$R_n(X, f(X)) = -\frac{1}{12} \sum_{k=1}^{n} f''(\eta_k) (\Delta X_k)^3,$$

where $X_{k-1} \leqslant \eta_k \leqslant X_k$, whence

$$|R_n(X, f(X))| \leqslant \frac{1}{12} \sup_{\eta} |f''(\eta)| \sum_{k=1}^{n} |\Delta X_k|^3 ,$$

where the supremum is taken over all η such that

$$\min (X_0, X_1, \ldots, X_n) \leqslant \eta \leqslant \max (X_0, X_1, \ldots, X_n).$$

We shall refer to formula (4.35) as the *discrete of Itô's formula* (DIF) see [148, 157, 158].

Note that the right-hand side of this formula contains the following three natural ingredients: "the discrete integral" $I_n(X, f(X))$, the quadratic covariation $\langle X, f(X) \rangle_n$, and the remainder term $R_n(X, f(X))$, which is so termed because it goes to zero in the limit transition to the continuous time.

In particular, observe that if $f(x) = a + bx$, then $R_n(X, f(X)) = 0$, and formula (4.35) takes the following form:

$$F(X_n) = F(X_0) + I_n(X, f(X)) + \frac{1}{2}\langle X, f(X) \rangle_n.$$

On the other hand, observe that if $X = \{X_n\}$ is a $\{\mathscr{F}_n\}$-martingale then $I_n(X, f(X))$ the "forward integral" is precisely the martingale transform of X by $f(X)$, since F is locally L^1-bounded.

Finally, Itô's formula can be obtained as a limit of the discrete version (4.35). Let $B = \{B_t\}_{0 \leqslant t \leqslant 1}$ be a standard Brownian motion and let $X_k = B_{k/n}, k = 0, 1, \ldots, n$. Then application of formula (4.35) leads to the following result:

$$F(B_1) = F(B_0) + \sum_{k=1}^{n} f\left(B_{(k-1)/n}\right) \Delta B_{k/n} + \frac{1}{2}\langle f(B_{\cdot/n}), B_{\cdot/n} \rangle_n + R_n\left(B_{\cdot/n}, f(B_{\cdot/n})\right).$$

Now, since

$$\mathbf{E}\left[\sum_{k=1}^{n} \left| B_{\frac{k}{n}} - B_{\frac{k-1}{n}} \right|^3 \right] = 2\sqrt{\frac{2}{\pi}} \frac{1}{\sqrt{n}},$$

then, by Markov's inequality,

$$\sum_{k=1}^{n} \left| B_{\frac{k}{n}} - B_{\frac{k-1}{n}} \right|^3 \to 0 \quad \text{as} \quad n \to \infty \tag{4.37}$$

in probability. Therefore, if f is twice differentiable and $|f''(x)| \leqslant C, x \in \mathbb{R}$, for some $C > 0$, then we obtain from (4.36), that $R_n(B \cdot /n, f(B \cdot /n)) \to 0$.

Now, again by Brownian motion theory, we obtain that for any Borel function $f \in L^2_{\text{loc}}$ there exists the limit (in probability) of "discrete integrals" $\sum_{k=1}^{n} f\left(B_{(k-1)/n}\right) \Delta B_{k/n}$. This limit is denoted by $\int_0^1 f(B_s) \, dB_s$ and called Itô's stochastic integral with respect to Brownian motion. For more details, see [79, 148] or [157].

4.4   Burkholder's and Davis's Inequalities

In this section, we are going to see generalizations of Khintchine's inequality (1.15) for sums of independent random variables and for martingales. Recall, from Chap. 1, that Khintchine's inequality (1.15) refers to Rademacher's functions, however, it can be written in a more general context as follows.

Theorem 4.14 (Khintchine's inequality) *Let* $\{\xi_k\}$ *be a sequence of independent Bernoulli random variables,*

$$\mathbf{P}\{\xi_k = 1\} = \mathbf{P}\{\xi_k = -1\} = 1/2,$$

and let $\{c_k\}$ *be a sequence of real numbers such that* $\{c_k\} \in \ell^2$. *Then, for any* $0 < p < \infty$ *there exists an universal constants* A_p, B_p *(independent of* $\{c_k\}$*) such that*

$$A_p \left(\sum_{k=1}^{n} |c_k|^2 \right)^{1/2} \leqslant \left\| \sum_{k=1}^{n} c_k \xi_k \right\|_p \leqslant B_p \left(\sum_{k=1}^{n} |c_k|^2 \right)^{1/2} \tag{4.38}$$

for any $n \geqslant 1$.

The next result generalizes Khintchine's inequality for random variables with an arbitrary distribution.

Theorem 4.15 (Marcinkiewicz-Zygmund inequality) *Let* $\{\xi_k\}$ *a sequence of independent random variables with* $\mathbf{E}[\xi_k] = 0$ *and* $\mathbf{E}[|\xi_k|^p] < \infty$, *for some* $1 \leqslant p < \infty$ *then there exist constants* A_p *and* B_p, *such that*

$$A_p \left\| \left(\sum_{k=1}^{n} \xi_k^2 \right)^{1/2} \right\|_p \leqslant \left\| \sum_{k=1}^{n} \xi_k \right\|_p \leqslant B_p \left\| \left(\sum_{k=1}^{n} \xi_k^2 \right)^{1/2} \right\|_p \tag{4.39}$$

for any $n \geq 1$, *where* A_p, B_p *depend only on* p.

For the proof of this result, we refer to the original article of Marcinkiewicz and Zygmund, [117, Theorem 13].

Note that in both cases the sequence $\{X_n\}$ considered $\{X_n = \sum_{i=1}^{n} c_j \xi_j\}$ and $\{X_n = \sum_{i=1}^{n} \xi_j\}$, are sums of independent centered random variables and therefore they are martingales.

A natural question is then whether this inequality can be extended to arbitrary martingales. We will use the quadratic function $S(X)$ to establish a general version of the inequality for general martingales; as we did for Paley's inequality (1.23) for Haar and Walsh functions. For this generalization, we are going to dedicate a good part of the section.

First, we establish an important distribution inequality for $S(X)$, which is analogous to Doob's maximal inequality (3.27). We have seen in Theorem 4.1, that the quadratic function $S(X)$ of a martingale L^1-bounded X is finite a.e. D. Burkholder strengthened Austin's result by proving that the operator $S(X)$ is weak- type $(1, 1)$.

Theorem 4.16 (L^1-weak inequality for $S(X)$) *Let* $X = \{X_n\}_n$ *be a* $\{\mathscr{F}_n\}$-*martingale or a non-negative* $\{\mathscr{F}_n\}$-*submartingale* L^1 *bounded, then for the quadratic function* $S(X)$ *of* X *there exists a constant* C *such that*

$$\lambda \mathbf{P}\{S(X) > \lambda\} \leqslant C\|X\|_1 \tag{4.40}$$

for any $\lambda > 0$.

Proof Let $\lambda > 0$ a fixed real number and define a stopping time

$$\mu = \inf\{n : |X_n| > \lambda\}.$$

Since $S_{\mu-1}(X) = S(X)$ on the set $\{\mu = \infty\} = \{X^* \leqslant \lambda\}$, we get by Tchebychev's inequality and Lemma 4.1,

$$\lambda \mathbf{P}\left\{S(X) > \lambda, X^* \leqslant \lambda\right\} \leqslant \lambda \mathbf{P}\left\{S_{\mu-1}(X) > \lambda\right\} \leqslant \frac{\lambda \left\|S_{\mu-1}(X)\right\|_2^2}{\lambda^2} \leqslant 2\|X\|_1. \tag{4.41}$$

Then by Doob's maximal inequality (Theorem 3.7)

$$\lambda \mathbf{P}\{S(X) > \lambda\} \leqslant \lambda \mathbf{P}\left\{X^* > \lambda\right\} + \lambda \mathbf{P}\left\{S(X) > \lambda; X^* \leqslant \lambda\right\} \leqslant \|X\|_1 + 2\|X\|_1 = 3\|X\|_1. \quad \square$$

Thus, we have proved (4.40) with $C = 3$, which is the original proof given by Burkholder in [30] and therefore the best constant C_0 in (4.40) satisfies $C_0 \leqslant 3$. Later in [31] he proved (4.40) with $C = 2$. Moreover, if

$$X_n = \sum_{k=1}^{n} c_k r_k(t),$$

where $\{r_k\}$ are the Rademacher functions, then $X = \{X_n\}$ is a martingale for which $S(X) = \left(\sum_{k=1}^{n} c_k^2\right)^{\frac{1}{2}} = \lambda$, then, by (4.40) and Khintchine inequality, Theorem 1.3,

$$\lambda \mathbf{P}\{S(X) \geqslant \lambda\} = \left(\sum_{k=1}^{n} c_k^2\right)^{\frac{1}{2}} \leqslant \sqrt{2} \int_0^1 \left|\sum_{k=1}^{n} c_k r_k(t)\right| dt = \sqrt{2}\|X\|_1.$$

as was proved by J. Sarek in [156]. Now it is clear that taking $c_1 = c_2 = 1$ and $c_k = 0, k \geqslant 3$ is an extremal sequence. Therefore, $C_0 \geqslant \sqrt{2}$, hence

$$\sqrt{2} \leqslant C_0 \leqslant 3.$$

Moreover, in [19] B. Bollobás improved the bounds for C_0,

$$\frac{3}{2} \leqslant C_0 \leqslant \sqrt{e}.$$

Finally, in [59] D. C. Cox proved inequality (4.40) with $C = \sqrt{e}$ and shows that this is the best possible constant.

Let us now discuss the famous Burkholder's inequality that generalizes for any martingales, the Khintchine, Marcinkiewicz-Zygmund, and Paley inequalities, for $p > 1$. For its proof we need the following interpolation technique, see also Proposition 3.10.

Lemma 4.2 *Let $X = \{X_n\}_n$ be a non-negative $\{\mathscr{F}_n\}$-submartingale. Then, for non-negative $n \in \mathbb{Z}$, fixed, the random variable $Y = \left(S_n(\theta X) \vee X_n^*\right)$ satisfies*

$$\lambda \mathbf{P}\{Y > \beta\lambda\} \leqslant 3 \int_{\{Y > \lambda\}} X_n d\mathbf{P}, \quad \lambda > 0 \tag{4.42}$$

with $\theta > 0$, $\beta = \left(1 + 2\theta^2\right)^{1/2}$, and

$$\|S_n(X)\|_p \leqslant 9 p^{1/2} q \, \|X_n\|_p, \tag{4.43}$$

with $\frac{1}{p} + \frac{1}{q} = 1$ and $1 < p < \infty$.

Proof First, in order to prove (4.42) observe that since $\beta > 1$,

$$\lambda \mathbf{P}\{Y > \beta\lambda\} \leqslant \lambda \mathbf{P}\left\{X_n^* > \lambda\right\} + \lambda \mathbf{P}\left\{S_n(\theta X) > \beta\lambda, X_n^* \leqslant \lambda\right\}.$$

For the first term of this expression, we have

$$\lambda \mathbf{P}\left\{X_n^* > \lambda\right\} \leqslant \int_{\{X_n^* > \lambda\}} X_n d\mathbf{P} \leqslant \int_{\{Y > \lambda\}} X_n d\mathbf{P}.$$

To prove that the second term is less or equal to twice the last integral, let us consider

$$Z_n = \chi_{\{S_n(\theta X) > \beta\lambda\}} X_n.$$

Then $Z = \{Z_n\}$ is a non-negative submartingale since

$$\chi_{\{S_n(\theta X) > \beta\lambda\}} \leqslant \chi_{\{S_{n+1}(\theta X) > \beta\lambda\}},$$

and then we have

$$\begin{aligned}
\mathbf{E}\left[Z_{n+1} \mid \mathscr{F}_n\right] &= \mathbf{E}\left[\chi_{\{S_{n+1}(\theta X) > \beta\lambda\}} X_{n+1} \mid \mathscr{F}_n\right] \\
&\geqslant \chi_{\{S_n(\theta X) > \beta\lambda\}} \mathbf{E}\left[X_{n+1} \mid \mathscr{F}_n\right] \geqslant \chi_{\{S_n(\theta X) > \beta\lambda\}} X_n = Z_n.
\end{aligned}$$

Now, we need to prove that

$$\left\{ S_n(Z) > \beta\lambda, Z_n^* \leqslant \lambda \right\} \supseteq \left\{ S_n(\theta X) > \beta\lambda, X_n^* \leqslant \lambda \right\}.$$

To prove the inclusion, let us consider the stopping time,

$$\tau = \inf\left\{ n : S_n(\theta X) > \lambda \right\}.$$

On the set $\mathbf{P}\left\{ S_n(\theta X) > \beta\lambda, X_n^* \leqslant \lambda \right\}$ we have $1 \leqslant \tau \leqslant n$, $Z_n^* \leqslant \lambda$ and

$$\beta^2\lambda^2 < S_n^2(\theta X) = S_{\tau-1}^2(\theta X) + \theta^2 \Delta X_\tau^2 + \theta^2 \sum_{\tau < k \leqslant n} \Delta X_n^2 \leqslant \lambda^2 + \theta^2 \Delta X_\tau^2 + \theta^2 S_n^2(Z).$$

Then, if $|\Delta X_\tau| \leqslant \lambda$,

$$\beta^2\lambda^2 < \lambda^2 + \theta^2\lambda^2 + \theta^2 S_n^2(Z)$$

which implies, by definition of β, that $S_n(Z) > \lambda$ if $|\Delta X_\tau| \leqslant \lambda$. From the fact that X is non-negative, we get

$$|\Delta X_k| \leqslant X_k, \quad \text{and} \quad X_{k-1} \leqslant X^*$$

so that in the set in question $|\Delta X_\tau| \leqslant X^* \leqslant \lambda$. Then, by the inclusion proved and using (4.41), we get

$$\lambda\mathbf{P}\left\{ S_n(\theta X) > \beta\lambda, X_n^* \leqslant \lambda \right\} \leqslant \lambda\mathbf{P}\left\{ S_n(Z) > \beta\lambda, Z_n^* \leqslant \lambda \right\} \leqslant 2\|Z_n\|_1 \leqslant 2 \int_{\{Y > \lambda\}} X_n d\mathbf{P}$$

which is the inequality needed. Then, by Proposition 3.10, we get

$$\theta\|S_n(X)\|_p = \|S_n(\theta X)\|_p \leqslant \|Y\|_p \leqslant 3\beta^p q \|X_n\|_p.$$

Finally, taking $\theta = p^{-1/2}$, we get $\beta^p = \left(1 + 2p^{-1}\right)^{p/2} < e < 3$ and therefore we obtain (4.43). $\qquad\square$

We are ready to state and prove the famous Burkholder's inequality (see [29, 30].)

Theorem 4.17 (Burkholder's inequality) *Given $1 < p < \infty$ there exist universal positive constants c_p and C_p, such that if $X = \{X_n\}$ is a $\{\mathscr{F}_n\}$-martingale, then*

$$c_p\|S(X)\|_p \leqslant \|X\|_p \leqslant C_p\|S(X)\|_p \tag{4.44}$$

or equivalently, for any n

$$c_p\|S_n(X)\|_p \leqslant \|X_n\|_p \leqslant C_p\|S_n(X)\|_p. \tag{4.45}$$

The proof shows that the optimal choice c_p satisfies $c_p^{-1} = O\left(p^{1/2}q\right)$ and analogously $C_p = O\left(q^{1/2}p\right)$ where $\frac{1}{p} + \frac{1}{q} = 1$ (in fact, we have, $c_p = \left[18p^{3/2}/(p-1)\right]^{-1}$ and $C_p = 18p^{3/2}/(p-1)^{1/2}$)).

Proof Fix n and define non-negative martingales Y and Z by

$$Y_k = \mathbf{E}\left[X_n^+ \mid \mathscr{F}_k\right] \text{ and } Z_k = \mathbf{E}\left[X_n^- \mid \mathscr{F}_k\right]$$

for $k \geqslant 1$. Then, $Y_n = X_n^+$, $Z_n = X_n^-$, and $X_k = Y_k - Z_k$, $1 \leqslant k \leqslant n$. Additionally

$$S_n(X) \leqslant S_n(Y) + S_n(Z).$$

Therefore by Lemma 4.2, we get

$$\|S_n(X)\|_p \leqslant \|S_n(Y)\|_p + \|S_n(Z)\|_p \leqslant 9p^{1/2}q\left(\|Y_n\|_p + \|Z_n\|_p\right) = 18p^{1/2}q\|X_n\|_p,$$

which implies the left-hand inequality (4.44).

The right-hand side inequality follows from the left-hand one using a simple argument by duality: we may assume that $\|S_n(X)\|_p < \infty$.

Since $|X_n| \leqslant n\Delta X^*$ and $\Delta X^* \leqslant S(X)$, and this implies X_n is in L^p. Let us assume that $\|X_n\|_p > 0$, then

$$\|X_n\|_p = \mathbf{E}[X_n W_n],$$

where the function $W_n = \frac{\text{sgn}(X_n)|X_n|^{p-1}}{\|X_n\|_p^{p-1}}$ satisfies $\|W_n\|_q = 1$.

Now, let $W = \{W_k\}$ be the martingale defined as $W_k = \mathbf{E}\left[W_n \mid \mathscr{F}_k\right], k \geqslant 1$. Then, by the orthogonality of the increments, and the inequalities of Cauchy-Schwarz and Hölder, we get

$$\|X_n\|_p = \mathbf{E}[X_n W_n] = \mathbf{E}\left[\sum_{k=1}^{n} \Delta X_n \Delta W_n\right] \leqslant \mathbf{E}\left[S_n(X) S_n(W)\right]$$

$$\leqslant \|S_n(X)\|_p \|S_n(W)\|_q \leqslant 18q^{1/2}p\|S_n(X)\|_p. \qquad \square$$

Burkholder's original proof uses explicitly Marcinkiewicz's interpolation theorem (see Zygmund [176]). The inequality (4.44) contains Paley's inequality (1.23) in the case that X is the sequence of 2^n the partial sums of a Walsh series, and the inequality obtained by Marcinkiewicz and Zygmund (4.39) in the case that ΔX, the difference sequence of X, is independent and satisfies $\mathbf{E}[\Delta X_n] = 0, n \geqslant 1$, in the latter case, the inequality is also true for $p = 1$. Actually, Paley's proof of his inequality (1.23) can be carried over to the general martingale case with only minor modifications as observed by R. Gundy. Therefore, Burkholder's proof given above may be viewed as an alternative to Paley's proof.

As it is already mentioned, Burkholder's inequality is satisfied for $p > 1$, while the Marcinkiewicz-Zygmund inequality is also satisfied in the case $p = 1$, what happen for the case $p = 1$ in general? As the following example shows that, in general, is false.

Example 4.2 Let $\xi_1, \xi_2, \ldots$ independent random variables with Bernoulli distribution

$$\mathbf{P}\{\xi_1 = 1\} = \mathbf{P}\{\xi_1 = -1\} = 1/2,$$

and let

$$X_n = \sum_{i=1}^{n\wedge\tau} \xi_j \quad \text{where } \tau = \inf\left\{k \geqslant 1 : \sum_{i=1}^{k} \xi_j = 1\right\}.$$

Then $X = \{X_n\}$ is a martingale with

$$\|X_n\|_1 = \mathbf{E}\left[|X_n|\right] = 2\mathbf{E}\left[X_n^+\right] \to 2 \text{ if } n \to \infty,$$

but

$$\|S_n(X)\|_1 = \mathbf{E}\left[S_n(X)\right] = E\left(\sum_{i=1}^{n\wedge\tau} 1\right)^{1/2} = \mathbf{E}[\sqrt{n \wedge \tau}] \to \infty, \text{ if } n \to \infty. \qquad \square$$

Thus, Burkholder's inequality in general, is not true for $p = 1$.

On the other hand, by Doob's inequality (Corollary 3.9), Burkholder's inequality is equivalent to the following one: there exist c_p and C_p such that if $X = \{X_n\}$ is a $\{\mathscr{F}_n\}$-martingale, then

$$c_p\|S(X)\|_p \leqslant \|X^*\|_p \leqslant C_p\|S(X)\|_p, \quad 1 < p < \infty \tag{4.46}$$

or equivalently, for all n

$$c_p\|S_n(X)\|_p \leqslant \|X_n^*\|_p \leqslant C_p\|S_n(X)\|_p, \quad 1 < p < \infty. \tag{4.47}$$

It is clear that for $p = 1$ the inequalities (4.44) and (4.46) are not equivalent since in that case Doob's inequality does not hold. The question is if for $p = 1$, (4.46) is still true. The answer is positive and is given in the next result (see [64]).

Theorem 4.18 (Davis) *Let $X = \{X_n\}$ is a L^2-$\{\mathscr{F}_n\}$-martingale, then there exist positive constants c, C such that*

$$c\|S(X)\|_1 \leqslant \|X^*\|_1 \leqslant C\|S(X)\|_1 \tag{4.48}$$

Proof First, by (3.7)

$$\|S(X)\|_2^2 = \lim_{n\to\infty} \|S_n(X)\|_2^2 = \lim_{n\to\infty} \|X_n\|_2^2 = \|X\|_2^2.$$

The most immediate connections between $S(X)$ are X^* are the following

$$\mathbf{P}\{S(X) > \lambda\} \leqslant \frac{\|X^*\|_2^2}{\lambda^2} \tag{4.49}$$

and

$$\mathbf{P}\left\{X^* > \lambda\right\} \leqslant \frac{\|S(X)\|_2^2}{\lambda^2}, \qquad\qquad (4.50)$$

since

$$\mathbf{P}\left\{X_n^* > \lambda\right\} = \mathbf{P}\left\{(X_n^2)^* > \lambda^2\right\} \leqslant \frac{\|X_n^2\|_1^2}{\lambda^2} = \frac{\|X_n\|_2^2}{\lambda^2}$$

and by the continuity of $\mathbf{P}$ we get

$$\mathbf{P}\left\{X^* > \lambda\right\} \leqslant \frac{1}{\lambda^2} \lim_{n\to\infty} \|X_n\|_2^2 = \frac{\|S(X)\|_2^2}{\lambda^2}.$$

On the other hand, by Tchebychev's inequality, we get

$$\mathbf{P}\left\{S_n(X) > \lambda\right\} \leqslant \frac{\|S_n(X)\|_2^2}{\lambda^2}$$

and then again by the continuity of $\mathbf{P}$,

$$\mathbf{P}\{S(X) > \lambda\} \leqslant \frac{1}{\lambda^2} \lim_{n\to\infty} \|S_n(X)\|_2^2 = \frac{1}{\lambda^2} \|X\|_2^2.$$

Moreover, as

$$\|X\|_2^2 = \sup_n \|X_n\|_2^2 = \sup_n \int_\Omega X_n^2 d\mathbf{P} \leqslant \|X^*\|_2^2,$$

we get

$$\mathbf{P}\{S(X) > \lambda\} \leqslant \frac{1}{\lambda^2} \|X^*\|_2^2.$$

Recall that by the L^1-weak inequality for $S(X)$ (Theorem 4.16)

$$\mathbf{P}\left\{S(X) > \lambda\right\} \leqslant \frac{3}{\lambda} \|X\|_1.$$

Let T be one of the operators $* : X \to X^*$ or $S : X \to S(X)$ and let H be the other one. The theorem will be proved if we prove that there exists a constant C such that for every martingale

$$\|H(X)\| \leqslant C \|T(X)\|,$$

since the roles of $T(X)$ and $H(X)$ can be interchanged. By Davis' decomposition, Theorem 4.6, we know X can be decomposed as the sum of two martingales $X = Y + Z$ and such that for $Z_k^\# = \Delta X_k \chi_{\{|\Delta X_k| > 2\Delta X_{k-1}^*\}}$ defined in that theorem we have

$$\sum_{k=1}^\infty |Z_k^\#| \leqslant 2\Delta X^* \leqslant 4 T(X)$$

for whatever T is chosen. Then,

$$\left\| \sum_{k=1}^{\infty} \left| \mathbf{E}\left[Z_k^{\#} \mid \mathscr{F}_{k-1} \right] \right| \right\|_1 = \mathbf{E}\left[\sum_{k=1}^{\infty} \left| \mathbf{E}\left[Z_k^{\#} \mid \mathscr{F}_{k-1} \right] \right| \right] \leqslant \mathbf{E}\left[\sum_{k=1}^{\infty} \mathbf{E}\left[|Z_k^{\#}| \mid \mathscr{F}_{k-1} \right] \right]$$

$$= \mathbf{E}\left[\sum_{k=1}^{\infty} |Z_k^{\#}| \right] \leqslant 4\mathbf{E}[T(X)] = 4\|T(X)\|_1.$$

Now as $X_n = Y_n + Z_n$, we obtain

$$\|H(X)\|_1 \leqslant \|H(Y)\|_1 + \|H(Z)\|_1 \leqslant \|H(Y)\|_1 + \left\| \sum_{k=1}^{\infty} |\Delta Z_k| \right\|_1$$

$$\leqslant \|H(Y)\|_1 + 8\|T(X)\|_1. \tag{4.51}$$

We need to estimate $\|H(Y)\|_1$. Let

$$\tau = \tau(\lambda) = \inf \left\{ n : \min\left(T_n(X), T_n(Z), \mathbf{E}\left[Z_{n+1}^{\#} \mid \mathscr{F}_n \right] \right) \geqslant \lambda \right\}$$

be a stopping time. Then,

$$|Y_\tau - Y_{\tau-1}| = \left| Y_\tau^{\#} - \mathbf{E}\left[Y_\tau^{\#} \mid \mathscr{F}_{\tau-1} \right] \right| = \left| Y_\tau^{\#} - \mathbf{E}\left[\Delta X_\tau - Z_\tau^{\#} \mid \mathscr{F}_{\tau-1} \right] \right|$$

$$= \left| \Delta X_\tau \chi_{\{|\Delta X_\tau| \leqslant 2\Delta X_{\tau-1}^*\}} + \mathbf{E}\left[Z_\tau^{\#} \mid \mathscr{F}_{\tau-1} \right] \right| \leqslant 5\lambda,$$

since $\left| \mathbf{E}\left[Z_\tau^{\#} \mid \mathscr{F}_{\tau-1} \right] \right| \leqslant \lambda$ and $\Delta X_{\tau-1}^* \leqslant 2|T_{\tau-1}(X)| \leqslant 2\lambda$. Therefore

$$T_\tau(Y) \leqslant T_{\tau-1}(Y) + |Y_\tau - Y_{\tau-1}| \leqslant 6\lambda. \tag{4.52}$$

Additionally,

$$T_\tau(Y) \leqslant T(Y) \leqslant T(X) + T(Z). \tag{4.53}$$

Now,

$$\mathbf{P}\{H(Y) > \lambda\} = \mathbf{P}\{\tau < \infty, H(Y) > \lambda\} + \mathbf{P}\{\tau = \infty, H(Y) > \lambda\}$$

$$\leqslant \mathbf{P}\{\tau < \infty\} + \mathbf{P}\{H_\tau(Y) > \lambda\},$$

and therefore

$$\int_0^{\infty} \mathbf{P}\{\tau < \infty\} d\lambda \leqslant \int_0^{\infty} \mathbf{P}\{T(X) > \lambda\} d\lambda + \int_0^{\infty} \mathbf{P}\{T(Y) > \lambda\} d\lambda$$

$$+ \int_0^{\infty} \mathbf{P}\{\sum_{k=1}^{\infty} |Z_k^{\#}| > \lambda\} d\lambda$$

$$\leqslant \|T(X)\|_1 + \|T(Y)\|_1 + \left\| \sum_{k=1}^{\infty} |Z_k^{\#}| \right\|_1$$

$$\leqslant 14\|T(X)\|_1,$$

since, by analogous argument as in (4.51) we have

$$\|T(Y)\|_1 \leqslant \|T(X)\|_1 + \|T(Z)\|_1 \leqslant 9\|T(X)\|_1.$$

It remains to analyze $\mathbf{P}\{H_\tau(Z) > \lambda\}$. By (4.49) and (4.50) we have

$$\mathbf{P}\{H(X) > \lambda\} \leqslant \frac{1}{\lambda^2}\|T(X)\|_2^2.$$

Then, by Tchebychev's inequality, we get, by (4.52) and (4.53)

$$
\begin{aligned}
\int_0^\infty \mathbf{P}\{H_\tau(Y) > \lambda\}\,d\lambda &\leqslant \int_0^\infty \frac{\|T_\tau(Y)\|_2^2}{\lambda^2}\,d\lambda \leqslant \int_0^\infty \|\min(T(X)+T(Z),6\lambda)\|_2^2 \frac{d\lambda}{\lambda^2} \\
&\leqslant \int_0^\infty \|(T(X)+T(Z))\chi_{\{T(X)+T(Z)\leqslant 6\lambda\}}\|_2^2 \frac{d\lambda}{\lambda^2} \\
&\qquad + \int_0^\infty (6\lambda)^2 \mathbf{P}\{T(X)+T(Z) > 6\lambda\}\frac{d\lambda}{\lambda^2} \\
&\leqslant \int_0^\infty \|(T(X)+T(Z))\chi_{\{T(X)+T(Z)\leqslant 6\lambda\}}\|_2^2 \frac{d\lambda}{\lambda^2} + 6\|T(X)+T(Z)\|_1.
\end{aligned}
$$

Let us estimate the first integral using Fubini's theorem

$$
\begin{aligned}
\int_0^\infty \|(T(X)+T(Z))\chi_{\{T(X)+T(Z)\leqslant 6\lambda\}}\|_2^2 \frac{d\lambda}{\lambda^2} &= \int_0^\infty \int_\Omega (T(X)+T(Z))^2 \chi_{\{T(X)+T(Z)\leqslant 6\lambda\}}\,d\mathbf{P}\frac{d\lambda}{\lambda^2} \\
&\leqslant \int_\Omega (T(X)+T(Z))^2 \left(\int_{\frac{T(X)+T(Z)}{6}}^\infty \frac{d\lambda}{\lambda^2}\right) d\mathbf{P} \\
&= 6\int |T(X)+T(Z)|\,d\mathbf{P} = 6\|T(X)+T(Z)\|_1.
\end{aligned}
$$

Then,

$$\int_0^\infty \mathbf{P}\{H_\tau(Y) > \lambda\}\,d\lambda \leqslant 12\|T(X)+T(Z)\|_1 + 12\cdot 8\|T(X)\|_1 \leqslant 108\|T(X)\|_1.$$

Finally,

$$
\begin{aligned}
\|H(Y)\|_1 &= \int_0^\infty \mathbf{P}\{H(Y) > \lambda\}\,d\lambda = \int_0^\infty \mathbf{P}\{\tau < \infty\}\,d\lambda + \int_0^\infty \mathbf{P}\{H_\tau(Y) > \lambda\}\,d\lambda \\
&\leqslant 14\|T(X)\|_1 + 108\|T(X)\|_1 = 122\|T(X)\|_1
\end{aligned}
$$

and therefore

$$\|H(X)\|_1 \leqslant 130\|T(X)\|_1. \qquad \square$$

In [31], D. L. Burkholder obtained the following (sharp) inequality that not only generalizes Paley's first inequality (1.23) but also the second one (1.26), and (1.27).

Theorem 4.19 (Burkholder) *For any* $1 < p < \infty$ *there exists a constant* c_p *such that for all* $\{\mathscr{F}_n\}$*-martingale,* X_n *and all multiplier sequences* $Y = \{Y_n\}_n$*, the transform* $(Y \circ X)$ *satisfies*

$$\|(Y \circ X)\|_p \leqslant c_p \|X\|_p, \tag{4.54}$$

in other words, for any $n \geqslant 1$

$$\left\|(Y \circ X)_n\right\|_p \leqslant c_p \|X_n\|_p. \tag{4.55}$$

It can be proved that the best constant for the inequality is $p^* - 1$ where $p^* = \max\{p, p/(p-1)\}$, see [31,36]. We will prove the inequality with $c_p = p^* - 1$, although we are not going to prove that it is the best constant.[3]

D. L. Burkholder gave a beautiful interpretation of this result in the introduction of that article: "*It must have been well known to Alexander Calder that it is possible to design a mobile that can be hung initially in a small room but which, if it is to move freely through all of its possible configurations, will have to be hung anew in an exceedingly large room. There is a close mathematical analog. To each possible configuration of a mobile made with string, rods, and weights, there corresponds a martingale with a similar arrangement of successive centers of gravity and this martingale is a transform of the martingale corresponding to the initial configuration. It is easy to see, either by looking first at mobiles or directly at martingales, that there do exist small martingales with large transforms.*"

Moreover, as also mentioned in the same paper, Theorem 4.19 is the martingale analog of the famous M. Riesz inequality about the L^p boundedness of conjugated function, which is the beginning of the study of the L^p-boundedness of singular integrals in harmonic analysis.

For the demonstration of Theorem 4.19, we need the following crucial lemma.

Lemma 4.3 *Given* $1 < p < \infty$*, let us define the function*

$$v(x, y) = \left|\frac{x+y}{2}\right|^p - (p^* - 1)^p \left|\frac{x-y}{2}\right|^p,$$

and let u *be a continuous function in* $\mathbb{R}^2$ *such that* $u(x, y) = u(y, x) = u(-x, -y)$ *defined for each* $p \geqslant 2$ *as*

$$u(x, y) = \begin{cases} \alpha_p x^p \left[1 - p^* \frac{x-y}{2x}\right] & \text{if } (x, y) \in A_1 \\ v(x, y) & \text{if } (x, y) \in A_2, \end{cases}$$

where $\alpha_p = p\left[p^*/(p^*-1)\right]^{1-p}$*, where*

$$A_1 = \left\{(x, y) \in \mathbb{R}^2 : x > 0, \ (1 - 2/p^*)x < y < x\right\}$$
$$A_2 = \left\{(x, y) \in \mathbb{R}^2 : x > 0, \ -x < y < (1 - 2/p^*)x\right\}$$

---

[3] Additionally, D. Burkholder obtained an analogous result for strongly subordinated submartingales, see [41, Theorem 1.1], with optimal constant.

Fig. 4.1 The regions A_1 and A_2

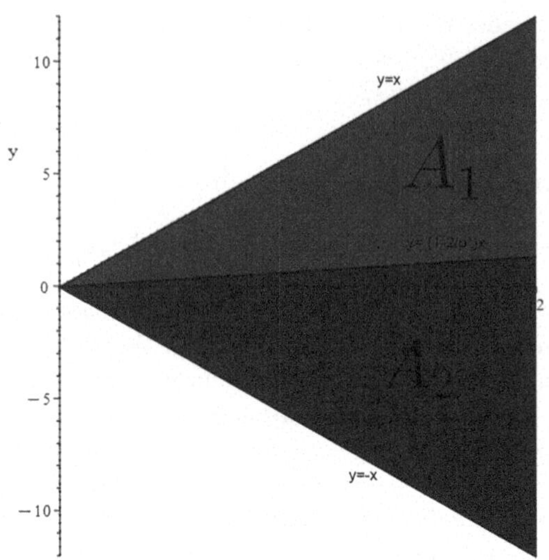

and if $1 < p < 2$, the regions A_1 and A_2 are exchanged. See (Figs. 4.1 and 4.2).
Then, u has the first derivatives u_x, u_y continuous and satisfies

$$u(x+h, y+k) \leqslant u(x, y) + u_x(x, y)h + u_y(x, y)k \quad if \ h \cdot k \leqslant 0. \tag{4.56}$$

Additionally,

$$u(x, y) \leqslant 0 \tag{4.57}$$

if $x \cdot y \leqslant 0$ and the strict inequality holds if $|x - y| \neq 2$. Finally, $v \leqslant u$ in $\mathbb{R}^2$.

The proof of this result will be given in full detail in an appendix at the end of the chapter. The function u is one of the so-called *Burkholder's functions*. The natural question is of course how to get this function? Where did it come from? As we will see in Chapter 5 this is the starting point of the Bellman function method.

Now, let us see the proof of Theorem 4.19.

Proof Here u and v are as in the previous lemma and let $Z = \{Z_n = (U_n, V_n)\}$ a martingale on $\mathbb{R}^2$ with $U_n = \sum_{k=1}^{n} (Y_k + 1) \, \Delta X_k$ and $V_n = \sum_{k=1}^{n} (Y_k - 1) \, \Delta X_k$.
Observe

$$U_1 \cdot V_1 = [(Y_1 + 1) \, \Delta X_1] \, [(Y_1 - 1) \, \Delta X_1] = \left(Y_1^2 - 1\right) \Delta X_1^2 \leqslant 0,$$

thus, by (4.57), $\mathbf{E}\left[u\left(Z_1\right)\right] \leqslant 0$.

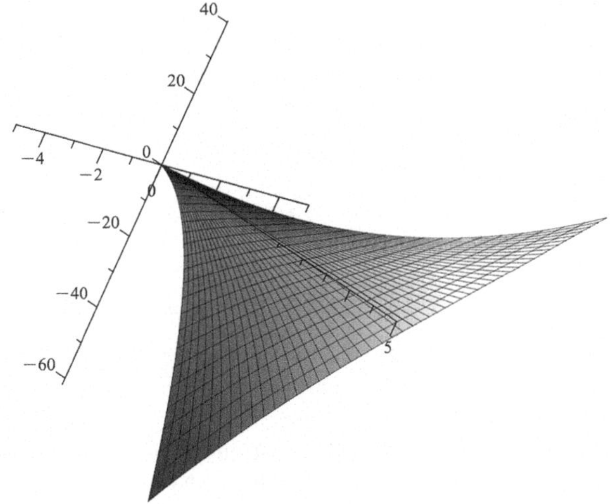

Fig. 4.2 A view of the function u on A_1 and A_2

Also, we want to see that

$$\mathbf{E}\left[u\left(Z_n\right)\right] \leqslant \cdots \leqslant \mathbf{E}\left[u\left(Z_1\right)\right] \leqslant 0.$$

We claim that

$$|u(x, y)| \leqslant C\left(|x|^p + |y|^p\right).$$

In fact, if $(x, y) \in A_1$, $0 < \frac{p^*(x-y)}{2x} < 1$ then

$$u(x, y) = \alpha_p x^p \left(1 - \frac{p^*(x - y)}{2x}\right)$$

then,

$$|u(x, y)| \leqslant \alpha_p |x|^p \leqslant \alpha_p \left(|x|^p + |y|^p\right).$$

If $(x, y) \in A_2$,

$$u(x, y) = \left|\frac{x + y}{2}\right|^p - (p^* - 1)^p \left|\frac{x - y}{2}\right|^p,$$

thus,

$$|u(x, y)| \leqslant \left|\frac{x + y}{2}\right|^p \leqslant |x + y|^p \leqslant 2^p (\max(|x|, |y|))^p \leqslant 2^p \left(|x|^p + |y|^p\right).$$

Analogously, it can be proved that

$$|u_x\left(Z_{n-1}\right)| \quad \text{and} \quad \left|u_y\left(Z_{n-1}\right)\right|$$

are bounded since $|u_x(x, y)|$ and $|u_y(x, y)|$ are bounded by $C(|x|^{p-1} + |y|^{p-1})$.

Therefore, the functions $u(Z_n), u(Z_{n-1}), u_x(Z_{n-1})(X_n - X_{n-1}), u_y(Z_{n-1})$ $(Y_n - Y_{n-1})$ are integrable. Then, since for $n \geqslant 2$, we have

$$(U_n - U_{n-1}) \cdot (V_n - V_{n-1}) = \left(Y_n^2 - 1\right) \Delta X_1^2 \leqslant 0$$

by (4.56) we would get

$$\mathbf{E}[u(Z_n)] \leqslant \mathbf{E}\left[u(Z_{n-1})\right]$$

if

$$\mathbf{E}\left[u_x(Z_{n-1})(X_n - X_{n-1})\right] = E\left[u_y(Z_{n-1})(Y_n - Y_{n-1})\right] = 0.$$

$u_x(Z_{n-1})$ and $u_y(Z_{n-1})$ are continuous functions of $\Delta X_1, \Delta X_2 \cdots \Delta$ $X_{n-1}, Y_1, Y_2, \ldots, Y_{n-1}$, and we have $U_n - U_{n-1} = (Y_n + 1)\Delta X_n$, $V_n - V_{n-1} = (Y_n - 1)\Delta X_n$.

Let $\Phi : \mathbb{R}^{2n-1} \to \mathbb{R}$ the continuous function associated to $u_x(Z_{n-1})$ and define the function $\Phi_j = \max(-j, \min(j, \Phi))$, then, by the dominated convergence theorem and the orthogonality condition, we get

$$\mathbf{E}\left[u_x(Z_{n-1})(Y_n - Y_{n-1})\right] = \mathbf{E}\left[\Phi(\Delta X_1, \ldots, \Delta X_{n-1}, Y_1, \ldots Y_{n-1})(Y_n + 1)\right]$$
$$= \lim_{j \to \infty} \mathbf{E}\left[\Phi_j(\Delta X_1, \ldots, \Delta X_{n-1}, Y_1, \ldots Y_{n-1})\right.$$
$$\times\left. ((Y_n + 1)\Delta X_n)\right] = 0.$$

In an analogous form, it can be proved that $\mathbf{E}\left[u_y(Z_{n-1})(Y_n - Y_{n-1})\right] = 0$. Then, we have

$$\mathbf{E}[u(Z_n)] \leqslant \mathbf{E}\left[u(Z_{n-1})\right] \leqslant \cdots \leqslant \mathbf{E}[u(Z_1)] \leqslant 0.$$

Finally since by Lemma 4.3 $v \leqslant u$ in $\mathbb{R}^2$, we conclude that

$$\mathbf{E}[v(Z_n)] = \mathbf{E}\left[\left|\frac{U_n + V_n}{2}\right|^p - (p^* - 1)^p \left|\frac{U_n - V_n}{2}\right|^p\right] \leqslant 0,$$

which as $\frac{U_n + V_n}{2} = (Y \circ X)_n$ and $\frac{U_n - V_n}{2} = X_n$ this give us (4.55). $\qquad\square$

A particular case of the inequality (4.55) is the ± 1 inequality: given $\{X_n\}$ a $\{\mathscr{F}_n\}$-martingale, and $\{\varepsilon_n\}$ any sequence of numbers in $\{-1, 1\}$, i.e., any random choice of sign, for $1 < p < \infty$ and $n \geqslant 1$,

$$\left\|\sum_{k=0}^{n} \varepsilon_k \Delta X_k\right\|_p \leqslant (p^* - 1)\left\|\sum_{k=0}^{n} \Delta X_k\right\|_p = (p^* - 1)\|X_n\|_p. \tag{4.58}$$

Theorem 1.1 of [36] establishes (4.55) with the optimal constant $c_p = p^* - 1$. The proof, which after some preliminary work reduces to the case when the predictable sequence $\{Y_n\} \in \{-1, 1\}$, rests on solving the non-linear partial differential equation

$$(p - 1)\left[yF_y - xF_x\right]F_{yy} - \left[(p - 1)F_y - xF_{xy}\right]^2 + x^2 F_{xx} F_{yy} = 0$$

for F non-constant and satisfying other conditions on a suitable domain of $\mathbb{R}^2$. Solving this equation leads to a system of five non-linear differential inequalities with boundary conditions. From this system, a function $u(x, y, t)$ is constructed in the domain

$$\Omega = \left\{ (x, y, t) \in \mathbb{R}^3 : \left| \frac{x - y}{2} \right|^p < t \right\}$$

with certain convexity properties from which, using the techniques of [32] D. Burkholder proves that

$$u(0, 0, 1) \, \|(Y \circ X)_n\|_p^p \leqslant \|X_n\|_p^p$$

for $1 < p \leqslant 2$ and shows that $u(0, 0, 1) = (p - 1)^p$. This, and duality, gives the bound $(p^* - 1)$ in (4.55).

Even today, the proofs in [36] seem extremely intricate. That D. L. Burkholder was able to show that these PDE's have a solution with the important properties needed for the martingale inequalities is impressive. A good reference of these ideas in a more formal way, which is what is called nowadays *Burkholder method* is [131, Chap. 2].

Proposition 4.3 *For $p > 1$, Burkholder's inequalities (4.44) and (4.54) are equivalents, i.e.,*

$$c_p \|S(X)\|_p \leqslant \|X\|_p \leqslant C_p \|S(X)\|_p$$

and

$$\|(Y \circ X)\|_p \leqslant c_p \|X\|_p$$

are equivalents.

Proof We will prove the equivalent inequalities (4.45) and (4.55). From Theorem 4.19 we know

$$\left\| (Y \circ X)_n \right\|_p \leqslant c_p \, \|X_n\|_p$$

for any n and any multiplier sequence $Y = \{Y_n\}$.

For $t \in [0, 1]$ fixed, let us consider the martingale transform $(Y \circ X)$ corresponding to the multiplier sequence $Y = \{r_n(t)\}_n$ where $\{r_n\}$ are the Rademacher functions and $t \in [0, 1]$ then from (4.54)

$$\left\| (Y \circ X)_n \right\|_p^p = \left\| \sum_{k=1}^{n} r_k(t) \Delta X_k \right\|_p^p \leqslant C_p^p \left\| \sum_{k=1}^{n} \Delta X_k \right\|_p^p = C_p^p \, \|X_n\|_p^p$$

but as $r_k^2 \equiv 1$, then X is the martingale transform of $(Y \circ X)$ with the same multiplier sequence Y and then, again by (4.54)

$$\|X_n\|_p^p \leqslant C_p^p \left\| \sum_{k=1}^{n} r_k(t) \Delta X_k \right\|_p^p = C_p^p \left\| (Y \circ X)_n \right\|_p^p$$

thus

$$C_p^{-p} \left\| (Y \circ X)_n \right\|_p^p \leqslant \| X_n \|_p^p \leqslant C_p^p \left\| (Y \circ X)_n \right\|_p^p.$$

Now, integrating on $t \in [0, 1]$ and using Fubini's theorem, we get

$$\int_0^1 \left\| \sum_{k=1}^n r_k(t) \Delta X_k \right\|_p^p dt = \mathbf{E}\left[\int_0^1 \left| \sum_{k=1}^n r_k(t) \Delta X_k \right|^p dt \right] = \left\| (Y \circ X)_n \right\|_p^p$$

and using Khinchine's inequality (1.3) we get

$$A_p^p \| S_n(X) \|_p^p = A_p^p \, \mathbf{E}\left[\sum_{k=1}^n (\Delta X_k)^2 \right]^{p/2} \leqslant \mathbf{E}\left[\int_0^1 \left| \sum_{k=1}^n r_k(t) \Delta X_k \right|^p dt \right] \leqslant B_p^p \| S_n(X) \|_p^p.$$

Therefore we get (4.44)

$$C_p^{-1} A_p \| S_n(X) \|_p \leqslant \| X_n \|_p \leqslant C_p B_p \| S_n(X) \|_p.$$

For the converse observe that the multiplier sequence is uniformly bounded by 1, then

$$S_n(Y \circ X) = \left[\sum_{k=1}^n (Y_k \Delta X_k)^2 \right]^{1/2} \leqslant \left[\sum_{k=1}^n (\Delta X_k)^2 \right]^{1/2} = S_n(X),$$

thus, by (4.44), we get (4.54)

$$\left\| (Y \circ X)_n \right\|_p \leqslant C_p \| S(Y \circ X) \|_p = C_p \| S(X) \|_p \leqslant C_p^{-1} C_p \| X \|_p. \qquad \square$$

Finally, Burkholder also obtained, for the case $p = 1$, a weak inequality for the martingale transforms, see [29, 36].

Theorem 4.20 (Burkholder) *There exists a constant C_1 such that for all $\{\mathscr{F}_n\}$-martingale, $X = X_n$ and all multiplier sequences $Y = \{Y_n\}_n$, the transform $(Y \circ X)$ satisfies*

$$\lambda \mathbf{P}\{(Y \circ X) > \lambda\} \leqslant C_1 \| X \|_1 \tag{4.59}$$

for any $\lambda > 0$.

Proof We will prove (4.59) with the optimal constant $C_1 = 2$,

$$\lambda \mathbf{P}\{(Y \circ X) > \lambda\} \leqslant 2 \| X \|_1.$$

To prove this we may assume that $X_1 \equiv 0$, define $Z = \{Z_n = (U_n, V_n)\}$ a martingale with values in $\mathbb{R}^2$ as

$$U_n = \sum_{k=2}^n (1 + Y_k) \Delta X_k \tag{4.60}$$

and

$$V_n = \sum_{k=2}^{n}(1 - Y_k)\Delta X_k \qquad (4.61)$$

starting at $x = 0$, $y = 0$.[4] Then

$$X_n = (U_n + V_n)/2 \text{ and } (Y \circ X)_n = (U_n - V_n)/2$$

and

$$(U_{n+1} - U_n)(V_{n+1} - V_n) = (1 - Y_{n+1}^2)\Delta X_{n+1}^2 \geqslant 0.$$

Now, set

$$D = \{(x, y) \in \mathbb{R}^2 : |x - y| < 2\}.$$

We may assume, without loss of generality that

$$|(Y \circ X)_\infty| \geqslant 1 \text{ a.e.} \qquad (4.62)$$

since if (4.62) is not satisfied, we can replace $(Y \circ X)$ by

$$(Y \circ X)^\tau = \{(Y \circ X)_{\tau \wedge n} : n \geqslant 1\},$$

where $\tau = \inf\{n : |(Y \circ X)_n| \geqslant 1\}$. On the set $\{\tau < \infty\}$,

$$\left|\left((Y \circ X)^\tau\right)_\infty\right| = \left|(Y \circ X)^\tau\right| \geqslant 1$$

and, on the set $\{\tau = \infty\}$, the inequality $|(Y \circ X)_n| < 1$ holds for all n but $\sup_n |(Y \circ X)_n| \geqslant 1$ a.e., so $(Y \circ X)^\tau$ satisfies (4.62) a.e.

Then, $Z_\infty = (U_\infty, V_\infty) \notin D$ a.e. since $|U_\infty - V_\infty|/2 = |(Y \circ X)_\infty| \geqslant 1$ a.e. and we also may assume that $Z_{n+1} = Z_n$ on the set $\{Z_n \notin D\}$. Then $\{(Y \circ X)_n^* \geqslant 1\} = \{Z_n \notin D\}$.

Let u be the continuous function defined on $\mathbb{R}^2$ by

$$u(x, y) = \begin{cases} 1 + xy & \text{if } |x| \vee |y| \leqslant 1 \\ |x + y| & \text{if } |x| \vee |y| > 1. \end{cases}$$

---

[4] This a particular case of what Burkholder defined as *zigzag martingales*: let $(x, y) \in \mathbb{R}^2$ consider $X = \{X_n\}$ a martingale starting at $(x + y)/2$ and $\{Y_n\}$ a multiplier sequence such that $Y_1 = 1$ and $Y_k \in \{-1, 1\}$, $k \geqslant 1$, then define the *zigzag martingale* on $\mathbb{R}^2$, $Z = \{Z_n = (U_n, V_n)\}$, starting at $Z_1 = (x, y)$, and

$$U_n = x + \sum_{k=2}^{n}(1 + Y_k)\Delta X_k, \ V_n = y + \sum_{k=2}^{n}(1 - Y_k)\Delta X_k, \ n > 1,$$

see [36, Sect. 3.]. Observe that $U_{n+1} - U_n = 0$ or $V_{n+1} - V_n = 0$.

It is elementary to check that, for $(x, y) \in \mathbb{R}^2$,

$$0 \leqslant u(x, y) - |x + y| \leqslant 1, \tag{4.63}$$

and that if $(x + h, y + k) \in \mathbb{R}^2$ with $h \cdot k \geqslant 0$, then

$$u(x + h, y + k) \geqslant u(x, y) + \varphi(x, y)h + \psi(x, y)k,$$

where $\varphi(x, y) = y$ and $\psi(x, y) = x$ if $|x| \vee |y| \leqslant 1$, and $\varphi(x, y) = \psi(x, y) = \mathrm{sgn}(x + y)$ elsewhere in $\mathbb{R}^2$. The functions φ and ψ are bounded and measurable so the martingale property of Z implies that

$$\mathbf{E}[\varphi\,(Z_n)\,(X_{n+1} - X_n)] = 0,$$

with a similar result for ψ. Therefore,

$$1 = u(0, 0) = \mathbf{E}[u\,(Z_1)] \leqslant \cdots \leqslant \mathbf{E}[u\,(Z_n)].$$

Then,

$$\mathbf{P}\left\{(Y \circ X)_n^* \geqslant 1\right\} = 1 - \mathbf{E}[\chi_D\,(Z_n)] \leqslant \mathbf{E}\left[u\,(Z_n) - \chi_D\,(Z_n)\right].$$

By the definition of u and (4.63), the latter integrand is not greater than $|U_n + V_n| = 2\,|X_n|$ and this gives

$$\mathbf{P}\left\{(Y \circ X)^* > 1\right\} \leqslant 2\|X\|_1.$$

This implies that

$$\lambda \mathbf{P}\left\{(Y \circ X)^* > \lambda\right\} \leqslant 2\|X\|_1$$

for all $\lambda > 0$, which implies (4.59). $\qquad\qquad\qquad\qquad\qquad\qquad\qquad\square$

Actually, in [36], Burkholder got a stronger version of inequality (4.59)

Theorem 4.21 (Burkholder) *Let $1 \leqslant p \leqslant 2$ there exists a constant c_p such that for all $\{\mathscr{F}_n\}$-martingale, $\{X_n\}$ and all multiplier sequences $Y = \{Y_n\}_n$, the transform $(X \circ Y)$ satisfies*

$$\sup_{\lambda > 0} \mathbf{P}\left\{(Y \circ X) > \lambda\right\} \leqslant \frac{2}{\Gamma(p + 1)}\|X\|_p^p \tag{4.64}$$

for any $\lambda > 0$.

This is a more complicated proof and since we will not use it, we will skip it; for the details see Theorem 1.3 of [36].

4.5   Generalizations of Burkholder's Inequality and the Good λ Inequalities

In this section, we will consider the generalizations of Burkholder's inequality. First, observe that Khintchine's inequality (1.15) which can be considered the source of these inequalities, is true for all $p > 0$. However, in all other inequalities, see for instance Theorem 4.17, the range $0 < p < 1$ is excluded. The question is what kind of restrictions are necessary to obtain the inequalities in that range?

We also want generalizations of Burkholder's inequality for functions beyond $\Phi(\lambda) = \lambda^p$, which we will discuss in the remaining of this section, and for a general class of martingale operators (that of course should include the operators $(\cdot)^*$ and S) to be discussed in detail in the next section. D. L. Burkholder and R. Gundy worked on these generalizations in their famous article *Extrapolation and interpolation of quasi-linear operators on martingales* published in Acta Mathematica [44], see also [30], and that will be the main reference of this section

Additionally, as we have mentioned before, if we have inequalities between distribution functions then using Marcinkiewicz's interpolation argument we could easily obtain integral inequalities, see for instance Lemma 4.2. However such inequalities between distribution functions do not always occur for all $\lambda > 0$ for the operators we are interested in, as the following example shows.

Example 4.3 Let $X_n = \sum_{k=1}^n \frac{r_k(t)}{k}$ where $\{r_k\}$ are the Rademacher functions; then $S(X) = \left(\sum_{k=1}^\infty \frac{1}{k^2}\right)^{1/2}$. By Khintchine's inequality, we know that

$$
A_p \left(\sum_{k=1}^\infty \frac{1}{k^2}\right)^{p/2} \leqslant \left\|\sum_{k=1}^\infty \frac{r_k}{k}\right\|_p \leqslant B_p \left(\sum_{k=1}^\infty \frac{1}{k^2}\right)^{p/2}
$$

but $m\{t \in [0, 1] : S(X) > \lambda\} = 0$, for any λ big enough.

Now, for any positive integer n

$$
m\left\{X^* > \lambda\right\} \geqslant m\left\{t \in [0, 1] : \left|\sum_{k=1}^n \frac{r_k(t)}{k}\right| > \lambda\right\} > 0
$$

for any $\lambda > 0$. Thus, we have a two-sided L^p inequality but the inequality

$$
m\left\{X^* > \lambda\right\} \leqslant Cm\{t \in [0, 1] : S(X) > \lambda\}
$$

is not possible if $\lambda > 0$ is big enough.

Despite that, D. L. Burkholder and R. Gundy in [44] gave a characterization for those numbers λ for which the inequalities between distribution functions hold which allows them, after justifying that such λ's are sufficient to integrate, to obtain integral formulas. These are the original formulation of the *good λ inequalities*: a given number $\lambda > 0$ is a "good λ" for a random variable Y if there exists $\alpha > 0$ and $\beta > 1$ such that

$$\mathbf{P}\{Y > \lambda\} \leqslant \alpha \mathbf{P}\{Y > \beta\lambda\}.$$

However, in [30], D. L. Burkholder obtained inequalities, that are more difficult than the previous ones (but valid for all λ) which are what nowadays is known as *good λ inequalities*. Thus, if it is possible to obtain *good λ inequalities,* one can also obtain the desired integral inequalities (see Theorem 4.22 below). Hence, the good λ inequalities are the correct translation, in probability, of Marcinkiewicz's interpolation argument.

- First, we will give the generalization of Burkholder inequality (4.44) for functions Φ beyond $\Phi(\lambda) = \lambda^p$, $p > 1$. Let us specify the type of functions for which we want generalization.

Definition 4.7 A non-trivial monotone non-decreasing continuous function Φ, defined on $[0, \infty]$ with $\Phi(0) = 0$ and satisfying a growth condition[5]

$$\Phi(2\lambda) \leqslant C\Phi(\lambda), \ \lambda > 0 \tag{4.65}$$

is said to be Young's function.

Observe that, trivially $\Phi(\lambda) = \lambda^p$ for $0 < p < \infty$ is Young's function. Additionally, from the definition we get the following inequalities.

(i) $\Phi(\lambda_1 \vee \lambda_2) \leqslant \Phi(\lambda_1) + \Phi(\lambda_2)$
(ii) $\Phi(\lambda_1 + \lambda_2) \leqslant \Phi(2\lambda_1) + \Phi(2\lambda_2) \leqslant C[\Phi(\lambda_1) + \Phi(\lambda_2)]$.

The following result, given by Burkholder in [30], explains in detail how the good λ inequalities are enough to get integral inequalities, and it will be the key, from now on, to get integral inequalities.

Theorem 4.22 *Suppose that X and Y are non-negative random variables over a probability space $(\Omega, \mathscr{F}, \mathbf{P})$ and assume there exist real numbers $\beta > 1, \delta > 0, \varepsilon > 0$, such that, for $\lambda > 0$*

$$\mathbf{P}\{Y > \beta\lambda, X \leqslant \delta\lambda\} \leqslant \varepsilon \mathbf{P}\{Y > \lambda\}. \tag{4.66}$$

Given Φ Young's function such that for all $\lambda > 0$

---

[5] In the theory of Orlicz spaces this condition is known as the Δ_2-condition.

$$\Phi(\beta\lambda) \leqslant \gamma\Phi(\lambda), \ \Phi\left(\delta^{-1}\lambda\right) \leqslant \eta\Phi(\lambda) \tag{4.67}$$

for some real numbers γ and η. Then, if $\gamma\varepsilon < 1$,

$$\mathbf{E}[\Phi(Y)] \leqslant \frac{\gamma\eta}{1-\gamma\varepsilon}\mathbf{E}[\Phi(X)]. \tag{4.68}$$

Observations 4.1 *The existence of γ and η satisfying (4.67) is guaranteed by (4.65). For instance, a possible choice for γ is C^K where K is a positive integer, such that $2^{K-1} < \beta \leqslant 2^K$ and C is the constant in (4.65).*

$$\Phi\left(2^{K-1}\lambda\right) \leqslant \Phi(\beta\lambda) \leqslant \Phi\left(2^K\lambda\right) \leqslant C\Phi\left(2^{K-1}\lambda\right) \leqslant \cdots \leqslant C^K\Phi(\lambda),$$

and analogously for η.

Proof First observe that we might assume that $\mathbf{E}[\Phi(X)] < \infty$ since otherwise (4.68) is trivially true. Also, we might assume that $\mathbf{E}[\Phi(Y)]$ is finite, since if Y satisfies (4.66), then $Y \wedge n$ also satisfies it, since for n big enough

$$
\begin{aligned}
\mathbf{P}\{Y \wedge n > \beta\lambda, X \leqslant \delta\lambda\} &= \mathbf{P}\{Y \wedge n > \beta\lambda, Y \leqslant n, X \leqslant \delta\lambda\} + \mathbf{P}\{Y \wedge n > \beta\lambda, Y > n, X \leqslant \delta\lambda\} \\
&= \mathbf{P}\{Y > \beta\lambda, Y \leqslant n, X \leqslant \delta\lambda\} + \mathbf{P}\{Y > \beta\lambda, Y > n, X \leqslant \delta\lambda\} \\
&\leqslant 2\mathbf{P}\{Y > \beta\lambda, X \leqslant \delta\lambda\} \leqslant 2\varepsilon\mathbf{P}\{Y > \lambda\}.
\end{aligned}
$$

Additionally, if (4.68) holds for $\{Y \wedge n, n \geqslant 1\}$, then it holds for Y. In fact, by the monotone convergence theorem, we have

$$\int_\Omega \Phi(Y)d\mathbf{P} = \lim_{n\to\infty}\int_\Omega \Phi(Y \wedge n)\,d\mathbf{P} \leqslant \frac{\eta}{1-\gamma\varepsilon}\int_\Omega \Phi(X)d\mathbf{P}.$$

Let us consider the Lebesgue-Stieltjes measure associated with Φ, denoted as μ_Φ, and defined by

$$\mu_\Phi([a, b)) = \int_a^b d\Phi(\lambda) = \Phi(b) - \Phi(a), 0 \leqslant a < b \leqslant \infty.$$

This is a positive and σ-finite measure on $\mathscr{B}[0, \infty)$. Observe that, if Z is a non-negative random variable on $(\Omega, \mathscr{F}, \mathbf{P})$, then

$$\Phi(Z) = \int_0^Z d\Phi(\lambda) = \int_0^\infty \chi_{\{Z>\lambda\}}d\Phi(\lambda).$$

By Fubini's theorem, we get then

$$\mathbf{E}[\Phi(Z)] = \int_{\Omega} \left(\int_0^{\infty} \chi_{\{Z>\lambda\}}(\lambda) d\Phi(\lambda) \right) d\mathbf{P}$$

$$= \int_0^{\infty} \left(\int_{\Omega} \chi_{\{Z>\lambda\}}(\omega) d\mathbf{P} \right) d\Phi(\lambda) = \int_0^{\infty} \mathbf{P}\{Z > \lambda\} d\Phi(\lambda).$$

Condition (4.66) implies that for any $\lambda > 0$,

$$\mathbf{P}\{Y > \beta\lambda\} = \mathbf{P}\{Y > \beta\lambda, X \leqslant \delta\lambda\} + \mathbf{P}\{Y > \beta\lambda, X > \delta\lambda\} \leqslant \varepsilon\mathbf{P}\{Y > \lambda\} + \mathbf{P}\{X > \delta\lambda\}.$$

Then, we get

$$\mathbf{E}\left[\Phi\left(\beta^{-1}Y\right)\right] = \int_0^{\infty} \mathbf{P}\left\{\beta^{-1}Y > \lambda\right\} d\Phi(\lambda) = \int_0^{\infty} \mathbf{P}\{Y > \beta\lambda\} d\Phi(\lambda)$$

$$\leqslant \int_0^{\infty} \varepsilon\mathbf{P}\{Y > \lambda\} d\Phi(\lambda) + \int_0^{\infty} \mathbf{P}\{X > \delta\lambda\} d\Phi(\lambda)$$

$$= \varepsilon\mathbf{E}[\Phi(Y)] + \mathbf{E}\left[\Phi\left(\delta^{-1}X\right)\right] \leqslant \varepsilon\mathbf{E}[\Phi(Y)] + \eta\mathbf{E}[\Phi(X)].$$

Moreover, as

$$\mathbf{E}[\Phi(Y)] = \mathbf{E}\left[\Phi\left(\beta\beta^{-1}Y\right)\right] \leqslant \gamma\mathbf{E}\left[\Phi\left(\beta^{-1}Y\right)\right]$$

$$\leqslant \gamma\varepsilon\mathbf{E}[\Phi(Y)] + \gamma\eta\mathbf{E}[\Phi(X)];$$

thus

$$(1 - \gamma\varepsilon)\mathbf{E}[\Phi(Y)] \leqslant \gamma\eta\mathbf{E}[\Phi(X)],$$

and given that $\mathbf{E}[\Phi(Y)] < \infty$, we get

$$\mathbf{E}[\Phi(Y)] \leqslant \frac{\gamma\eta}{1 - \gamma\varepsilon}\mathbf{E}[\Phi(X)]. \qquad \square$$

Observations 4.2

- *If the inequality*

$$\mathbf{P}\{Y > \lambda\} \leqslant C\mathbf{P}\{X > \delta\lambda\},$$

  holds for any λ then (4.68) is immediate, for any Φ Young's function

$$\mathbf{E}[\Phi(Y)] = \int_{\Omega} \Phi(Y)d\mathbf{P} = \int_0^{\infty} \mathbf{P}\{Y > \lambda\}d\Phi(\lambda) \leqslant C \int_0^{\infty} \mathbf{P}\{X/\delta > \lambda\}d\Phi(\lambda)$$

$$= C\mathbf{E}\left[\Phi\left(\frac{X}{\delta}\right)\right] \leqslant C'\mathbf{E}[\Phi(X)].$$

- *Now, if λ is a "good λ" for Y in the original sense and (4.66) holds, then*

$$\mathbf{P}\{Y > \beta\lambda\} = \mathbf{P}\{Y > \beta\lambda, X \leqslant \delta\lambda\} + \mathbf{P}\{Y > \beta\lambda, X > \delta\lambda\}$$
$$\leqslant \varepsilon\mathbf{P}\{Y > \lambda\} + \mathbf{P}\{X > \delta\lambda\}$$
$$\leqslant \alpha\varepsilon\mathbf{P}\{Y > \beta\lambda\} + \mathbf{P}\{X > \delta\lambda\}$$

and therefore, if $\alpha\varepsilon < 1$, we get

$$\mathbf{P}\{Y > \beta\lambda\} \leqslant C\mathbf{P}\{X > \delta\lambda\},$$

with $C = \frac{1}{1-\alpha\varepsilon}$ and we are back to the previous case.

The following example, due to Marcinkiewicz and Zygmund [117], demonstrates that with no additional conditions on the martingale, Burkholder's inequalities (4.44) and (4.54) are false for the range $0 < p < 1$.

Example 4.4 Let j be a positive integer, and $\xi_j = \{\xi_{j1}, \xi_{j2}, \ldots\}$, a sequence of independent random variables with distribution

$$\mathbf{P}\left\{\xi_{jk} = 1\right\} = 1 - (j+1)^{-1}, \mathbf{P}\left\{\xi_{jk} = -j\right\} = (j+1)^{-1}.$$

Let $(Y \circ X)_j = \left((Y \circ X)_{j1}, (Y \circ X)_{i2}, \ldots\right)$ a martingale defined as

$$(Y \circ X)_{jn} = \sum_{k=1}^{n} \xi_{jk}.$$

Clearly, $(Y \circ X)_j$ is a martingale since it is the sum of independent random variables with $\mathbf{E}\left[\xi_{jk}\right] = 0$. Now, let us see that

$$\lim_{m \to \infty} \lim_{j \to \infty} \frac{\| (Y \circ X)_j^m \|_p}{\| S((Y \circ X)_j^m) \|_p} = \infty.$$

If $j > 1$, $\mathbf{P}\left\{\xi_{jk} = -j\right\} < \mathbf{P}\left\{\xi_{jk} = 1\right\}$, therefore

$$\left\| (Y \circ X)_j^m \right\|_p = \sup_{1 \leqslant n < \infty} \left\| (Y \circ X)_{jn}^m \right\|_p = \sup_{1 \leqslant n < \infty} \left\| \sum_{k=1}^n \chi_{\{m>k\}} \xi_{jk} \right\|_p$$

$$= \sup_{1 \leqslant n < \infty} \left\| \sum_{k=1}^{n \wedge m} \xi_{jk} \right\|_p = \sup_{1 \leqslant n < \infty} \left(\int_\Omega \left| \sum_{k=1}^{n \wedge m} \xi_{jk} \right|^p d\mathbf{P} \right)^{1/p}$$

$$\geqslant \sup_{1 \leqslant n < \infty} \left(\int_{\{\xi_{jk}=1 \text{ for all } k\}} \left(\sum_{k=1}^{n \wedge m} \xi_{jk} \right)^p d\mathbf{P} \right)^{1/p} = \sup_{1 \leqslant n < \infty} (n \wedge m) = m.$$

Analogously,

$$\left\| S\left((Y \circ X)_j^m \right) \right\|_p = \left\| \left(\sum_{k=1}^m \xi_{jk}^2 \right)^{1/2} \right\|_p = \left(\int_\Omega \left| \sum_{k=1}^m \xi_{jk}^2 \right|^{p/2} d\mathbf{P} \right)^{1/p}$$

$$= \left(\int_{\{\xi_{jk}=1 \text{ for all } k\}} \left| \sum_{k=1}^m \xi_{jk}^2 \right|^{p/2} d\mathbf{P} + \int_{\{\xi_{jk}=1 \text{ for all } k\}^c} \left| \sum_{k=1}^m \xi_{jk}^2 \right|^{p/2} d\mathbf{P} \right)^{1/p}$$

$$\leqslant \left[m^{p/2} \left(1 - \frac{1}{j+1} \right)^m + \left(m j^2 \right)^{p/2} \left(1 - \left(1 - \frac{1}{j+1} \right) \right)^m \right]^{1/p}$$

$$= m^{1/2} \left[\left(1 - \frac{1}{j+1} \right)^m + \left(\frac{j^p}{(j+1)^m} \right) \right]^{1/p}.$$

Thus,

$$\lim_{m \to \infty} \lim_{j \to \infty} \frac{\| (Y \circ X)_j^m \|_p}{\| S((Y \circ X)_j^m) \|_p} \geqslant \lim_{m \to \infty} \lim_{j \to \infty} \frac{m}{m^{1/2} \left[\left(1 - \frac{1}{j+1} \right)^m + \left(\frac{j^p}{(j+1)^m} \right) \right]^{1/p}}$$

$$\geqslant \lim_{m \to \infty} \lim_{j \to \infty} \frac{m^{1/2}}{\left[\left[\left(1 - \frac{1}{j+1} \right)^m + \left(\frac{j^p}{(j+1)^m} \right) \right] \right]^{1/p}} = \infty.$$

with $0 < p < 1$, hence, we cannot have an inequality of type

$$\| (Y \circ X) \|_p \leqslant C \| S((Y \circ X)) \|_p.$$

The problem in this example is due to the fact that the martingale can have very large jumps. The crucial idea of good λ inequalities is that when a martingale crosses the λ level, it cannot go too far, i.e., the excursion is bounded above by a multiple of λ. This idea appears already in R. Gundy's article from the late 1960s, see [87].

As we have already mentioned a possible generalization of Burkholder's inequality is to try to extend it to be true for all of Young's functions. From the example, we know that for this to be possible we have to assume additional conditions on the martingale, i.e., to restrict the set of martingales under consideration.

Definition 4.8 Let $X = \{X_n\}$ be a $\{\mathscr{F}_n\}$-martingale, X is called regular if

$$\mathbf{E}\left[(\Delta X_k)^2 \mid \mathscr{F}_{k-1}\right] = 1, \tag{4.69}$$

and there exists $\delta > 0$ such that

$$\mathbf{E}\left[|\Delta X_k| \mid \mathscr{F}_{k-1}\right] \geq \delta \tag{4.70}$$

for any $k \geqslant 1$.

This definition is due to Richard Gundy in [87], who called it the Marcinkiewicz-Zygmund (MZ) condition. Under this name, it is referred to in the literature, see for instance [55, 63, 112]. Gundy considers this condition to study a.e. convergence problems for martingales transform of regular martingales which will be studied in the last section of this chapter.

Additionally, in [87] Gundy proves that condition (4.70) is equivalent to

$$\mathbf{P}\left\{|\Delta X_k| > \lambda \mid \mathscr{F}_{k-1}\right\} \geqslant \gamma > 0 \tag{4.71}$$

for some constants $\lambda > 0$, $\gamma > 0$ uniformly for $k \geqslant 1$.

The proof of the equivalence is an immediate consequence of the following lemma, which is due, essentially, to Paley and Zygmund [135].

Lemma 4.4 *(Paley and Zygmund). Let $\mathscr{F}$ be a σ-field, $Y \geqslant 0$ a random variable, and $0 < \eta < 1$ a fixed constant. If $\mathbf{E}[Y \mid \mathscr{F}] \geqslant \alpha > 0$ and $\mathbf{E}\left[Y^2 \mid \mathscr{F}\right] \leqslant \beta$ on a set F, then*

$$\mathbf{P}\{Y \geqslant \eta\alpha \mid \mathscr{F}\} \geqslant (1 - \eta)^2 \alpha^2 / \beta$$

a.e. on F.

Proof Since, trivially

$$\mathbf{E}\left[Y \chi_{\{Y < \eta\alpha\}} \mid \mathscr{F}\right] < \eta\alpha,$$

we have, by the Cauchy-Schwarz inequality,

$$
\begin{aligned}
(\alpha - \eta\alpha)^2 &\leqslant \left(\mathbf{E}[Y \mid \mathscr{F}] - \mathbf{E}\left[Y \chi_{\{Y < \eta\alpha\}} \mid \mathscr{F}\right]\right)^2 = \left(\mathbf{E}\left[Y \chi_{\{Y \geqslant \eta\alpha\}} \mid \mathscr{F}\right]\right)^2 \\
&\leqslant \mathbf{E}\left[Y^2 \chi_{\{Y \geqslant \eta\alpha\}} \mid \mathscr{F}\right] \mathbf{E}\left[\chi_{\{Y \geqslant \eta\alpha\}} \mid \mathscr{F}\right] \leqslant \beta \mathbf{P}\left[\chi_{\{Y \geqslant \eta\alpha\}} \mid \mathscr{F}\right],
\end{aligned}
$$

where these inequalities are understood to hold a.e. on F. $\square$

Now let us prove the equivalence between (4.70) and (4.71). Suppose (4.70) holds. Let $\beta = 1 = \mathbf{E}\left[\Delta X_k^2 \mid \mathscr{F}_{k-1}\right]$ and $\alpha = \delta$, $\eta = \frac{1}{2}$. Then by Lemma 4.4,

$$\mathbf{P}\left\{|\Delta X_k| \geqslant \delta/2 \mid \mathscr{F}_{k-1}\right\} \geqslant \delta^2 / 4 > 0$$

so that (4.71) holds. Now suppose (4.71) holds for some $\lambda > 0$ and $\gamma > 0$. Since the relation $|\Delta X_k| \geqslant |\Delta X_k| \, \chi_{\{|\Delta X_k| \geqslant \lambda\}}$ holds trivially, then

$$\mathbf{E}\left[|\Delta X_k| \mid \mathscr{F}_{k-1}\right] \geqslant \mathbf{E}\left[|\Delta X_k| \, \chi_{\{|\Delta X_k| \geqslant \lambda\}} \mid \mathscr{F}_{k-1}\right] \geqslant \lambda \mathbf{P}\{|\Delta X_k| \geqslant \lambda \mid \mathscr{F}_{k-1}\} \geqslant \lambda\gamma > 0$$

a.e., so that (4.70) holds with $\lambda\gamma = \delta$.

Also, we will need later the following technical result, see Zygmund [176, Chap. V], see also [44, Lemma 2.3].

Lemma 4.5 *Let Y a non-negative random variable on $(\Omega, \mathscr{F}, \mathbf{P})$, α, β, p_1, p_2 positive real numbers with $p_1 < p_2$, and $A \in \mathscr{F}$ such that*

$$\int_A Y^{p_1} d\mathbf{P} \geqslant \alpha^{p_1} \mathbf{P}(A)$$

and for all $\eta > 0$,

$$\eta^{p_2} \mathbf{P}\{Y > \eta, A\} \leqslant \beta^{p_2} \mathbf{P}(A)$$

then

$$\mathbf{P}\{Y > \theta\alpha, A\} = \left[\left(1 - \theta^{p_1}\right) \frac{p_2 - p_1}{p_2} \left(\frac{\alpha}{\beta}\right)^{p_1}\right]^{\frac{p_2}{p_2 - p_1}} \mathbf{P}(A),$$

with $0 < \theta < 1$.

Proof

- It is enough to prove it for $A = \Omega$, $p_1 = 1$ and $p_2 = p > 1$, since if $P(A) > 0$, taking the conditional probability with respect to A,

$$\mathbf{P}(\cdot \mid A) = \frac{\mathbf{P}(\cdot \cap A)}{\mathbf{P}(A)},$$

  we can write the inequalities of the lemma as follows:

$$\int_A Y^{p_1} d\mathbf{P}(\cdot \mid A) \geqslant \alpha^{p_1}$$

$$\eta^{p_2} \mathbf{P}\{Y > \eta \mid A\} \leqslant \beta^{p_2}$$

$$\mathbf{P}\{Y > \theta\alpha \mid A\} \geqslant \left[\left(1 - \theta^{p_1}\right) \frac{p_2 - p_1}{p_2} \left(\frac{\alpha}{\beta}\right)^{p_1}\right]^{\frac{p_2}{p_2 - p_1}}.$$

- That it is enough to consider $p_1 = 1$ and $p_2 = p > 1$, is also immediate, since instead of writing the lemma as before we can take the following parameters

$$Y' = Y^{1/p_1}, \eta' = \eta^{1/p_1}, \theta' = \theta^{1/p_1}, \alpha' = \alpha^{1/p_1}, \beta' = \beta^{1/p_1},$$

  then we have $p_1 = 1$ and $p = \frac{p_2}{p_1}$.

For any positive real number B, by hypothesis and using Fubini's theorem we get

$$\alpha \leqslant \int_\Omega Y d\mathbf{P} = \int_\Omega \left(\int_0^Y d\eta \right) d\mathbf{P} = \int_\Omega \left(\int_0^\infty \chi_{\{Y > \eta\}} d\eta \right) d\mathbf{P}$$

$$= \int_0^\infty \left(\int_\Omega \chi_{\{Y > \eta\}} d\mathbf{P} \right) d\eta = \int_0^\infty \mathbf{P}\{Y > \eta\} d\eta$$

$$\leqslant \int_0^{\theta\alpha} d\eta + \int_{\theta\alpha}^{B\alpha} \mathbf{P}\{Y > \eta\} d\eta + \int_{B\alpha}^\infty \mathbf{P}\{Y > \eta\} d\eta$$

$$\leqslant \int_0^{\theta\alpha} d\eta + \int_{\theta\alpha}^{B\alpha} \mathbf{P}\{Y > \theta\alpha\} d\eta + \int_{B\alpha}^\infty \mathbf{P}\{Y > \eta\} d\eta,$$

since in the second integral $\theta\alpha \leqslant \eta \leqslant B\alpha$, and then $\{Y > \eta\} \subset \{Y > \theta\alpha\}$. In addition, by hypothesis

$$\int_0^{\theta\alpha} d\eta + \int_{\theta\alpha}^{B\alpha} \mathbf{P}\{Y > \theta\alpha\} d\eta + \int_{B\alpha}^\infty \mathbf{P}\{Y > \eta\} d\eta$$

$$\leqslant \theta\alpha + B\alpha\mathbf{P}\{Y > \theta\alpha\} + \int_{B\alpha}^\infty \beta^p \eta^{-p} d\eta$$

$$\theta\alpha + B\alpha\mathbf{P}\{Y > \theta\alpha\} + \frac{1}{p-1} \left(\frac{\beta}{\alpha} \right)^p \frac{\alpha}{B^{p-1}}.$$

Taking $B := \left[\frac{1}{1-\theta} \frac{p}{p-1} \left(\frac{\beta}{\alpha} \right)^p \right]^{\frac{1}{p-1}}$ we get

$$\theta\alpha + \beta\alpha\mathbf{P}\{Y > \theta\alpha\} + \frac{1}{p-1} \left(\frac{\beta}{\alpha} \right)^p \frac{\alpha}{B^{p-1}} = \theta\alpha + B\alpha\mathbf{P}\{Y > \theta\alpha\}$$

$$+ \frac{1}{p-1} \left(\frac{\beta}{\alpha} \right)^p \frac{\alpha}{\frac{1}{1-\theta} \frac{p}{(p-1)} \left(\frac{\beta}{\alpha} \right)^p}$$

$$= \theta\alpha + B\alpha\mathbf{P}\{Y > \theta\alpha\} + \frac{\alpha}{p}(1 - \theta)$$

then

$$\alpha \leqslant \theta\alpha + B\alpha\mathbf{P}\{Y > \theta\alpha\} + \frac{\alpha}{p}(1 - \theta).$$

From there we get

$$\mathbf{P}\{Y > \theta\alpha\} \geqslant \frac{1}{B} \left[1 - \theta - \frac{1}{p}(1 - \theta) \right] = \frac{(1 - \theta)}{B} \left[1 - \frac{1}{p} \right]$$

$$= (1 - \theta) \left(\frac{p - 1}{p} \right) \frac{1}{\left[\frac{1}{1-\theta} \left(\frac{p}{p-1} \right) \left(\frac{\beta}{\alpha} \right)^p \right]^{\frac{1}{p-1}}}$$

$$= \left[(1 - \theta) \left(\frac{p - 1}{p} \right) \left(\frac{\alpha}{\beta} \right) \right]^{\frac{p}{p-1}}.$$

$\square$

For regular martingales , we get the following result

Lemma 4.6 *Let $X = \{X_n\}$ be a regular $\{\mathscr{F}_n\}$-martingale and let $\{Y_n\}$ be a multiplier sequence. Then, for any $\lambda > 0$*

$$\mathbf{P}\left\{Y^* > \lambda\right\} \leqslant C\mathbf{P}\left\{c\Delta\left(Y \circ X\right)^* > \lambda\right\},$$

where C is a constant which depends only on δ.

Proof Let $\tau = \inf\{k : |Y_k| > \lambda\}$ and $A_k = \{\tau = k\}, k \geqslant 1$, then $A_k \in \mathscr{F}_{k-1}$ (since Y_k is predictable) and by the regularity of X we get

$$\int_{A_k} |\Delta X_k| \, d\mathbf{P} \geqslant \delta\mathbf{P}\left(A_k\right) \text{ and } \int_{A_k} \Delta X_k^2 d\mathbf{P} = \mathbf{P}\left(A_k\right).$$

Now, by Tchebychev's inequality we have

$$\mathbf{P}\left\{|\Delta X_k| > \eta, A_k\right\} \leqslant \frac{1}{\eta^2}\int_{A_k}(\Delta X_k)^2 d\mathbf{P} = \frac{1}{\eta^2}\mathbf{P}\left(A_k\right).$$

Therefore, by Lemma 4.5, by taking $p_1 = 1, p_2 = 2, \alpha = \delta, \beta = 1$ we get

$$\mathbf{P}\left\{|\Delta X_k| > \theta\delta, A_k\right\} \geqslant \frac{(1-\theta)^2\delta^2}{4}\mathbf{P}\left(A_k\right).$$

Thus,

$$\mathbf{P}\{Y^* > \lambda\} = \sum_{k=1}^{\infty}\mathbf{P}\left(A_k\right) \leqslant \frac{4}{(1-\theta)^2\delta^2}\sum_{k=1}^{\infty}\mathbf{P}\left\{|\Delta X_k| > \theta\delta, A_k\right\}$$

$$\leqslant C\sum_{k=1}^{\infty}\mathbf{P}\left\{|\Delta\left(Y \circ X\right)_k| > \lambda\theta\delta, A_k\right\} \leqslant \mathbf{P}\left\{c\Delta\left(Y \circ X\right)^* > \lambda\right\}. \quad \Box$$

The following result is a particular case of Theorem 4.30, see also Theorems 2.1 and 2.2 of [44], and it has its own interest. It shows how the jumps of a regular martingale can be controlled. The regularity hypothesis cannot be weakened without the result being false (see Example 8.3 in [44]).

Proposition 4.4 *Let $X = \{X_n\}$ be a regular $\{\mathscr{F}_n\}$-martingale and $(Y \circ X)$ a martingale transform of X by a multiplier sequence $Y = \{Y_n\}$ uniformly bounded by a constant b. If*

$$\tau = \inf\left\{n : |(Y \circ X)_n| > b\right\} \quad or \quad \tau = \inf\{n : S_n\left(Y \circ X\right) > b\}$$

then

$$\| ((Y \circ X)_\tau)^* \|_2^2 \leqslant cb^2\mathbf{P}\{Y^* > 0\} \leqslant cb^2, \tag{4.72}$$

with c depending only on the constant δ of the regularity conditions.

Finally, we are ready to state a generalization of Burkholder's inequality (4.44) for Young's functions.

Theorem 4.23 (Burkholder-Gundy) *Let Φ be Young's function, $X = \{X_n\}$ be a regular $\{\mathscr{F}_n\}$-martingale, $Y = \{Y_n\}$ a multiplier sequence. Then, there exist universal constants c, C such that*

$$c\mathbf{E}[\Phi(S(Y \circ X))] \leqslant \mathbf{E}\left[\Phi\left((Y \circ X)^*\right)\right] \leqslant C\mathbf{E}[\Phi(S(Y \circ X))]. \tag{4.73}$$

The choice of c and C depends only on Φ and the regularity of X.

By Theorem 4.22 it is sufficient to prove the following good-λ inequalities for the operators S and $*$.

Proposition 4.5 *Let $X = \{X_n\}$ be a regular $\{\mathscr{F}_n\}$-martingale, $Y = \{Y_n\}$ multiplier sequence. Then, if $\beta > 1$ and $0 < \delta < \beta - 1$, we have for any $\lambda > 0$*

$$\mathbf{P}\left\{(Y \circ X)^* > \beta\lambda, S(Y \circ X) \vee Y^* \leqslant \delta\lambda\right\} \leqslant \frac{C\delta^2}{(\beta - \delta - 1)^2}\mathbf{P}\left\{(Y \circ X)^* > \lambda\right\},$$
$$\tag{4.74}$$

and

$$\mathbf{P}\left\{S(Y \circ X) > \beta\lambda, (Y \circ X)^* \vee Y^* \leqslant \delta\lambda\right\} \leqslant \frac{c\delta^2}{(\beta^2 - \delta^2 - 1)}\mathbf{P}\{S(Y \circ X) > \lambda\}.$$
$$\tag{4.75}$$

Proof We will prove (4.74), the proof of (4.75) is proved similarly. Let us consider the following $\{\mathscr{F}_n\}$-stopping times.

$$\mu = \inf\left\{n : |(Y \circ X)_n| > \lambda\right\}$$
$$\nu = \inf\left\{n : |(Y \circ X)_n| > \beta\lambda\right\}$$
$$\sigma = \inf\{n : S_n(Y \circ X) > \delta\lambda \text{ or } |Y_{n+1}| > \delta\lambda\}.$$

Let $Z = \{Z_n\}$ the martingale X started at μ and stopped at $\nu \wedge \sigma$:

$$Z_n = [{}^\mu(Y \circ X)^{\nu \wedge \sigma}]_n = \sum_{k=1}^{n}\left[\chi_{\{\mu < k \leqslant \nu \wedge \sigma\}}Y_k\right]\Delta X_k = \sum_{k=1}^{n}U_k\Delta X_k.$$

Then Z is a martingale transform of X by the multiplier sequence $U = \{U_k\}$, which is uniformly bounded by $\delta\lambda$, by the definition of σ. Consider the stopping time

$$\tau = \inf\{n : S_n(Z) > \delta\lambda\},$$

then we can apply Proposition 4.4 and conclude

$$\|Z_\tau^*\|_2^2 \leqslant c\delta^2\lambda^2\mathbf{P}\left\{U^* > 0\right\} \leqslant c\delta^2\lambda^2\mathbf{P}\{\mu < \infty\} \leqslant c\delta^2\lambda^2\mathbf{P}\left\{(Y \circ X)^* > \lambda\right\}.$$

Let us consider the set

$$E = \left\{ (Y \circ X)^* > \beta\lambda, \, S(Y \circ X) \vee Y^* \leqslant \delta\lambda \right\}.$$

On E, since $S(Z) \leqslant S(Y \circ X)$, we have $\tau = \infty$ and $\sigma = \infty$. By Proposition 3.7 (iv) we have $Z_n = \left[{}^\mu (Y \circ X)^\nu \right]_n = (Y \circ X)_n^\nu - (Y \circ X)_n^\mu$. Obviously, $(Y \circ X)_n^\mu = (Y \circ X)_n^{\mu-1} + \Delta (Y \circ X)_n^\mu$, then $|(Y \circ X)_n^\mu| \leqslant |(Y \circ X)_n^{\mu-1}| + |Y_n^\mu|$, therefore as $\sigma = \infty$

$$\sup_n \left| (Y \circ X)_n^\mu \right| \leqslant \lambda + \delta\lambda = (1 + \delta)\lambda,$$

and

$$\sup_n | (Y \circ X)_n^\nu | \geqslant \beta\lambda,$$

thus we get

$$Z^* \geqslant \beta\lambda - (1 + \delta)\lambda = (\beta - \delta - 1)\lambda,$$

and therefore, as by Tchebychev's inequality, we get

$$
\begin{aligned}
\mathbf{P}\left\{ (Y \circ X)^* > \beta\lambda, \, S(Y \circ X) \vee Y^* \leqslant \delta\lambda \right\} &\leqslant \mathbf{P}\{Z^* \geqslant (\beta - \delta - 1)\lambda, \, \tau = \infty\} \\
&\leqslant \mathbf{P}\{Z_\tau^* \geqslant (\beta - \delta - 1)\lambda\} \\
&\leqslant \frac{1}{(\beta - \delta - 1)^2\lambda^2} \left\| Z_\tau^* \right\|_2^2 \\
&\leqslant \frac{1}{(\beta - \delta - 1)^2\lambda^2} C\delta^2\lambda^2 \mathbf{P}\left((Y \circ X)^* > \lambda \right). \quad \square
\end{aligned}
$$

Let us now consider the proof of Theorem 4.23.

Proof By Theorems 4.22 and (4.74) we have then

$$\mathbf{E}\left[\Phi \left((Y \circ X)^* \right) \right] \leqslant C\mathbf{E}\left[\Phi \left(S(Y \circ X) \vee Y^* \right) \right] \leqslant C \left(\mathbf{E}\left[\Phi \left(S(Y \circ X) \right) \right] + \mathbf{E}\left[\Phi \left(Y^* \right) \right] \right).$$

As a consequence of Lemma 4.6 we have

$$\mathbf{E}\left[\Phi \left(Y^* \right) \right] \leqslant C\mathbf{E}\left[\Phi \left(c\Delta (Y \circ X)^* \right) \right] \leqslant C\mathbf{E}\left[\Phi \left(\Delta (Y \circ X)^* \right) \right] \leqslant C\mathbf{E}\left[\Phi \left(S(Y \circ X) \right) \right].$$

We obtain the right inequality of (4.73); for the left inequality, the argument is analogous using (4.75). $\square$

For the case $\Phi(\lambda) = \lambda^p$, taking $\beta = 1 + \frac{1}{p}$ and $\delta = cp^{-1/2}$ in Theorem 4.22, we obtain for $1 < p < \infty$

$$\|S(X)\|_p \leqslant Cp^{1/2} \left\| X^* \right\|_p \leqslant Cp^{1/2}q\|S(X)\|_p$$

which gives another proof of (4.46).

- Another way to control the jumps of a martingale to obtain integral inequalities, without asking for regularity, is as follows

Theorem 4.24 *Let Φ be Young's function and let $X = \{X_n\}$ be a $\{\mathscr{F}_n\}$-martingale, such that*

$$|\Delta X_n| \leqslant W_n \in \mathscr{F}_{n-1}, \tag{4.76}$$

that is, X has bounded $\mathscr{F}_{n-1}$-increments. Then there exists a constant c, depending only on Φ, such that

$$\mathbf{E}\left[\Phi\left(X^*\right)\right] \leqslant c\mathbf{E}[\Phi(S(X))] + c\mathbf{E}\left[\Phi\left(W^*\right)\right] \tag{4.77}$$

and

$$\mathbf{E}[\Phi(S(X))] \leqslant c\mathbf{E}\left[\Phi\left(X^*\right)\right] + c\mathbf{E}\left[\Phi\left(W^*\right)\right]. \tag{4.78}$$

In order to prove this result, again by Theorem 4.22, it is enough to prove a pair of good λ inequalities, contained in the following result.

Proposition 4.6 *Let $X = \{X_n\}$ be a $\{\mathscr{F}_n\}$-martingale, such that $|\Delta X_n| \leqslant W_n \in \mathscr{F}_{n-1}$ then, if $\beta > 1$ and $0 < \delta < \beta - 1$, we have, for all $\lambda > 0$*

$$\mathbf{P}\left\{X^* > \beta\lambda,\, S(X) \vee W^* \leqslant \delta\lambda\right\} \leqslant \frac{2\delta^2}{(\beta - \delta - 1)^2}\mathbf{P}\left\{X^* > \lambda\right\} \tag{4.79}$$

and

$$\mathbf{P}\left(S(X) > \beta\lambda,\, X^* \vee W^* \leqslant \delta\lambda\right\} = \frac{2\delta^2}{(\beta^2 - \delta^2 - 1)}\mathbf{P}\{S(X) > \lambda\}. \tag{4.80}$$

Proof In order to prove (4.79) let us consider $\lambda > 0$ and the following stopping times,

$$\mu = \inf\{n : |X_n| > \lambda\}$$
$$\nu = \inf\{n : |X_n| > \beta\lambda\}$$
$$\sigma = \inf\{n : S_n(X) > \delta\lambda \text{ or } W_{n+1} > \delta\lambda\}.$$

Let $Z = \{Z_n\}$ be the martingale X started at μ and stoped at $\nu \wedge \sigma$ as defined in (3.16), i.e.,

$$Z_n = \left({}^{\mu}X^{\nu\wedge\sigma}\right)_n = \sum_{k=1}^{n} \chi_{\{\mu < k \leqslant \nu\wedge\sigma\}}\Delta X_k.$$

Z is a transform of X such that $S(Z) = 0$ in $\{\mu = \infty\} = \{X^* \leqslant \lambda\}$ and moreover

$$S^2(Z) \leqslant 2\delta^2\lambda^2.$$

To prove the latter statement, we see that in the set $\{0 < \sigma < \infty\}$

$$S^2(Z) \leqslant S_\sigma^2(X) = S_{\sigma-1}^2(X) + \Delta X_\sigma^2 \leqslant \delta^2\lambda^2 + W_\sigma^2 \leqslant 2\delta^2\lambda^2$$

and in the case $\{\sigma = \infty\}$ is immediate, since

$$S^2(Z) \leqslant S^2(X) \leqslant \delta^2\lambda^2 \leqslant 2\delta^2\lambda^2.$$

Therefore,

$$\|Z\|_2^2 = \sup_n \|Z_n\|_2^2 = \sup_n \mathbf{E}\left[S_n^2(Z)\right] = \mathbf{E}\left[S^2(Z)\right] = \|S(Z)\|_2^2$$

$$= \int_{\{X^* \leqslant \lambda\}} S^2(Z)d\mathbf{P} + \int_{\{X^* > \lambda\}} S^2(Z)d\mathbf{P} = \int_{\{X^* > \lambda\}} S^2(Z)d\mathbf{P}$$

$$\leqslant 2\delta^2\lambda^2\mathbf{P}\left\{X^* > \lambda\right\}.$$

Now, by a similar argument as the one given in the proof of Proposition 4.74, let us consider the set

$$E = \left\{X^* > \beta\lambda, S(X) \vee W^* \leqslant \delta\lambda\right\}.$$

On E, as $\beta > \delta$, $\sigma = \infty$ and then by Proposition 3.7 (iv) we have $Z_n = (^\mu X^\nu)_n = X_n^\nu - X_n^\mu$. Hence, as trivially, $X_n^\mu = X_n^{\mu-1} + \Delta X_n^\mu$, then $|X_n^\mu| \leqslant |X_n^{\mu-1}| + W_n^\mu$, therefore

$$\sup_n \left|X_n^\mu\right| \leqslant \lambda + \delta\lambda = (1+\delta)\lambda,$$

and

$$\sup_n |X_n^\nu| \geqslant \beta\lambda,$$

thus we get

$$Z^* \geqslant \beta\lambda - (1+\delta)\lambda = (\beta - \delta - 1)\lambda$$

on $\{\sigma = \infty\}$. So, by Tchebychev's inequality

$$\mathbf{P}\left\{X^* > \beta\lambda, S(X) \vee W^* \leqslant \delta\lambda\right\} \leqslant \mathbf{P}\left\{Z^* > (\beta - \delta - 1)\lambda, \sigma = \infty\right\} \leqslant$$

$$\mathbf{P}\left\{Z^* > (\beta - \delta - 1)\lambda\right\}$$

$$\leqslant \frac{1}{(\beta + \delta - 1)^2\lambda^2}\|Z^*\|_2^2 = \frac{1}{(\beta + \delta - 1)^2\lambda^2}\|Z\|_2^2$$

$$\leqslant \frac{2\delta^2}{(\beta - \delta - 1)^2}\mathbf{P}\left\{X^* > \lambda\right\}.$$

Then, we have obtained the inequality (4.79). The proof of the inequality (4.80) is analogous. $\square$

- Now, if we ask for non-negativity for the martingale, we have the following result

Theorem 4.25 *Let Φ be Young's function and $X = \{X_n\}$ a non-negative $\{\mathscr{F}_n\}$-martingale. Then, there exists a constant C, depending only on Φ, such that*

$$\mathbf{E}[\Phi(S(X))] \leqslant C\mathbf{E}\left[\Phi\left(X^*\right)\right]. \tag{4.81}$$

As usual, to prove this theorem, again by Theorem 4.22, it suffices to obtain the following inequality of the good λ inequality.

Proposition 4.7 *Let $X = \{X_n\}$ be a non-negative $\{\mathscr{F}_n\}$-martingale. Then, if $\beta > 1$ and $0 < \delta < \beta - 1$, we have for all $\lambda > 0$*

$$\mathbf{P}\left\{S(X) > \beta\lambda, \, X^* \leqslant \delta\lambda\right\} \leqslant \frac{2\delta^2}{\left(\beta^2 - \delta^2 - 1\right)}\mathbf{P}\{S(X) > \lambda\}. \tag{4.82}$$

Proof Consider the stopping time defined by $\mu = \inf\{n : X_n > \delta\lambda\}$, then

$$\mathbf{P}\left\{S(X) > \beta\lambda, \, X^* \leqslant \delta\lambda\right\} \leqslant \mathbf{P}\left\{S_{\mu-1}(X) > \beta\lambda\right\}.$$

Note that

$$S_{\mu-1}^2(X) = \sum_{k=1}^{\infty} \chi_{\{\mu>k\}}\Delta X_k^2,$$

where $\chi_{\{\mu>k\}}\Delta X_k^2 = \chi_{\{X_k\leqslant\delta\lambda\}}\Delta X_k^2 \leqslant \delta^2\lambda^2$.

Now, let ν be the stopping time defined as

$$\nu = \inf\{n : \sum_{k=1}^{n} \chi_{\{\mu>k\}}\Delta X_k^2 > \lambda^2\}.$$

Then, for all $t > 0$

$$\mathbf{P}\left\{S_{\mu-1}^2(X) > \lambda^2 + t + \delta^2\lambda^2\right\} \leqslant \mathbf{P}\left\{S_{\mu-1}^2(X) > \lambda^2 + t\right\}$$

$$\leqslant \sum_{n=1}^{\infty}\mathbf{P}\left\{\sum_{k=n+1}^{\infty} \chi_{\{\mu>k\}}\Delta X_k^2 > t, \nu = n\right\}.$$

Integrating with respect to t we get

$$\int_{(1+\delta^2)\lambda^2}^{\infty} \mathbf{P}\left\{S_{\mu-1}^2(X) > s\right\} ds \leqslant \sum_{n=1}^{\infty} \mathbf{E}\left[\sum_{k=n+1}^{\infty} \chi_{\{\mu>k,\nu=n\}} \Delta X_k^2\right]. \tag{4.83}$$

On the other hand, for n fixed let us consider the martingale Y defined as

$$Y_k = \chi_{\{\mu>n,\nu=n\}} X_{n+k-1}$$

for $k \geqslant 1$. Then Y is a non-negative martingale relative to the filtration $\{\mathscr{F}_{n+k-1}\}_k$. Using Lemma 4.1 we get

$$\mathbf{E}\left[\sum_{k=n+1}^{\infty} \chi_{\{\mu>k,\nu=n\}} \Delta X_k^2\right] \leqslant \mathbf{E}\left[\sum_{k=n+1}^{\infty} \chi_{\{Y_k^* \leqslant \delta\lambda\}} \Delta Y_k^2\right] \tag{4.84}$$

$$\leqslant 2\delta\lambda\|Y\|_1 = 2\delta\lambda\mathbf{E}[Y_1] \leqslant 2\delta^2\lambda^2\mathbf{P}\{\nu=n\}.$$

Therefore, putting together (4.83) and (4.84)

$$\int_{(1+\delta^2)\lambda^2}^{\infty} \mathbf{P}\left\{S_{\mu-1}^2(X) > s\right\} ds \leqslant 2\delta^2\lambda^2 \sum_{n=1}^{\infty} \mathbf{P}\{\nu=n\} = 2\delta^2\lambda^2\mathbf{P}\{\nu < \infty\}$$

$$= 2\delta^2\lambda^2\mathbf{P}\left\{S_{\mu-1}^2(X) > \lambda^2\right\}.$$

Thus

$$\left(\beta^2 - \delta^2 - 1\right)\lambda^2\mathbf{P}\left\{S_{\mu-1}(X) > \beta\lambda\right\} \leqslant \int_{(1+\delta^2)\lambda^2}^{\beta^2\lambda^2} \mathbf{P}\left\{S_{\mu-1}^2(X) > s\right\} ds$$

$$\leqslant \int_{(1+\delta^2)\lambda^2}^{\infty} \mathbf{P}\left\{S_{\mu-1}^2(X) > s\right\} ds$$

$$\leqslant 2\delta 2\lambda^2\mathbf{P}\left\{S_{\mu-1}^2(X) > \lambda^2\right\} = 2\delta^2\lambda^2\mathbf{P}\{S(X) > \lambda\}. \qquad \square$$

- An alternative to generalize Burkholder's inequality is not to restrict the set of martingales but the set of functions Φ; observe that if $p > 1$, $\Phi(\lambda) = \lambda^p$ is convex.

  First, we need to discuss some results about convex functions. For simplicity, we assume that the convex function Φ can be written as

$$\Phi(\lambda) = \int_0^{\lambda} \phi(\gamma) d\gamma,$$

with ϕ a strictly increasing and non-negative function on $[0, \infty)$.[6]

---

[6] In general, this representation is true, but we can only say that ϕ is monotonically increasing, see Wheeden and Zygmund [170, Chap. 7].

Observe that Φ is Young's function (see (4.65)) since, using the change-of-variables $\gamma = \frac{1}{2}u$, we get

$$\Phi(2\lambda) = \int_0^{2\lambda} \phi(\gamma)d\gamma = 2\int_0^{\lambda} \phi(\frac{u}{2})du < 2\int_0^{\lambda} \phi(u)du = 2\Phi(\lambda),$$

as ϕ is strictly increasing. Additionally, let us consider Ψ the convex function conjugate in the sense of Young to Φ, i.e., it can be written as

$$\Psi(\lambda) = \int_0^{\lambda} \psi(\gamma)d\gamma,$$

where $\psi = \phi^{-1}$, is the inverse function of ϕ.
The functions Φ and Ψ satisfy the following relations:

(i) $\lambda\phi(\lambda) = \Phi(\lambda) + \Psi(\phi(\lambda))$.
(ii) $\int_0^{\gamma} \lambda d\Psi(\lambda) = \Phi(\psi(\gamma))$ and $\int_0^{\lambda} \gamma d\Phi(\gamma) = \Psi(\phi(\lambda))$.
(iii) Young's inequality: $\lambda\gamma \leqslant \Phi(\lambda) + \Psi(\gamma)$.
(iv) $\Phi\left(\frac{\lambda}{\alpha}\right) \leqslant \frac{1}{\alpha}\Phi(\lambda)$, and $\Psi\left(\frac{\lambda}{\alpha}\right) \leqslant \frac{1}{\alpha}\Psi(\lambda)$, for all $\alpha \geqslant 1$.
(v) Observe that, by definition, for any $\lambda > 0$,

$$1 = \frac{\int_0^{\lambda} \phi(\gamma)d\gamma}{\Phi(\lambda)} < \frac{\lambda\phi(\lambda)}{\Phi(\lambda)}.$$

Moreover, if C is the constant in the Δ-condition (4.65), taking $\rho = \sup_{\lambda>0}\left(\frac{\lambda\phi(\lambda)}{\Phi(\lambda)}\right)$, we get, using the mean value theorem,

$$1 < \rho \leqslant C - 1 < \infty$$

as

$$\lambda\phi(\xi) = \Phi(2\lambda) - \Phi(\lambda) \leqslant (C-1)\Phi(\lambda), \quad \text{for some } \xi \in (\lambda, 2\lambda).$$

(vi) For any $\alpha > 1$

$$\Phi(\alpha\lambda) \leqslant \alpha^{\rho}\Phi(\lambda), \quad \text{and} \quad \Psi(\lambda) \leqslant (\rho - 1)\Phi(\Psi(\lambda)).$$

Detailed proofs of these assertions and more can be found in [109].
Now, we will consider an interesting inequality for this kind of functions.

Proposition 4.8 *Let Ψ and ρ as above, and $\{X_n\}$ a $\{\mathscr{F}_n\}$-martingale, then we have*

$$\mathbf{E}\left[\Psi\left(X_n^*\right)\right] \leqslant \rho\mathbf{E}\left[\Psi\left(\rho X_n\right)\right]. \tag{4.85}$$

Proof Using Fubin's theorem and Doob's maximal inequality (3.27), Theorem 3.7 we get

$$
\begin{aligned}
\mathbf{E}\left[\Phi\left(\Psi\left(X_n^*\right)\right)\right] &= \mathbf{E}\left[\int_0^{X_n^*} \lambda \psi(\lambda)d\lambda\right] = \mathbf{E}\left[\int_0^\infty \lambda \chi_{\{X_n^*\geqslant\lambda\}}\psi(\lambda)d\lambda\right] \\
&= \int_0^\infty \lambda \mathbf{P}\{X_n^* \geqslant \lambda\}\psi(\lambda)d\lambda \leqslant \int_0^\infty \left(\int_{\{X_n^*\geqslant\lambda\}} X_n\, d\mathbf{P}\right)\psi(\lambda)d\lambda \\
&= \int_0^\infty \left(\int_\Omega \chi_{\{X_n^*\geqslant\lambda\}} X_n\, d\mathbf{P}\right)\psi(\lambda)d\lambda = \mathbf{E}\left[X_n\left(\int_0^{X_n^*}\psi(\lambda)d\lambda\right)\right] \\
&= \mathbf{E}\left[X_n\Psi\left(X_n^*\right)\right],
\end{aligned}
$$

using property (ii) for Ψ. Thus,

$$
\mathbf{E}\left[\Phi\left(\Psi\left(X_n^*\right)\right)\right] \leqslant \mathbf{E}\left[X_n\Psi\left(X_n^*\right)\right].
$$

Now, by Young's inequality and (iv)

$$
\begin{aligned}
\mathbf{E}\left[\Phi\left(\Psi\left(X_n^*\right)\right)\right] &\leqslant \mathbf{E}\left[\rho X_n\frac{1}{\rho}\Psi\left(X_n^*\right)\right] \leqslant \mathbf{E}\left[\Phi\left(\frac{1}{\rho}\Psi\left(X_n^*\right)\right)\right] + \mathbf{E}\left[\Psi\left(\rho X_n\right)\right] \\
&\leqslant \mathbf{E}\left[\frac{1}{\rho}\Phi\left(\Psi\left(X_n^*\right)\right)\right] + \mathbf{E}\left[\Psi\left(\rho X_n\right)\right].
\end{aligned}
$$

Thus,

$$
\frac{\rho-1}{\rho}\mathbf{E}\left[\Phi\left(\Psi\left(X_n^*\right)\right)\right] \leqslant \mathbf{E}\left[\Psi\left(\rho X_n\right)\right].
$$

Finally, using (vi) we obtain

$$
\mathbf{E}\left[\Psi\left(X_n^*\right)\right] \leqslant (\rho-1)\mathbf{E}\left[\Phi\left(\Psi\left(X_n^*\right)\right)\right] \leqslant \rho\mathbf{E}\left[\Psi\left(\rho X_n\right)\right]. \qquad \square
$$

Note that in the case of $\Psi(\lambda) = \lambda^p$, $p > 1$, (4.85) is precisely Doob's inequality (3.31).

To establish the generalization of Burkholder's inequality for convex functions Φ, we need the following result [45], which has interest by itself.

Proposition 4.9 (Burkholder-Davis-Gundy) *Let Φ be a convex Young's function and let $\{\xi_n\}$ a sequence of non-negative random variables on $(\Omega, \mathscr{F}, \mathbf{P})$, then*

$$
\mathbf{E}\left[\Phi\left(\sum_{k=1}^\infty \mathbf{E}\left[\xi_k \mid \mathscr{F}_{k-1}\right]\right)\right] \leqslant C\mathbf{E}\left[\Phi\left(\sum_{k=1}^\infty \xi_k\right)\right]. \tag{4.86}
$$

Proof We will give A. Garsia's proof [81]. We will prove (4.86) with $C = p^{2p}$. For n fix let us define

$$Z_n = \sum_{k=1}^{n} \xi_k, \ W_n = \sum_{k=1}^{n} E\left[\xi_k \mid \mathscr{F}_{k-1}\right] \ \text{ and for } 1 \leqslant k \leqslant n, \ X_k = E\left[\phi\left(W_n\right) \mid \mathscr{F}_k\right]$$

then we have

$$\mathbf{E}\left[W_n \phi(W_n)\right] = \sum_{n=1}^{n} \mathbf{E}\left[\xi_k X_{k-1}\right] \leqslant \mathbf{E}\left[Z_n X_n^*\right].$$

Then, by Young's inequality and (4.85) we get

$$\mathbf{E}\left[W_n \phi\left(W_n\right)\right] \leqslant \mathbf{E}\left[\Phi(\rho^2 Z_n)\right] + \mathbf{E}\left[\Psi\left(\frac{X_n^*}{\rho^2}\right)\right] \leqslant \mathbf{E}\left[\Phi\left(\rho^2 Z_n\right)\right] + \rho\mathbf{E}\left[\Psi\left(\frac{\phi\left(W_n\right)}{\rho}\right)\right].$$

Now, using (i) and (iv) we obtain

$$\mathbf{E}\left[\Phi\left(W_n\right)\right] + \mathbf{E}\left[\Psi\left(\phi\left(W_n\right)\right)\right] = \mathbf{E}\left[W_n \phi\left(W_n\right)\right] \leqslant \mathbf{E}\left[\Phi\left(\rho^2 Z_n\right)\right] + \mathbf{E}\left[\Psi\left(\phi\left(W_n\right)\right)\right],$$

and finally (4.86) is obtained using (v), since

$$\mathbf{E}\left[\Phi\left(W_n\right)\right] \leqslant \mathbf{E}\left[\Phi\left(\rho^2 Z_n\right)\right] \leqslant \rho^{2\rho}\mathbf{E}\left[\Phi\left(Z_n\right)\right]. \qquad \square$$

Now, let us consider the generalization of Burkholder's inequality to convex functions.

Theorem 4.26 (Burkholder-Davis-Gundy): *Let Φ be a convex Young's function and let $X = \{X_n\}$ be a $\{\mathscr{F}_n\}$-martingale. Then there exist universal constants c and C such that*

$$c\mathbf{E}[\Phi(S(X))] \leqslant \mathbf{E}\left[\Phi\left(X^*\right)\right] \leqslant C\mathbf{E}[\Phi(S(X))]. \qquad (4.87)$$

The choice of c and C depends only on Φ.

Proof We will prove only the right-hand side of the inequality (4.87); the proof of the left-hand side is analogous. Unlike the previous proofs, in this one, the good λ inequalities technique will not be used, but the Davis decomposition (Theorem 4.6). Thus, we know that we can then write the martingale X as the sum of two martingales Y and Z, $X = Y + Z$. Also note that

$$X^* \leqslant Y^* + Z^* \leqslant Y^* + \sum_{k=1}^{\infty} |\Delta Z_k|, \qquad (4.88)$$

and

$$S(Y) \leqslant S(X) + S(Z) \leqslant S(X) + \sum_{k=1}^{\infty} |\Delta Z_k|. \qquad (4.89)$$

Then, by (4.88) and the additional property ii) of Φ, we have

$$\mathbf{E}\left[\Phi\left(X^*\right)\right] \leqslant C\mathbf{E}\left[\Phi\left(Y^*\right)\right] + C\mathbf{E}\left[\Phi\left(\sum_{k=1}^{\infty}|\Delta Z_k|\right)\right].$$

Now, given that Y has $\mathscr{F}_{n-1}$-bounded increments, then by (4.77) we get

$$\mathbf{E}\left[\Phi\left(Y^*\right)\right] \leqslant c\mathbf{E}[\Phi(S(Y))] + c\mathbf{E}\left[\Phi\left(4\Delta X^*\right)\right].$$

Therefore, from the above inequalities, and the fact that $\Delta X^* \leqslant S(X)$, we obtain

$$\mathbf{E}\left[\Phi\left(X^*\right)\right] \leqslant C\mathbf{E}[\Phi(S(X))] + C\mathbf{E}\left[\Phi\left(\sum_{k=1}^{\infty}|\Delta Z_k|\right)\right].$$

Now, by the definition of Z_n, and using (4.13)

$$\sum_{k=1}^{\infty}|\Delta Z_k| \leqslant \sum_{k=1}^{\infty}|Z_k^{\#}| + \sum_{k=1}^{\infty}\mathbf{E}\left[|Z_k^{\#}| \mid \mathscr{F}_{k-1}\right] \leqslant 2\Delta X^* + \sum_{k=1}^{\infty}\mathbf{E}\left[|Z_k^{\#}| \mid \mathscr{F}_{k-1}\right],$$

then, using the Delta condition of Φ, we get

$$\mathbf{E}\left[\Phi\left(X^*\right)\right] \leqslant C\mathbf{E}[\Phi(S(X))] + C\mathbf{E}\left[\Phi\left(\sum_{k=1}^{\infty}\mathbf{E}\left[|Z_k^{\#}| \mid \mathscr{F}_{k-1}\right]\right)\right].$$

Now, using (4.86), (4.13) and again as $\Delta X^* \leqslant S(X)$, we have

$$\mathbf{E}\left[\Phi\left(X^*\right)\right] \leqslant C\mathbf{E}[\Phi(S(X))] + C\mathbf{E}\left[\Phi\left(\sum_{k=1}^{\infty}|Z_k^{\#}|\right)\right] \leqslant C\mathbf{E}[\Phi(S(X))]. \qquad \square$$

- Let us now consider the case that Young's function Φ is concave. In this case, we get immediately (4.65) with $C = 2$,

$$\Phi(2\lambda) \leqslant 2\Phi(\lambda),$$

  using the fact that for a concave function, its increment quotients are non-increasing.
  Moreover, if Φ is concave this implies that there exists a non-negative, non-increasing ϕ function on $(0, \infty)$, such that:

$$\Phi(t) = \int_0^t \phi(\lambda)d\lambda = \int_0^t \phi(\lambda)d\lambda - t\phi(\infty) + t\phi(\infty) = \int_0^t (\phi(\lambda) - \phi(\infty))d\lambda + t\phi(\infty),$$

where $\phi(\infty) = \lim_{t\to\infty} \phi(t)$. Let us suppose that $\phi(\infty) = 0$, then integrating by parts, using that ϕ is non-negative, non-increasing and $\phi(\infty) = 0$.

$$
\begin{aligned}
\Phi(t) &= \lim_{a\to 0, b\to\infty} \int_a^b \phi(\lambda)d(\lambda \wedge t) \\
&= \lim_{a\to 0, b\to\infty} \left[\phi(b)(t \wedge b) - \phi(a)(t \wedge a) - \int_a^b (t \wedge \lambda)d\phi(\lambda) \right] \\
&= \phi(\infty)t - \phi(0)0 - \int_0^\infty (t \wedge \lambda)d\phi(\lambda) = -\int_0^\infty (t \wedge \lambda)d\phi(\lambda) \\
&= \int_0^\infty (t \wedge \lambda)(-d\phi(\lambda)).
\end{aligned}
\tag{4.90}
$$

Note that $-d\phi(\lambda)$ defines a positive measure and the formula holds also for $t = 0$ and $t = \infty$.

Let us consider now the concave version of Proposition 4.9. Observe that in order to prove this result, the trivial trick of considering the convex function $-\Phi$ does not work since we are working only with positive functions.

Proposition 4.10 *Let Φ be a concave Young's function and let $\{\xi_n\}$ be a sequence of non-negative random variables on $(\Omega, \mathscr{F}, \mathbf{P})$ then*

$$
\mathbf{E}\left[\Phi\left(\sum_{k=1}^\infty \xi_k \right) \right] \leqslant 2\mathbf{E}\left[\Phi\left(\sum_{k=1}^\infty \mathbf{E}\left[\xi_k \mid \mathscr{F}_{k-1} \right] \right) \right].
\tag{4.91}
$$

Proof For $n > 0$ fix let us define

$$
Z_n = \sum_{k=1}^n \xi_k, \quad W_n = \sum_{k=1}^n E\left[\xi_k \mid \mathscr{F}_{k-1} \right], \quad Z = Z_\infty, \quad \text{and} \quad W = W_\infty.
$$

Then, we will prove that

$$
\mathbf{E}[Z \wedge \lambda] \leqslant 2\mathbf{E}[W \wedge \lambda], \quad \lambda > 0.
\tag{4.92}
$$

To see this, let us consider the stopping time $\tau = \inf\{n \geqslant 0 : W_{n+1} > \lambda\}$ and notice that $W_\tau \leqslant W \wedge \lambda$. In view of

$$
Z \wedge \lambda \leqslant Z_\tau + \lambda\chi_{\{\tau < \infty\}}
$$

the inequality (4.92) follows from

$$
\begin{aligned}
\mathbf{E}[Z_\tau] = \mathbf{E}\left[\sum_{k=1}^\infty \chi_{\{\tau \geqslant k\}}\xi_k \right] &= \mathbf{E}\left[\sum_{k=1}^\infty \chi_{\{\tau \geqslant k\}}\mathbf{E}\left[\xi_k \mid \mathscr{F}_{k-1} \right] \right] \\
&= \mathbf{E}[W_\tau] \leqslant \mathbf{E}[W \wedge \lambda]
\end{aligned}
$$

and

$$\mathbf{E}[\lambda \chi_{\{\tau < \infty\}}] = \lambda \mathbf{P}\{W > \lambda\} \leqslant \mathbf{E}[W \wedge \lambda].$$

Inequality (4.92) is a special case of (4.91) but actually implies (4.91). Therefore, by Fubini's theorem and the representation of Φ considered above (4.90)

$$\mathbf{E}[\Phi(Z)] = -\int_0^\infty \mathbf{E}[Z \wedge \lambda] d\phi(\lambda)$$

$$\leqslant -2 \int_0^\infty \mathbf{E}[W \wedge \lambda] d\phi(\lambda) = 2E\Phi(W). \qquad \square$$

Now, we are going to introduce another martingale operator over $\mathcal{M}$, the set of all $\{\mathcal{F}_n\}$-martingales defined on $(\Omega, \mathcal{F}, \mathbf{P})$.

Definition 4.9 Let $X = \{X_n\}$ a $\{\mathcal{F}_n\}$-martingale. Let us define, for each $n \geqslant 1$ the operator $s_n(X)$ the conditional quadratic function of X, of order n as

$$s_n(X) = \left(\sum_{k=1}^n \mathbf{E}\left[\Delta X_k^2 \mid \mathcal{F}_{k-1} \right] \right)^{1/2} \tag{4.93}$$

and the operator $s(X)$ the conditional quadratic function

$$s(X) = \left(\sum_{k=1}^\infty \mathbf{E}\left[\Delta X_k^2 \mid \mathcal{F}_{k-1} \right] \right)^{1/2}. \tag{4.94}$$

Theorem 4.27 (Bernstein's inequality for martingales) *Let $\{X_n\}$ be a $\{\mathcal{F}_n\}$-martingale, with $X_0 = 0$, with bounded increments $|\Delta X_k| \leqslant K$. Then for all constants $t > 0$, and $n \geqslant 1$*

$$\mathbf{P}\{X_n^* \geq t, \ (s_n(X))^2 \leqslant \nu\} \leq \exp\left(-\frac{t^2}{2(\nu + Kt/3)} \right). \tag{4.95}$$

Proof Consider the exponential process

$$Z_k = \exp\left(\lambda X_k - \frac{\lambda^2}{2}(s_k(X))^2 \right),$$

where $\lambda > 0$ is a parameter that will be optimized later. Clearly Z_k is non-negative, let us show that Z_k is a supermartingale, By definition

$$Z_k = Z_{k-1} \exp\left(\lambda \Delta X_k - \frac{\lambda^2}{2} \mathbf{E}[\Delta X_k^2 \mid \mathcal{F}_{k-1}] \right).$$

Now, taking conditional expectation:

$$\mathbf{E}[Z_k \mid \mathscr{F}_{k-1}] = Z_{k-1}\mathbf{E}\left[\exp\left(\lambda \Delta X_k - \frac{\lambda^2}{2}\mathbf{E}[\Delta X_k^2 \mid \mathscr{F}_{k-1}]\right) \mid \mathscr{F}_{k-1}\right].$$

Since $\mathbf{E}[\Delta X_k \mid \mathscr{F}_{k-1}] = 0$, if we expand $\exp(\lambda \Delta X_k)$ using Taylor theorem of order 2,

$$\mathbf{E}[\exp(\lambda \Delta X_k) \mid \mathscr{F}_{k-1}] \leq 1 + \frac{\lambda^2}{2}\mathbf{E}[\Delta X_k^2 \mid \mathscr{F}_{k-1}].$$

Using the fact that $|\Delta X_k| \leqslant K$ we have then

$$
\begin{aligned}
Z_k &= Z_{k-1}\exp\left(\lambda \Delta X_k - \frac{\lambda^2}{2}\mathbf{E}[\Delta X_k^2 \mid \mathscr{F}_{k-1}]\right)\\
&\leqslant Z_{k-1}\left(1 + \frac{\lambda^2}{2}\mathbf{E}[\Delta X_k^2 \mid \mathscr{F}_{k-1}]\right)\exp\left(-\frac{\lambda^2}{2}\mathbf{E}[\Delta X_k^2 \mid \mathscr{F}_{k-1}]\right)\\
&\leqslant Z_{k-1}\left(1 + \frac{\lambda^2}{2}K^2\right)\exp\left(-\frac{\lambda^2}{2}K^2\right) \leqslant Z_{k-1},
\end{aligned}
$$

as $(1+x)e^{-x} \leqslant 1$, for $x \geqslant 0$. Thus, $\{Z_k\}$ is a non-negative supermartingale. Then, by Doob's maximal inequality for non-negative supermartingales (3.28)

$$\mathbf{P}\left\{\max_{k\leq n} Z_k \geq z,\ (s_n(X))^2 \leqslant v\right\} \leqslant \mathbf{P}\left\{\max_{k\leq n} Z_k \geq z\right\} \leqslant \frac{\mathbf{E}[Z_0]}{z}.$$

Now, set $Z_0 = 1$ and choose $z = e^{\lambda t}$, since

$$
\begin{aligned}
\left\{\max_{k\leq n} Z_k \geq e^{\lambda t},\ (s_n(X))^2 \leqslant v\right\} &\supset \left\{\max_{k\leq n} e^{\lambda X_k}e^{-\frac{\lambda^2}{2}v} \geqslant e^{\lambda t},\ (s_n(X))^2 \leqslant v\right\}\\
&= \left\{\max_{k\leq n} X_k \geqslant t + \frac{\lambda}{2}v,\ (s_n(X))^2 \leqslant v\right\},
\end{aligned}
$$

we obtain

$$\mathbf{P}\left\{\max_{k\leq n} X_k \geqslant t + \frac{\lambda}{2}v,\ (s_n(X))^2 \leqslant v\right\} \leqslant \mathbf{P}\left\{\max_{k\leq n} Z_k \geq e^{\lambda t}\right\},$$

or equivalently,

$$
\begin{aligned}
\mathbf{P}\left\{\max_{k\leq n} X_k \geqslant t,\ (s_n(X))^2 \leqslant v\right\} &\leqslant \mathbf{P}\left\{\max_{k\leq n} Z_k \geq e^{\lambda t - \frac{\lambda^2}{2}v}\right\}\\
&\leqslant \exp\left(-\lambda e^{\lambda t} + \frac{\lambda^2}{2}v\right).
\end{aligned}
$$

Finally, minimizing the exponent $-\lambda e^{\lambda t} + \frac{\lambda^2}{2}v$, it can be proved that the minimizer is[7]

$$\lambda = \frac{t}{v + Kt/3}.$$

Thus, using this value of λ we get

$$\mathbf{P}\left\{X_n^* \geqslant t, \ (s_n(X))^2 \leqslant v\right\} \leqslant \exp\left(-\frac{t^2}{2(v + Kt/3)}\right). \qquad \square$$

Although Burkholder's inequality is extremely useful, it does not give particularly good estimates in the case of martingales with bounded increments. Bernstein's inequality gives better estimates for that case. It belongs to an extensive family of exponential inequalities and concentration inequalities which are another tool to get inequalities for martingales but giving rates of convergence. For more about those inequalities we refer to the book of Bercu, Delyon, and Rio [14] .

Let us see now another inequality for the conditional quadratic function $s(x)$.

Lemma 4.7 *Let $0 < p \leqslant 2$, $X = \{X_n\}$ be a $\{\mathscr{F}_n\}$-martingale, and $(Y \circ X)$ be a martingale transform of X by a multiplier sequence $Y = \{Y_1, Y_2, \ldots\}$, then*

$$\left\|(Y \circ X)^*\right\|_p \leqslant c_p \|s((Y \circ X))\|_p.$$

The choice of c_p depends only on p.

Proof We may assume that $\|s((Y \circ X))\|_p < \infty$. If $p = 2$, by (3.31) and the orthogonality of the difference sequence $Y_k \Delta X_k$ of $(Y \circ X)$, we have that

$$\left\|(Y \circ X)^*\right\|_2 \leqslant 2\|(Y \circ X)\|_2 = 2\|S((Y \circ X))\|_2 = 2\|s((Y \circ X))\|_2.$$

Now let $0 < p < 2$. Since

$$s_n((Y \circ X)) = \left[\sum_{k=1}^{n} \mathbf{E}\left[(Y_k \Delta X_k)^2 \mid \mathscr{F}_{k-1}\right]\right]^{\frac{1}{2}}$$

is $\mathscr{F}_{n-1}$-measurable,

$$\tau = \inf\left\{0 \leqslant n < \infty : s_{n+1}((Y \circ X)) > \lambda\right\}$$

is a stopping time. Let $Z = (Y \circ X)^\tau$. Then $s(Z) = s_\tau((Y \circ X)) \leqslant \lambda$, $s(Z) \leqslant s((Y \circ X))$, and

$$\mathbf{P}\left\{(Y \circ X)^* > \lambda\right\} \leqslant \mathbf{P}\{s((Y \circ X)) > \lambda\} + \mathbf{P}\left\{(Y \circ X)^* > \lambda, s((Y \circ X)) \leqslant \lambda\right\}.$$

---

[7] For details see, for instance, [165, Chap. 2].

Since $Z = (Y \circ X)$ on $\{\tau = \infty\} = \{s((Y \circ X)) \leqslant \lambda\}$, the last probability is equal to

$$\mathbf{P}\left\{Z^* > \lambda, s((Y \circ X)) \leqslant \lambda\right\} \leqslant \mathbf{P}\left\{Z^* > \lambda\right\} \leqslant \lambda^{-2}\|Z\|_2^2$$

$$= \lambda^{-2} \int_{\{s((Y \circ X)) > \lambda\}} s(Z)^2 d\mathbf{P} + \lambda^{-2} \int_{\{s((Y \circ X)) \leqslant \lambda\}} s(Z)^2 d\mathbf{P}$$

$$\leqslant \mathbf{P}\{s((Y \circ X)) > \lambda\} + \lambda^{-2} \int_{\{s((Y \circ X)) \leqslant \lambda\}} s((Y \circ X))^2 d\mathbf{P}.$$

Therefore,

$$\|(Y \circ X)^*\|_p^p = p \int_0^\infty \lambda^{p-1} \mathbf{P}\left\{(Y \circ X)^* > \lambda\right\} d\lambda$$

$$\leqslant 2\|s((Y \circ X))\|_p^p + p \int_\Omega s((Y \circ X))^2 \int_{s((Y \circ X))}^\infty \lambda^{p-3} d\lambda d\mathbf{P}$$

$$= [2 + p/(2 - p)]\|s((Y \circ X))\|_p^p = c_p^p \|s((Y \circ X))\|_p^p. \qquad \square$$

On the other hand,

Theorem 4.28 *Let $X = \{X_n\}$ be a $\{\mathscr{F}_n\}$-martingale and let Φ be a concave Young's function. Then,*

$$\mathbf{E}\left[\Phi\left((X^*)^2\right)\right] \leqslant 5\mathbf{E}\left[\Phi\left(s^2(X)\right)\right]. \tag{4.96}$$

Proof Let $\lambda > 0$ and define

$$\tau = \inf\left\{n \geqslant 0 : s_{n+1}^2(X) > \lambda\right\}$$

observe that τ is a stopping time since $s_{n+1}(X) \in \mathscr{F}_n$.

Let us consider X^τ, the stopped martingale by τ, then, we get

- if $\tau < \infty$, $\lambda \chi_{\{\tau < \infty\}} = \lambda$, and then $(X^*)^2 \wedge \lambda \leqslant \left[(X^\tau)^*\right]^2 + \lambda$, and
- if $\tau = \infty$ $(X^\tau)^* = X^*$ and then $(X^*)^2 \wedge \lambda \leqslant \left[(X^\tau)^*\right]^2$.

Moreover, by Doob's inequality (3.31)

$$\left\|(X^\tau)^*\right\|_2^2 \leqslant 4\left\|X^\tau\right\|_2^2.$$

But, by orthogonality

$$\|X\|_2^2 = \left\|s\left(X^\tau\right)\right\|_2^2 = \|s_\tau(X)\|_2^2.$$

In addition, observe that

$$s_\tau^2(x) \leqslant s^2(x) \wedge \lambda$$

since $s_\tau^2(X) \leqslant \lambda$, by definition of τ, and $s_\tau^2(X) \leqslant s^2(X)$, then

$$\left\| s_\tau^2(X) \right\|_2^2 \leqslant \left\| s^2(X) \wedge \lambda \right\|_2^2.$$

Then we obtain

$$\mathbf{E}\left[\left((X^*)^2 \wedge \lambda \right) \right] \leqslant \mathbf{E}\left[\left((X^\tau)^* \right)^2 \right] + \mathbf{E}\left[\lambda \chi_{\{\tau < \infty\}} \right] \leqslant 4\mathbf{E}\left[s^2(X) \wedge \lambda \right] + \mathbf{E}\left[\lambda X_{\{\tau < \infty\}\}} \right].$$

Now, by definition of τ, we get

$$
\begin{aligned}
\mathbf{E}\left[\lambda \chi_{\{\tau < \infty\}} \right] = {} & \lambda \mathbf{P}\{\tau < \infty\} = \lambda \mathbf{P}\left\{ s^2(X) > \lambda \right\} \\
\leqslant {} & \lambda \mathbf{P}\left\{ s^2(X) > \lambda \right\} + \int_{\{s^2(X) \leqslant \lambda\}} \left(s^2(X) \wedge \lambda \right) d\mathbf{P} \\
= {} & \int_{\{s^2(X) > \lambda\}} \left(s^2(X) \wedge \lambda \right) d\mathbf{P} + \int_{\{s^2(X) \leqslant \lambda\}} \left(s^2(X) \wedge \lambda \right) d\mathbf{P} = \mathbf{E}\left[s^2(X) \wedge \lambda \right].
\end{aligned}
$$

Thus, we have

$$\mathbf{E}\left[(X^*)^2 \wedge \lambda \right] \leqslant 5\mathbf{E}\left[s^2(X) \wedge \lambda \right].$$

Finally, by Fubini's Theorem and (4.90) we have

$$
\begin{aligned}
\mathbf{E}\left[\Phi\left((X^*)^2 \right) \right] &= \mathbf{E}\left[-\int_0^\infty \left((X^*)^2 \wedge \lambda \right) d\phi(\lambda) \right] = -\int_0^\infty \mathbf{E}\left[(X^*)^2 \wedge \lambda \right] d\phi(\lambda) \\
&\leqslant -\int_0^\infty 5\mathbf{E}\left[s^2(X) \wedge \lambda \right] d\phi(\lambda) \\
&= 5\mathbf{E}\left[-\int_0^\infty \left(s^2(X) \wedge \lambda \right) d\phi(\lambda) \right] = 5\mathbf{E}\left[\Phi\left(s^2(X) \right) \right]. \qquad \square
\end{aligned}
$$

In the case that $\Phi(\lambda) = \lambda^p$, $0 < p \leqslant 1$, we obtain

$$\mathbf{E}\left[(X^*)^{2p} \right] \leqslant 5\mathbf{E}\left[s^{2p}(X)s \right].$$

Now, if $0 < p' = 2p \leqslant 2$, this can be written as

$$\left\| X^* \right\|_{p'} \leqslant 5^{1/p'} \| s(X) \|_{p'}. \tag{4.97}$$

An inequality like the one in the previous theorem is not true for a general Φ.

Theorem 4.29 *Let Φ be Young's function and let $\{X_n\}$ be a $\{\mathscr{F}_n\}$-martingale, then*

$$\mathbf{E}\left[\Phi\left(X^* \right) \right] \leqslant C\mathbf{E}[\Phi(s(X))] + C\mathbf{E}\left[\Phi\left(\Delta X^* \right) \right]. \tag{4.98}$$

Again, to establish this inequality, by Theorem 4.22 we need a good λ inequality.

Proposition 4.11 *Let $X = \{X_n\}$ be a $\{\mathscr{F}_n\}$-martingale, then if $\beta > 1$ and $0 < \delta < \beta - 1$, we have for $\lambda > 0$*

$$\mathbf{P}\left\{X^* > \beta\lambda,\; s(X) \vee \Delta X^* \leqslant \delta\lambda\right\} \leqslant \frac{\delta^2}{(\beta - \delta - 1)^2}\mathbf{P}\left\{X^* > \lambda\right\}. \qquad (4.99)$$

Proof The proof is analogous to the one of (4.77), so we are going to skip details. Let $\lambda > 0$ and consider

$$\mu = \inf\{n : |X_n| > \lambda\},$$
$$\nu = \inf\{n : |X_n| > \beta\lambda\},$$
$$\sigma = \inf\{n : s_{n+1}(X) > \delta\lambda \text{ or } |\Delta X_n| > \delta\lambda\}$$

and consider $Z = \{Z_n\}$ be the martingale started at μ and stopped at $\nu \wedge \sigma$, as defined in (3.16), i.e.,

$$Z_n = \left({}^\mu X^{\nu\wedge\sigma}\right)_n = \sum_{k=1}^{n} \chi_{\{\mu < k \leqslant \nu\wedge\sigma\}}\Delta X_k,$$

then

$$
\begin{aligned}
\mathbf{P}\left\{X^* > \beta\lambda,\; s(X) \vee \Delta X^* \leqslant \delta\lambda\right\} &\leqslant \mathbf{P}\{\mu \leqslant \nu < \infty, \sigma = \infty\} \\
&\leqslant \mathbf{P}\left\{Z^* > (\beta - \delta - 1)\lambda\right\} \\
&\leqslant \frac{1}{(\beta - \delta - 1)^2\lambda^2}\|Z\|_2^2 \\
&= \frac{1}{(\beta - \delta - 1)^2\lambda^2}\|s(Z)\|_2^2 \\
&\leqslant \frac{\delta^2\lambda^2}{(\beta - \delta - 1)^2\lambda^2}\mathbf{P}\left\{X^* > \lambda\right\}. \qquad \square
\end{aligned}
$$

4.6 Burkholder, Davis, Gundy for a General Class of Operators

D. L. Burkholder and R. Gundy in their famous and fundamental article in Acta Matematica *Extrapolation and interpolation of quasi-linear operators on martingales* [44] obtained generalization of Burkholder, Davis and Gundy inequalities, (4.44),(4.48) and (4.73), for a general class of operators defined on the class, $\mathcal{E}$ of all martingales transforms of regular martingales.

However, the proofs in that article, while containing all the basic ideas for good inequalities, do not use them to their full potential, so the proofs were very complicated. Precisely with the formulation of good lambda inequalities, and the technique implied by them, Theorem 4.22, these proofs can be greatly simplified. In [30] D. L. Burkholder illustrates this technique for the classical operators $*$, S and s. The main goal of this section is to give the proofs of the Burkholder-Gundy theorems (Theorems 3.2 and 4.2 of [44]) for general operators, using good λ inequalities, see [123]. Also, we shall study some local properties for martingales, see [124].

In the article [87] where Gundy's decomposition appears, Theorem 4.5, a class of operators on martingales is also introduced (see also [88]).

Definition 4.10 An operator T is said to be in the class $\mathscr{B}$ if

(i) Its domain is a family of $\{\mathscr{F}_n\}$-martingales closed under addition.
(ii) Its range is a family of random variables, such that,

(1) T is quasi-linear, i.e.,

$$|T(X + Y)| \leqslant C(|T(X)| + |T(Y)|).$$

(2) $\mathbf{P}\{|\,T(X)\,| > 0\} \leqslant \mathbf{P}\{X^* > 0\}$.
(3) The operator T satisfies the following inequalities in norm

(a) $\|T(X)\|_2 \leqslant C\|X\|_2$
(b) $\|T(X)\|_1 \leqslant C \left\|\sum_{k=1}^{\infty} |\Delta X_k|\right\|_1$.

It is easy to see that the operators $(\,)^*$, S and s belong to the class $\mathscr{B}$.

Using his decomposition, Gundy [87] proved that indeed every operator T over martingales belonging to the class $\mathscr{B}$, satisfies a weak inequality

$$\mathbf{P}\{|T(X)| > \lambda\} \leqslant \frac{C}{\lambda}\|X\|_1,$$

see [87].

However, the class of operators on which D. L. Burkholder and R. Gundy worked in [44], and which is the one that most of the time we are going to consider here, is wider than the one in the previous definition.

Definition 4.11 Let us consider an operator T from $\mathscr{M}$ (or $\mathscr{M}(X)$) into the non-negative $\mathscr{F}$-measurable functions. The operator T satisfies condition $\mathfrak{B}$ if, for $\gamma \geqslant 1$,

(B1) T is quasi-linear:

$$T(X + Y) \leqslant \gamma(TX + TY);$$

(B2) T is local: $TX = 0$ on the set $\{s(X) = 0\}$; where

$$s(X) = \left(\sum_{k=1}^{\infty} \mathbf{E}\left[(\Delta X_k)^2 \mid \mathscr{F}_{k-1} \right] \right)^{1/2}$$

is the conditional quadratic function, see Definition 4.9.
(B3) T is symmetric: $T(-X) = TX$.

Observe that the operators $T(X) = S(X)$, $T(X) = s(X)$, and $T(X) = X^*$ satisfy condition $\mathfrak{B}$.

Non-negativity and symmetry are not essential: if T does not satisfy these conditions, it can be replaced, without loss of generality for our results, by $\tilde{T}X = |TX| \vee |T(-X)|$. Note that (B1) and (B3) imply that $T(X - Y) \leqslant \gamma(TX + TY)$.

We adopt the following notation:

$T_n(X) = T(X)^n$, $1 \leqslant n \leqslant \infty$, where, as before, T^n is the martingale stopped at n,

$$T^*(X) = \sup_{1 \leqslant n < \infty} T_n X.$$

In some cases, $T = T^*$; for example, S and s have this property. However, $TX = \limsup_{n \to \infty} |X_n|$ does not since it can happen that $TX < X^* = T^*X$.

Observe that, if T satisfies condition $\mathfrak{B}$, then so do T^*.

In what follows we need a more strict condition than being a regular martingale, Definition 4.8.

Definition 4.12 Let $X = \{X_n\}$ be a $\{\mathscr{F}_n\}$-martingale. Let $0 < \delta \leqslant 1$ and $\varrho \geqslant 2$. We say that X satisfies condition **A** if, for all $k \geqslant 1$

(A1) $\mathbf{E}\left[|\Delta X_k| \mid \mathscr{F}_{k-1} \right] \geqslant \delta$,
(A2) $\mathbf{E}\left[(\Delta X_k)^2 \mid \mathscr{F}_{k-1} \right] = 1$,
(A3) $\mathbf{E}\left[|\Delta X_k|^{\varrho} \mid \mathscr{F}_{k-1} \right] \leqslant c$, for some fixed c.

Note that these conditions are redundant in some cases. If (A2) holds and $\varrho = 2$, (A3) imposes no extra restriction and the constant c in (A3), may be taken to be 1. If (A2) and (A3) hold with $\varrho > 2$, then (A1) holds, which follows from Hölder's inequality.

Finally, we need additional restrictions for the operators.

Definition 4.13 The operator T satisfies condition **L** if, for all $\lambda > 0$ and $X \in \mathcal{M}$, if there exist $0 < p_1 < p_2 \leqslant \varrho$, and $0 < s_1 < \varrho$ where ϱ is the same as in (A3) such that

(L1) $\lambda^{s_1} \mathbf{P}\{T(X) > \lambda\} \leqslant c \, \|X^*\|_{s_1}^{s_1}$;
(L2) T satisfies condition **R**

(R1) $\lambda^{p_1} \mathbf{P}\{X^* > \lambda\} \leqslant c\|T(X)\|_{p_1}^{p_1}$;

(R2) $\lambda^{p_2} \mathbf{P}\{T(X) > \lambda\} \leqslant c\,\|X^*\|_{p_2}^{p_2}$.

(L3) $T_n f$ is $\mathscr{F}_n$-measurable, $n \geqslant 1$.

It can be proved that the $(L1)$ part of condition L is not needed to obtain the results.

Example 4.5

(i) The operator $T(X) = X^*$ trivially satisfies condition **L**.

(ii) The operator $T(X) = S(X)$ also satisfies condition **L** with, $s_1 = 1$, $p_1 = 1$ and $p_2 = 2$, since condition (L3) is trivially satisfied and by (4.40) and (3.34) we get

$$\lambda \mathbf{P}\{S(X) > \lambda\} \leqslant C\|X\|_1 \leqslant C\|X^*\|_1,$$

so (L1) holds, by Tchebychev inequality. From the right-hand side of (4.48) we get

$$\lambda \mathbf{P}\{X^* > \lambda\} \leqslant C\|X^*\|_1 \leqslant C\|S(X)\|_1,$$

which is (R1), and finally by Tchebychev inequality, the left-hand side of (4.44) and (3.34)

$$\lambda^2 \mathbf{P}\{S(X) > \lambda\} \leqslant \|S(X)\|_2 \leqslant C\|X\|_2 \leqslant C\|X^*\|_2,$$

we get (R2).

(iii) Consider the operator

$$T(X) = \limsup_{n \to \infty} \left({}^n X\right)^*.$$

Then,

$$T_n(X) = \limsup_{m \to \infty} \left({}^m X^n\right)^* = 0$$

for all positive integers n. On the other hand,

$$T(X) = \infty \quad \text{on } \left\{X^* = \infty\right\}.$$

In what follows, we will need the following technical result.

Lemma 4.8 *If T be an operator satisfying condition $\mathfrak{B}$ and τ is a stopping time, then*

$$\gamma^{-1} T_\tau(X) \leqslant T(X)^\tau \leqslant \gamma T_\tau(X).$$

Proof Since for $0 \leqslant n \leqslant \infty$; on the set $\{\tau = n\}$,

$$T(X)^\tau = T\left(X^n + \left(X^\tau - X^n\right)\right) \leqslant \gamma \left[T(X^n) + T\left(X^\tau - X^n\right)\right] = \gamma T_n(X) = \gamma T_\tau(X).$$

The left-hand inequality is obtained similarly. $\square$

The next lemma shows how the variability of X^* and, more generally, TX is controlled by the variability of $s(X)$.

Lemma 4.9 *Let $X = \{X_n\}$ be a $\{\mathscr{F}_n\}$-martingale satisfying condition **A**, and $(Y \circ X)$ be a martingale transform of X by a multiplier sequence $Y = \{Y_1, Y_2, \ldots\}$, uniformly bounded by a constant b, $\lambda > 0$, and $A \in \mathscr{F}_m$ for some $m \geqslant 0$ such that*

$$A \subset \{s_m((Y \circ X)) = 0, \lambda \leqslant s((Y \circ X)) \leqslant 2\lambda\}, \tag{4.100}$$

$$\mathbf{P}\left(\{(Y \circ X)^* > c\lambda\} \cap A\right) \geqslant c\mathbf{P}(A), \tag{4.101}$$

$$\int_A \left|(Y \circ X)^*\right|^p d\mathbf{P} \leqslant c\lambda^p \mathbf{P}(A), \quad 0 < p \leqslant \rho. \tag{4.102}$$

*If T is any operator satisfying conditions $\mathfrak{B}$ and **R**, then*

$$\mathbf{P}\left(\{T(Y \circ X) > c\lambda\} \cap A\right) \geqslant c\mathbf{P}(A). \tag{4.103}$$

*The choice of the constant in (4.101) depends only on δ; that of (4.102) only on p and the constant on (A3); and that of (4.103) only on the parameters of **A**, $\mathfrak{B}$, and **R**.*

Proof We may assume in the proof that

$$s((Y \circ X)) = 0 \text{ on } A^c. \tag{4.104}$$

Consider $Z = {}^m(Y \circ X)^\nu$ with ν the stopping time defined by

$$\nu = m \text{ on } A^c, \quad \nu = \infty \text{ on } A.$$

Then $s(Z) = 0$ on A^c, $Z = (Y \circ X)$ on A, so that $(Y \circ X)$ satisfies not only (4.100) but also (4.104). If Z satisfies the conclusions of the lemma, then so does $(Y \circ X)$. This is clear for the first two conditions, (4.101) and (4.102). For the third, note that $\nu = \infty$ and $s\left((Y \circ X)^m\right) = 0$ on the set $\{TZ > C\lambda\} \cap A$, so that, by Lemma 4.8

$$C\lambda < TZ = T\left[(Y \circ X)^\nu - (Y \circ X)^m\right] \leqslant \gamma \left[T(Y \circ X)^\nu + T(Y \circ X)^m\right]$$
$$= \gamma T(Y \circ X)^\nu \leqslant \gamma^2 T_\nu(Y \circ X) = \gamma^2 T(Y \circ X).$$

Therefore, letting $c = \gamma^{-2}C$, we have that

$$\mathbf{P}\left(\{T(Y \circ X) > c\lambda\} \cap A\right) \geqslant \mathbf{P}\left(\{TZ > c\lambda\} \cap A\right)$$

and we may assume (4.104). Note that, under condition (A2),

$$s((Y \circ X)) = \left(\sum_{k=1}^{\infty} Y_k^2\right)^{\frac{1}{2}}, \tag{4.105}$$

so that, by Cauchy-Schwarz inequality,

$$s((Y \circ X))S((Y \circ X)) \geqslant \sum_{k=1}^{\infty} Y_k^2 |\Delta X_k|.$$

Therefore, by (A1), (A2), and (4.100),

$$2\lambda \int_A S((Y \circ X))d\mathbf{P} \geqslant \int_A s((Y \circ X))S((Y \circ X))d\mathbf{P} \geqslant \int_A \sum_{k=m+1}^{\infty} Y_k^2 |\Delta X_k|\, d\mathbf{P}$$

$$\geqslant \delta \int_A \sum_{k=m+1}^{\infty} Y_k^2 d\mathbf{P} = \delta \int_A s((Y \circ X))^2 d\mathbf{P} \geqslant \delta\lambda^2 \mathbf{P}(A).$$

Also, we have that

$$\int_A S((Y \circ X))^2 d\mathbf{P} = \int_A \sum_{k=m+1}^{\infty} Y_k^2 \Delta X_k^2 d\mathbf{P} = \int_A \sum_{k=m+1}^{\infty} Y_k^2 d\mathbf{P}$$

$$= \int_A s((Y \circ X))^2 d\mathbf{P} \leqslant 4\lambda^2 \mathbf{P}(A).$$

The conditions of Lemma 4.5 are satisfied by $Y = S((Y \circ X))$, $\alpha = 2^{-1}\delta\lambda$, $\beta = 2\lambda$, $p_1 = 1$, $p_2 = 2$, $\theta = 2^{-1}$, so that

$$\mathbf{P}\left\{S((Y \circ X)) > 2^{-2}\delta\lambda, A\right\} \geqslant 2^{-8}\delta^2 \mathbf{P}(A).$$

Therefore, by (4.104), (4.40), and (3.34),

$$\int_A (Y \circ X)^* d\mathbf{P} = \left\|(Y \circ X)^*\right\|_1 \geqslant \|(Y \circ X)\|_1 \geqslant c\lambda \mathbf{P}\{S((Y \circ X)) > c\lambda\} \geqslant c\lambda \mathbf{P}(A),$$

$$\int_A \left|(Y \circ X)^*\right|^2 d\mathbf{P} \leqslant \left\|(Y \circ X)^*\right\|_2^2 \leqslant 4\|(Y \circ X)\|_2^2 = 4\|s((Y \circ X))\|_2^2$$

$$= 4 \int_A s((Y \circ X))^2 d\mathbf{P} \leqslant 16\lambda^2 \mathbf{P}(A).$$

Another application of Lemma 4.5 now gives (4.101).

If $2 \leqslant p \leqslant \varrho$, then by (4.44), (4.104), (4.105), and the inequality

$$\mathbf{E}\left[|\Delta X_k|^p \mid \mathscr{F}_{k-1}\right] \leqslant c,$$

which follows from (A3), we have that, using Hölder's inequality for $p/2$ and $p/(p-2)$,

$$\int_A \left|(Y \circ X)^*\right|^p d\mathbf{P} \leqslant \left\|(Y \circ X)^*\right\|_p^p \leqslant c\|S((Y \circ X))\|_p^p = c\int_\Omega \left(\sum_{k=1}^\infty Y_k^2 \Delta X_k^2\right)^{p/2} d\mathbf{P}$$

$$\leqslant c\int_\Omega s((Y \circ X))^{p-2}\left(\sum_{k=1}^\infty Y_k^2 |\Delta X_k|^p\right) d\mathbf{P}$$

$$= c\int_A s((Y \circ X))^{p-2}\left(\sum_{k=1}^\infty Y_k^2 |\Delta X_k|^p\right) d\mathbf{P}$$

$$\leqslant c\lambda^{p-2}\left(\sum_{k=m+1}^\infty \int_A Y_k^2 \, d\mathbf{P}\right) \leqslant c\lambda^{p-2}\int_A s((Y \circ X))^2 d\mathbf{P} \leqslant c\lambda^p \mathbf{P}(A).$$

If $0 < p \leqslant 2$, then by Lemma 4.7,

$$\int_A \left|(Y \circ X)^*\right|^p d\mathbf{P} \leqslant \left\|(Y \circ X)^*\right\|_p^p \leqslant c\|s((Y \circ X))\|_p^p = c\int_A s((Y \circ X))^p d\mathbf{P} \leqslant c\lambda^p \mathbf{P}(A).$$

This proves (4.102).

Using (4.104), (B2), (R1), and (4.101), we have that

$$\int_A |T(Y \circ X)|^{p_1} d\mathbf{P} = \|T(Y \circ X)\|_{p_1}^{p_1} \geqslant c\lambda^{p_1} \mathbf{P}\left\{(Y \circ X)^* > c\lambda\right\} \geqslant c\lambda^{p_1} \mathbf{P}(A).$$

By (R2), (4.104), and (4.102), for $\eta > 0$,

$$\eta^{p_2} \mathbf{P}\left\{T(Y \circ X) > \eta\right\} \leqslant c\left\|(Y \circ X)^*\right\|_{p_2}^{p_2} = c\int_A \left|(Y \circ X)^*\right|^{p_2} d\mathbf{P} \leqslant c\lambda^{p_2} \mathbf{P}(A).$$

Therefore, (4.103) follows after another application of Lemma 4.5. This completes the proof. $\qquad\square$

The following two results provide upper bounds for stopped martingales, see Theorems 2.1 and 2.2 of [44].

Theorem 4.30 *Let $X = \{X_n\}$ be a $\{\mathscr{F}_n\}$-martingale satisfying condition* **A**, $0 < p \leqslant \varrho$, *let T be an operator satisfying conditions* $\mathfrak{B}$, **R**, *and condition* (L3), *and $(Y \circ X)$ be a martingale transform of X by a multiplier sequence $Y = \{Y_k\}$, uniformly bounded by a constant b and τ is the stopping time defined by*

$$\tau = \inf\left\{n : T_n(Y \circ X) > b\right\}.$$

Then, there exists a constant c such that

$$\left\| \left((Y \circ X)^{\tau} \right)^* \right\|_p \leqslant cb[\mathbf{P}\{s\,(Y \circ X) > 0\}]^{1/p} \leqslant cb.$$

The choice of c depends only on p and the parameters of $\mathbf{A}$, $\mathfrak{B}$, and $\mathbf{R}$.

Proof Let N be a positive integer. We must show that there exists $c > 0$ such that

$$\left\| \left((Y \circ X)^{\tau \wedge N} \right)^* \right\|_p \leqslant cb[\mathbf{P}\,\{s\,((Y \circ X)) > 0\}]^{1/p},$$

with c not depending on N.

Let $\lambda = 2\gamma^2 b/\beta$ where $\beta < 1$. Note that $Y^* \leqslant b < \lambda$. Define the multiplier sequence W as

$$W_k = Y_k, \quad 1 \leqslant k \leqslant N, \quad W_k = \lambda \chi_{\{s((Y \circ X)) > 0\}}, \quad k > N.$$

Let $(W \circ X)$ be the corresponding transform:

$$(W \circ X)_n = \sum_{k=1}^{n} W_k \Delta X_k, \ n \geqslant 1.$$

Then $(Y \circ X)^j = (W \circ X)^j$, $1 \leqslant j \leqslant N$, and on the set $\{s\,((Y \circ X)) > 0\}$, by condition (A2) and the fact that $\{W_k\}$ is $\mathscr{F}_{k-1}$ measurable

$$s_n\,((W \circ X)) = \left(\sum_{k=1}^{n} \mathbf{E}\left[(\Delta(W \circ X)_k)^2 \mid \mathscr{F}_{k-1} \right] \right)^{1/2}$$

$$= \left(\sum_{k=1}^{n} \mathbf{E}\left[W_k^2 (\Delta X_k)^2 \mid \mathscr{F}_{k-1} \right] \right)^{1/2} = \left(\sum_{k=1}^{n} W_k^2 \right)^{1/2} \to \infty$$

as $n \to \infty$. Define a sequence of stopping times as follows. Let

$$\mu_0 = \inf\{n \geqslant 0 : s_{n+1}\,((Y \circ X)) > 0\}.$$

μ_0 is a stopping time. Note that $\{\mu_0 < \infty\} = \{s\,((Y \circ X)) > 0\}$ and $s\,((W \circ X)^{\mu_0}) = 0$. Recursively, if $j \geqslant 1$ and assuming that we have already defined the stopping time μ_{j-1}, we defined μ_j as

$$\mu_j = \inf\left\{n : s\left(^m (W \circ X)^n\right) > \lambda\right\}$$

on the set where $\mu_{j-1} = m, m \geqslant 0$, and let $\mu_j = \infty$ on the set where $\mu_{j-1} = \infty$. Then μ_j, is a stopping time satisfying

$$\mu_{j-1} < \mu_j < \infty \text{ on } \{s\,((Y \circ X)) > 0\}$$
$$\mu_j = \infty \text{ on } \{s\,((Y \circ X)) = 0\}.$$

Let $Z_j = \{Z_{j1}, Z_{j2,\ldots}\}$ denote the martingale transform $\{^{\mu_j-1}(W \circ X)_k^{\mu_j}\}_k$ for each $j \geqslant 1$.

Since $W^* \leqslant \lambda$ and $s\left(^m(W \circ X)^n\right) \leqslant s\left(^m(W \circ X)^{n-1}\right) + |W_n|$, we have that

$$\lambda < s\left(Z_j\right) \leqslant 2\lambda, \quad j \geqslant 1 \quad \text{on} \quad \{s\left((Y \circ X)\right) > 0\}.$$

Also, $Z_j^* = 0$ on $\{s\left((Y \circ X)\right) = 0\}$. Now let

$$\sigma = \inf\left\{j : T_{\mu_j}^*\left(W \circ X\right) > b\right\}$$

$\mu_\infty = \infty$, and $\nu = \mu_\sigma$. Note that $\sigma \geqslant 1$, and by Lemma 4.8 and (B2),

$$T_{\mu_0}^*\left(W \circ X\right) \leqslant \gamma T^*\left(W \circ X\right)^{\mu_0} = 0,$$

since $s\left((W \circ X)^{\mu_0}\right) = 0$. Also, if $n \leqslant N$, then $\tau \wedge N \leqslant \nu$, since on the set $\{\nu = n\}$, we have that $b < T_n^*\left(W \circ X\right) = T^*\left(W \circ X\right)^n = T^*\left(Y \circ X\right)^n = T_n^*\left(Y \circ X\right)$ and $\tau \leqslant n$.

Therefore,

$$\left((Y \circ X)^{\tau \wedge N}\right)^* = \left((W \circ X)^{\tau \wedge N}\right)^* \leqslant \left((W \circ X)^\nu\right)^* \leqslant \sum_{j=1}^\infty \chi_{\{\sigma \geqslant j\}} Z_j^*,$$

since on $\{\sigma = j\}$,

$$\left((W \circ X)^{\mu_j}\right)^* = \left((W \circ X)^{\mu_0} +^{\mu_0}(W \circ X)^{\mu_1} + \cdots +^{\mu_j-1}(W \circ X)^{\mu_j}\right)^*$$
$$= \left(0 + Z_1 + \cdots + Z_j\right)^* \leqslant Z_1^* + \cdots + Z_j^*, \quad j \geqslant 1.$$

Consequently,

$$\left\|\left((Y \circ X)^{\tau \wedge N}\right)^*\right\|_p^{p \wedge 1} \leqslant \sum_{j=1}^\infty \|\chi_{\{\sigma \geqslant j\}} Z_j^*\|_p^{p \wedge 1}$$

for $0 < p < \infty$. From now on assume that $0 < p \leqslant \varrho$.

Now, fix $j \geqslant 1$ and let

$$A_m = \left\{\mu_{j-1} = m, T_m^*\left(W \circ X\right) \leqslant b\right\}, \quad m \geqslant 0.$$

Clearly,

$$\{\sigma \geqslant j, s\left((Y \circ X)\right) > 0\} = \left\{T_{\mu_{j-1}}^*\left(W \circ X\right) \leqslant b, s\left((Y \circ X)\right) > 0\right\} = \bigcup_{m=0}^\infty A_m.$$

Note that

$$A_m \subset \left\{ s_m\left(Z_j\right) = 0, \lambda \leqslant s\left(Z_j\right) \leqslant 2\lambda \right\},$$

and $A_m \in \mathscr{F}_m$ by (L3). Applying Lemma 4.9 to Z_j, we have that

$$\left\| \chi_{\{\sigma \geqslant j\}} Z_j^* \right\|_p^p = \left\| \chi_{\{\sigma \geqslant j, s((Y \circ X)) > 0\}} Z_j^* \right\|_p^p = \sum_{m=0}^{\infty} \int_{A_m} \left| Z_j^* \right|^p d\mathbf{P}$$

$$\leqslant c\lambda^p \sum_{m=0}^{\infty} \mathbf{P}(A_m) = c\lambda^p \mathbf{P}\{\sigma \geqslant j, s((Y \circ X)) > 0\}.$$

On the set where $T_{\mu_j}^*(W \circ X) \leqslant b$ and $\mu_j < \infty$, using Lemma 4.8,

$$TZ_j = T\left[(W \circ X)^{\mu_j} - (W \circ X)^{\mu_{j-1}}\right] \leqslant \gamma \left[T(W \circ X)^{\mu_j} + T(W \circ X)^{\mu_{j-1}}\right]$$

$$\leqslant \gamma^2 \left[T_{\mu_j}(W \circ X) + T_{\mu_{j-1}}(W \circ X)\right] \leqslant 2\gamma^2 T_{\mu_j}^*(W \circ X) \leqslant 2\gamma^2 b = \beta\lambda.$$

Therefore, applying Lemma 4.9 to Z_j a second time, we obtain

$$\mathbf{P}\{\sigma \geqslant j+1, s((Y \circ X)) > 0\} = \mathbf{P}\left\{ T_{\mu_j}^*(W \circ X) \leqslant b, s((Y \circ X)) > 0 \right\}$$

$$= \sum_{m=0}^{\infty} \mathbf{P}\left\{ \mu_{j-1} = m, T_{\mu_j}^*(W \circ X) \leqslant b \right\}$$

$$\leqslant \sum_{m=0}^{\infty} \mathbf{P}\left(\{TZ_j \leqslant \beta\lambda\} \cap A_m\right) \leqslant (1-\beta) \sum_{m=0}^{\infty} \mathbf{P}(A_m)$$

$$= (1-\beta)\mathbf{P}\{\sigma \geqslant j, s((Y \circ X)) > 0\}.$$

By induction, for all $j \geqslant 1$,

$$\mathbf{P}\{\sigma \geqslant j, s((Y \circ X)) > 0\} \leqslant (1-\beta)^{j-1}\mathbf{P}\{\sigma \geqslant 1, s((Y \circ X)) > 0\}$$
$$= (1-\beta)^{j-1}\mathbf{P}\{s((Y \circ X)) > 0\}.$$

Accordingly, for $0 < p \leqslant 1$,

$$\left\| \left((Y \circ X)^{\tau \wedge N}\right)^* \right\|_p^p \leqslant \sum_{j=1}^{\infty} \left\| \chi_{\{\sigma \geqslant j\}} Z_j^* \right\|_p^p \leqslant \sum_{j=1}^{\infty} c\lambda^p (1-\beta)^{j-1}\mathbf{P}\{s((Y \circ X)) > 0\}$$

$$= c\lambda^p \mathbf{P}\{s((Y \circ X)) > 0\},$$

and for $1 \leqslant p \leqslant \varrho$,

$$\left\| \left((Y \circ X)^{\tau \wedge N}\right)^* \right\|_p \leqslant \sum_{j=1}^{\infty} \left\| \chi_{\{\sigma \geqslant j\}} Z_j^* \right\|_p \leqslant \sum_{j=1}^{\infty} \left[c\lambda^p (1-\beta)^{j-1}\mathbf{P}\{s((Y \circ X)) > 0\}\right]^{1/p}$$

$$= c\lambda[\mathbf{P}\{s((Y \circ X)) > 0\}]^{1/p}. \qquad \square$$

Corollary 4.31 *Let $X = \{X_n\}$ be a $\{\mathscr{F}_n\}$-martingale satisfying condition $\mathbf{A}$ and $0 < p \leqslant \varrho$. Let $\{(Y \circ X)_n\}$ be the martingale transform of X by a multiplier sequence $Y = \{Y_1, Y_2, \ldots\}$, uniformly bounded by a constant b and τ is the stopping time defined by*

$$\tau = \inf \{n : |(Y \circ X)_n| > b\},$$

then

$$\left\| \left((Y \circ X)^\tau\right)^* \right\|_p \leqslant cb[\mathbf{P}\{s\left((Y \circ X)\right) > 0\}]^{1/p} \leqslant cb.$$

The choice of c depends only on p and the parameters of $\mathbf{A}$.

Proof It follows immediately from Theorem 4.30: let T be the operator defined by $TX = X^*$. In this case $T_n X = (X^n)^*$ and $\mathfrak{B}$, $\mathbf{R}$, and $L3$ are satisfied with $\gamma = 1$, $p_1 = 1$, $p_2 = 2$, $c = 1$. $\qquad\square$

Condition $\mathbf{A}$ cannot be substantially weakened. See Example 8.3 in [44].

Lemma 4.10 *Let $X = \{X_n\}$ be a $\{\mathscr{F}_n\}$-martingale satisfying conditions (A1) and (A2). Then for all $\lambda > 0$ and all martingale transform $(Y \circ X) \in \mathscr{M}(X)$,*

$$\mathbf{P}\left\{Y^* > \lambda\right\} \leqslant c\mathbf{P}\left\{c(Y \circ X)^* > \lambda\right\}$$

with the choice of c depending only on δ.

Proof Let $\tau = \inf \{k : |Y_k| > \lambda\}$ and $A_k = \{\tau = k\}, k \geqslant 1$. Then $A_k \in \mathscr{F}_{k-1}$ and, by (A1),

$$\int_{A_k} |\Delta X_k|\, d\mathbf{P} \geqslant \delta\mathbf{P}(A_k),$$

and by (A2),

$$\int_{A_k} (\Delta X_k)^2\, d\mathbf{P} = \mathbf{P}(A_k)$$

hence by Lemma 4.5,

$$\mathbf{P}\{|\Delta X_k| > c, A_k\} \geqslant c\mathbf{P}(A_k).$$

Therefore,

$$\mathbf{P}\left\{Y^* > \lambda\right\} = \sum_{k=1}^{\infty} \mathbf{P}(A_k) \leqslant c \sum_{k=1}^{\infty} \mathbf{P}\{|\Delta X_k| > c, A_k\}$$

$$\leqslant c \sum_{k=1}^{\infty} \mathbf{P}\{|\Delta(Y \circ X)_k| > c\lambda, A_k\} \leqslant c\mathbf{P}\left\{c(Y \circ X)^* > \lambda\right\}. \qquad\square$$

4.7   Burholder-Gundy Inequalities for General Operators

The main result of this section is the extension of Theorem 4.23 for more general operators T under the conditions discussed above.

Remember that Φ is Young's function, see Definition 4.7, i.e., non-trivial monotone non-decreasing continuous function defined on $[0, \infty]$ with $\Phi(0) = 0$ that satisfies the *delta condition*,

$$\Phi(2\lambda) \leqslant C\Phi(\lambda), \ \lambda > 0. \tag{4.106}$$

Theorem 4.32 (Burkholder-Gundy) *Given Young's function Φ, a $\{\mathscr{F}_n\}$-martingale $X = \{X_n\}$ satisfying condition* **A**, *a multiplier sequence $Y = \{Y_n\}$, and an operator T from $\mathscr{M}$ into the non-negative $\mathscr{F}$-measurable functions that satisfies conditions* $\mathfrak{B}$ *and* **L**, *then:*

(i) There exist a constant C such that

$$\mathbf{E}\left[\Phi\left((Y \circ X)^*\right)\right] \leqslant C\mathbf{E}\left[\Phi\left(T^*(Y \circ X)\right)\right] + C\mathbf{E}[\Phi\left(\Delta(Y \circ X)^*\right)] \tag{4.107}$$

for the martingale transform $(Y \circ X)$ of X; where, as before, $\Delta(Y \circ X)$ is the sequence of differences of $(Y \circ X)$ and $\Delta(Y \circ X)^* = \sup_n |\Delta(Y \circ X)_n|$.
(ii) There exist a constant C such that

$$\mathbf{E}\left[\Phi\left(T^*(Y \circ X)\right)\right] \leqslant C\mathbf{E}\left[\Phi\left((Y \circ X)^*\right)\right] + C\mathbf{E}\left[\Phi\left(\Delta^*\right)\right] \tag{4.108}$$

for the martingale transform $(Y \circ X)$ of X, where $\Delta_n = T\left(^{n-1}(Y \circ X)^n\right)$, $n \geqslant 1$, and $\Delta^* = \sup_n |\Delta_n|$.

As we have already mentioned, see the comments before Theorem 4.22, the way that D. Burkholder and R. Gundy proved those inequalities originally in [44] was by observing that certain inequalities between distribution functions are true for sufficiently many λ's, called them *good lambdas*, and they allow to prove the integral inequality (observe that it is needed to justify that integrating over those λ's is enough!). This is precisely the *good lambda condition* defined by them.

Nevertheless, in [30] D. L. Burkholder was able to obtain the inequality (4.73) for the maximal operator $(\ )^*$ and the quadratic function operator S, see Theorem 4.73, using weaker inequalities (but valid for all λ's) that simplify dramatically the proof. These inequalities are what now are called *good-λ inequalities*, see for instance (4.74) and (4.75). In the same paper, D. L. Burkholder suggested that this could be done for general operators T but, as far as we know, it was not done by them. This technique will be used to get a simpler proof for general operators T and for more general functions Φ than the ones used in [44].

Thus, again by Theorem 4.22 it is enough to prove the following good-λ inequalities. For inequality (4.107) we need

Proposition 4.12 *If $\beta > 1, 0 < \delta < \beta - 1$, and let X be a $\{\mathscr{F}_n\}$-martingale that satisfies condition $\mathbf{A}$ and a operator T that satisfies conditions $\mathfrak{B}$ and $\mathbf{L}$, then*

$$\mathbf{P}\left\{(Y \circ X)^* > \beta\lambda, T^*(Y \circ X) \vee \Delta(Y \circ X)^* \vee Y^* \leqslant \delta\lambda\right\} \leqslant \frac{c\gamma^p\delta^p}{(\beta - \delta - 1)^p}\mathbf{P}\left\{(Y \circ X)^* > \lambda\right\}$$

$$(4.109)$$

for $0 < p < \rho$.

Proof Let $\lambda > 0$ and

$$\mu = \inf\{n : |(Y \circ X)_n| > \lambda\},$$
$$v = \inf\{n : |(Y \circ X)_n| > \beta\lambda\} \text{ and}$$
$$\sigma = \inf\{n : T_n(Y \circ X) > \delta\lambda \text{ or } \Delta(Y \circ X)_n > \delta\lambda \text{ or } |Y_{n+1}| > \delta\lambda\}$$

be stopping times. Now let $(W \circ X)$ be the martingale $(Y \circ X)$ starting at μ and stoped at $v \wedge \sigma$, thus $(W \circ X) = {}^\mu(Y \circ X)^{v \wedge \sigma}$, where

$$(W \circ X)_n = \sum_{k=1}^{n} W_k \Delta X_k \text{ with } W_k = \chi_{\{\mu < k \leqslant v \wedge \sigma\}} Y_k.$$

The sequence $\{W_k\}_n$ is, by construction, uniformly bounded by $\delta\lambda$.

Let us consider $\lambda > 0$ and the stopping time $\tau = \inf\{n : T_n(W \circ X) > 2\gamma\delta\lambda\}$, then, by Theorem 4.30, it can be obtained that

$$\left\|\left((W \circ X)^\tau\right)^*\right\|_p^p \leqslant C(\gamma\delta\lambda)^p\mathbf{P}\{s((W \circ X)) > 0\} \leqslant C(\gamma\delta\lambda)^p\mathbf{P}\left\{W^* > 0\right\}$$
$$\leqslant C(\gamma\delta\lambda)^p\mathbf{P}\{\mu < \infty\} = C(\gamma\delta\lambda)^p\mathbf{P}\left\{(Y \circ X)^* > \lambda\right\}.$$

Then, by the condition (B1) and Lemma 4.8, we have

$$T^*(W \circ X) = T^*\left({}^\mu(Y \circ X)^{v \wedge \sigma}\right) = T^*\left((Y \circ X)^{v \wedge \sigma} - (Y \circ X)^\mu\right)$$
$$\leqslant \gamma\left(T^*(Y \circ X)^{v \wedge \sigma} + T^*(Y \circ X)^\mu\right)$$
$$\leqslant 2\gamma T^*(Y \circ X).$$

Now observe that

(i) On the set $\{(Y \circ X)^* > \beta\lambda, T^*(Y \circ X) \vee \Delta(Y \circ X)^* \vee Y^* \leqslant \delta\lambda\}$, we have

$$T^*(W \circ X) \leqslant 2\gamma\delta\lambda, \Delta(Y \circ X) \leqslant \Delta(Y \circ X)^* \leqslant \delta\lambda, \mu < v < \infty,$$

$\tau = \infty$ and $\sigma = \infty$.

(ii) On the set $\{\mu < \nu < \infty\}$,

$$(W \circ X)^* = \sup_{1 \leqslant n < \infty} |(W \circ X)_n| = \sup_{1 \leqslant n < \infty} |^{\mu}(Y \circ X)_n^{\nu}| = \sup_{1 \leqslant n < \infty} |(Y \circ X)_n^{\nu} - (Y \circ X)_n^{\mu}|$$

$$\geqslant \sup_{1 \leqslant n < \infty} |(Y \circ X)_n^{\nu}| - \sup_{1 \leqslant n < \infty} |(Y \circ X)_n^{\mu}| = |(Y \circ X)_{\nu}| - |(Y \circ X)_{\mu}|$$

$$> (\beta - \delta - 1)\lambda$$

since $|(Y \circ X)_{\mu}| \leqslant |(Y \circ X)_{\mu-1}| + |\Delta(Y \circ X)_{\mu}| \leqslant \lambda + \delta\lambda$ and $|(Y \circ X)_{\nu}| > \beta\lambda$.

Therefore by Tchebychev's inequality, we obtain

$$\mathbf{P}\left\{(Y \circ X)^* > \beta\lambda, T^*(Y \circ X) \vee \Delta(Y \circ X)^* \vee Y^* \leqslant \delta\lambda\right\} \leqslant \mathbf{P}\{\mu < \nu < \infty, \tau = \infty, \sigma = \infty\}$$

$$\leqslant \mathbf{P}\left\{(W \circ X)^* > (\beta - \delta - 1)\lambda, \tau = \infty\right\} = \mathbf{P}\left\{((W \circ X)^{\tau})^* > (\beta - \delta - 1)\lambda, \tau = \infty\right\}$$

$$\leqslant \mathbf{P}\left\{((W \circ X)^{\tau})^* > (\beta - \delta - 1)\lambda\right\} \leqslant \frac{1}{(\beta - \delta - 1)^p \lambda^p} \left\|((W \circ X)^{\tau})^*\right\|_p^p$$

$$\leqslant \frac{C(\gamma\delta\lambda)^p}{(\beta - \delta - 1)^p \lambda^p} \, \mathbf{P}\left\{(Y \circ X)^* > \lambda\right\} = \frac{C(\gamma\delta)^p}{(\beta - \delta - 1)^p} \, \mathbf{P}\left\{(Y \circ X)^* > \lambda\right\}.$$

Let us prove now inequality (4.107).

Proof Assume that, besides the previous conditions, Φ satisfies[8]

$$\Phi(\lambda_1 \vee \lambda_2) \leqslant \Phi(\lambda_1) + \Phi(\lambda_2),$$

then by Theorem 4.22, Proposition 4.12 and Lemma 4.6 we have

$$\mathbf{E}\left[\Phi\left((Y \circ X)^*\right)\right] \leqslant C\mathbf{E}\left[\Phi\left(T^*(Y \circ X) \vee \Delta^*(Y \circ X) \vee Y^*\right)\right]$$
$$\leqslant C\mathbf{E}\left[\Phi\left(T^*(Y \circ X)\right)\right] + C\mathbf{E}\left[\Phi\left(\Delta(Y \circ X)^*\right)\right] + C\mathbf{E}\left[\Phi\left(Y^*\right)\right]$$
$$\leqslant C\mathbf{E}\left[\Phi\left(T^*(Y \circ X)\right)\right] + C\mathbf{E}\left[\Phi\left(\Delta(Y \circ X)^*\right)\right].$$

$\square$

Now, for (4.108) we need the following good-λ inequality

Proposition 4.13 *Let $\beta > 1, 0 < \delta < \beta\gamma^{-4} - 1$, and let X be a $\{\mathscr{F}_n\}$-martingale that satisfies condition* **A** *and an operator T that satisfies conditions* $\mathfrak{B}$ *and* **L**, *then:*

$$\mathbf{P}\left\{T^*(Y \circ X) > \beta\lambda, (Y \circ X)^* \vee \Delta^* \vee Y^* \leqslant \delta\lambda\right\} \leqslant \frac{C\delta^p}{(\beta\gamma^{-4} - \delta - 1)^p} \mathbf{P}\left\{T^*(Y \circ X) > \lambda\right\}$$

$$\tag{4.110}$$

for $0 < p \leqslant \rho$.

---

[8] This condition is immediately satisfied if Φ is convex.

Proof Let $\lambda > 0$ and

$$\mu = \inf \{n : T_n(Y \circ X) > \lambda\}$$
$$v = \inf \{n : T_n(Y \circ X) > \beta\lambda\} \text{ and}$$
$$\sigma = \inf \{n : |(Y \circ X)_n| > \delta\lambda \text{ or } |\Delta_n| > \delta\lambda \text{ or } |Y_{n+1}| > \delta\lambda\}$$

be stopping times. Analogously as in Proposition 4.12, let

$$(W \circ X) = {}^{\mu}(Y \circ X)^{v \wedge \sigma}$$

be the martingale $(Y \circ X)$ starting at μ and stopped at $v \wedge \sigma$, with $W = \{W_n\} = \{\chi_{\{\mu < k \leqslant v \wedge \sigma\}} Y_k\}$ the multiplier sequence. By construction W is uniformly bounded by $\delta\lambda$.

Let us consider $\lambda > 0$ and the stopping time $\tau = \inf\{n : |(Y \circ X)| > \delta\lambda\}$. Then, by Theorem 4.30 it can be proved that

$$\left\| \left((W \circ X)^{\tau}\right)^* \right\|_p^p \leqslant C(\delta\lambda)^p \mathbf{P}\{s((W \circ X)) > 0\} \leqslant C(\delta\lambda)^p \mathbf{P}\left\{W^* > 0\right\}$$
$$\leqslant C(\delta\lambda)^p \mathbf{P}\{\mu < \infty\} = C(\delta\lambda)^p \mathbf{P}\left\{T^*(Y \circ X) > \lambda\right\}.$$

Observe that

(i) On the set $\{\mu = n\}$,

$$T_{\mu}(Y \circ X) = T_n(Y \circ X) = T(Y \circ X)^n = T\left((Y \circ X)^{n-1} + {}^{n-1}(Y \circ X)^n\right)$$
$$\leqslant \gamma \left(T\left((Y \circ X)^{n-1}\right) + T\left({}^{n-1}(Y \circ X)^n\right)\right)$$
$$= \gamma \left(T\left((Y \circ X)^{n-1}\right) + \Delta_n\right) \leqslant \gamma \left(\lambda + \Delta^*\right).$$

(ii) On the set $\{T^*(Y \circ X) > \beta\lambda, (Y \circ X)^* \vee \Delta^* \vee Y^* \leqslant \delta\lambda\}$,

$$\mu < v < \infty, \tau = \infty, \sigma = \infty.$$

(iii) On the set $\{\mu < v < \infty, \sigma = \infty\}$, by Lemma 4.8

$$\beta\lambda < T_v(Y \circ X) \leqslant \gamma T(Y \circ X)^v = \gamma T\left((Y \circ X)^{\mu} + {}^{\mu}(Y \circ X)^v\right)$$
$$= \gamma T\left((Y \circ X)^{\mu} + (W \circ X)\right) + \gamma^2 \left(T(Y \circ X)^{\mu} + T(W \circ X)\right)$$
$$\leqslant \gamma^2 \left(\gamma T_{\mu}(Y \circ X) + \gamma T(W \circ X)\right) = \gamma^3 \left(T_{\mu}(Y \circ X) + T(W \circ X)\right)$$
$$\leqslant \gamma^3 \left(\gamma \left(\lambda + \Delta^*\right) + \gamma T(W \circ X)\right) = \gamma^4 \left(\lambda + \Delta^* + T(W \circ X)\right).$$

Therefore we obtain

$$T(W \circ X) > \left(\beta\gamma^{-4} - \delta - 1\right)\lambda$$

Thus by condition (R2):

$$\mathbf{P}\left\{T^*(Y \circ X) > \beta\lambda, (Y \circ X)^* \vee \ \Delta^* \vee Y* \leqslant \delta\lambda\right\}$$
$$\leqslant \mathbf{P}\{\mu < \nu < \infty, \sigma = \infty, \tau = \infty\}$$
$$\leqslant \mathbf{P}\left\{T(W \circ X)^\tau > \left(\beta\gamma^{-4} - \delta - 1\right)\lambda, \tau = \infty\right\}$$
$$\leqslant \mathbf{P}\left\{T(W \circ X)^\tau > \left(\beta\gamma^{-4} - \delta - 1\right)\lambda\right\}$$
$$\leqslant \frac{C}{\left(\beta\gamma^{-4} - \delta - 1\right)^P \lambda^P}\left\|\left((W \circ X)^\tau\right)^*\right\|_P^P$$
$$\leqslant \frac{C(\delta\lambda)^P}{\left(\beta\gamma^{-4} - \delta - 1\right)^P \lambda^P}\mathbf{P}\left\{T^*(Y \circ X) > \lambda\right\}$$
$$= \frac{C\delta^P}{\left(\beta\gamma^{-4} - \delta - 1\right)^P}\mathbf{P}\left\{T^*(Y \circ X) > \lambda\right\}.$$

Now, let us prove inequality (4.108).

Proof By Theorem 4.22, Proposition 4.13 and Lemma 4.6 we have

$$\mathbf{E}\left[\Phi\left(T^*(Y \circ X)\right)\right] \leqslant \mathbf{CE}\left[\Phi\left((Y \circ X)^* \vee \Delta^* \vee Y^*\right)\right]$$
$$\leqslant \mathbf{CE}\left[\Phi\left(((Y \circ X)^*)\right)\right] + \mathbf{CE}\left[\Phi\left(\Delta^*\right)\right] + \mathbf{CE}\left[\Phi\left(Y^*\right)\right]$$
$$\leqslant \mathbf{CE}\left[\Phi\left((Y \circ X)^*\right)\right] + \mathbf{CE}\left[\Phi\left(\Delta^*\right)\right] + \mathbf{CE}\left[\Phi\left(\Delta(Y \circ X)^*\right)\right]$$
$$\leqslant \mathbf{CE}\left[\Phi\left((Y \circ X)^*\right)\right] + \mathbf{E}\left[\Phi\left(\Delta^*\right)\right]$$

since,

$$\Delta(Y \circ X)^* \leqslant 2(Y \circ X)^*,$$

as

$$\Delta(Y \circ X)^* = \sup_n |\Delta(Y \circ X)_n| = \sup_n \left|\sum_{k=1}^n \Delta(Y \circ X)_k - \sum_{k=1}^{n-1}\Delta(Y \circ X)_k\right|$$
$$\leqslant \sup_n \left[\left|\sum_{k=1}^n \Delta(Y \circ X)_k\right| + \left|\sum_{k=1}^{n-1}\Delta(Y \circ X)_k\right|\right]$$
$$\leqslant 2\sup_n \left|\sum_{k=1}^n \Delta(Y \circ X)_k\right| = 2(Y \circ X)^*.$$

$\square$

Finally, imposing additional conditions on Φ we have

Corollary 4.33 *Given Young's function Φ, a $\{\mathscr{F}_n\}$-martingale $X = \{X_n\}$ satisfying condition **A**, a multiplier sequence $Y = \{Y_n\}$, and an operator T from $\mathscr{M}$ into the non-negative $\mathscr{F}$-measurable functions satisfies conditions $\mathfrak{B}$ and **L**.*

(i) If we assume that Young's function Φ satisfies the condition

$$\mathrm{R}\Phi: \quad \mathbf{E}\left[\Phi\left(\Delta(Y \circ X)^*\right)\right] \leqslant C\mathbf{E}\left[\Phi\left(T^*(Y \circ X)\right)\right]$$

for all $(Y \circ X)$ martingale transforms of a $\{\mathscr{F}_n\}$-martingale X, then,

$$\mathbf{E}\left[\Phi\left((Y \circ X)^*\right)\right] \leqslant C\mathbf{E}\left[\Phi\left(T^*(Y \circ X)\right)\right]. \tag{4.111}$$

(ii) If we assume that Young's function Φ satisfies the condition

$$L\Phi: \quad \mathbf{E}\left[\Phi\left(\Delta^*\right)\right] \leqslant C\mathbf{E}\left[\Phi\left((Y \circ X)^*\right)\right]$$

for all $(Y \circ X)$ martingale transforms of a $\{\mathscr{F}_n\}$-martingale X, then,

$$\mathbf{E}\left[\Phi\left(T^*(Y \circ X)\right)\right] \leqslant C\mathbf{E}\left[\Phi\left((Y \circ X)^*\right)\right]. \tag{4.112}$$

Finally, even though the inequalities for the operators S and s are essentially analogous, there are significant differences between them. Under a boundedness condition on the increments, we know that the double inequality occurs for the S operator, Theorem 4.24, but, as we will see in the next result, only the left inequality occurs for s, that is to say, there exists an analog of (4.77) for s, but not of (4.78). Moreover, the double inequality in L_p-norm is valid for S in the interval $0 < p < \infty$, as a consequence of Theorem 4.24, but for s, it only occurs in the range $0 < p \leqslant 2$.

Theorem 4.34

(i) Let $2 \leqslant p < \infty$, $X = \{X_n\}$ be a $\{\mathscr{F}_n\}$-martingale, $X \in \mathscr{M}$,

$$\|s(X)\|_p \leqslant c\|X\|_p.$$

The choice of c depends only on p.
(ii) Let $0 < p \leqslant 2$, $X = \{X_n\}$ be a $\{\mathscr{F}_n\}$-martingale, $X \in \mathscr{M}$,

$$\left\|X^*\right\|_p \leqslant C\|s(X)\|_p.$$

The choice of C depends only on p.
(iii) Let $0 < p \leqslant 2$, $X = \{X_n\}$ be a $\{\mathscr{F}_n\}$-martingale, and $(Y \circ X) \in \mathscr{M}(X)$ be a martingale transform of X by a multiplier sequence $Y = \{Y_k\}$. If $(A1)$ and $(A2)$ are satisfied, then,

$$\|s((Y \circ X))\|_p \leqslant c\left\|(Y \circ X)^*\right\|_{p^*}.$$

The choice of c depends only on p and δ.

(iv) Let $2 \leqslant p \leqslant \varrho$, $X = \{X_n\}$ be a $\{\mathscr{F}_n\}$-martingale, and $(Y \circ X) \in \mathscr{M}(X)$ be a martingale transform of X by a multiplier sequence $Y = \{Y_k\}$, if condition $\mathbf{A}$ holds, then,

$$\left\| (Y \circ X)^* \right\|_p \leqslant C \| s((Y \circ X)) \|_p.$$

The choice of C depends only on p, δ, and the constant of condition (A3).

Proof (i) Let us prove, by induction, that if j is a positive integer and $X \in \mathscr{M}$, then

$$\| s(X) \|_{2j} \leqslant j^{1/2} \| S(X) \|_{2j}. \tag{4.113}$$

This is trivially true if $j = 1$. Let $j > 1$ and suppose that $\| S(X) \|_{2j} < \infty$. Then, considering $Y_k^2 = \mathbf{E}\left[\Delta X_k^2 \mid \mathscr{F}_{k-1} \right]$, we have

$$\left\| Y_k^2 \right\|_j \leqslant \left\| \Delta X_k^2 \right\|_j \leqslant \| s(X) \|_{2j}^2 < \infty$$

by the norm diminishing property of conditional expectations, and therefore all of the following integrals are finite. Let n be a positive integer and

$$K = \{1, \ldots, n\} \times \cdots \times \{1, \ldots, n\}(j \text{ factors }).$$

If $k = (k_1, \ldots, k_j) \in K$, let $|k| = \max(k_1, \ldots, k_j)$. Then letting

$$A_i = \{k \in K : |k| = k_i\}$$
$$K \subset A_1 \cup \cdots \cup A_j$$

we have that and

$$\int_\Omega \left| \sum_{k=1}^n Y_k^2 \right|^j d\mathbf{P} \leqslant \sum_{i=1}^j \sum_{k \in A_i} \int_\Omega Y_{k_1}^2 \cdots Y_{k_j}^2 \, d\mathbf{P}$$

$$= \sum_{i=1}^j \sum_{k \in A_i} \int_\Omega \Delta X_{k_i}^2 \prod_{\substack{m=1 \\ m \neq i}}^j Y_{k_m}^2 \, d\mathbf{P} \leqslant \sum_{i=1}^j \int_\Omega \sum_{k \in K} \Delta X_{k_i}^2 \prod_{\substack{m=1 \\ m \neq i}}^j Y_{k_m}^2 \, d\mathbf{P}$$

$$= j \int_\Omega \left| \sum_{k=1}^n Y_k^2 \right|^{j-1} \left| \sum_{k=1}^n \Delta X_k^2 \right| d\mathbf{P}$$

$$\leqslant j \left[\int_\Omega \left| \sum_{k=1}^n Y_k^2 \right|^j d\mathbf{P} \right]^{(j-1)/j} \left[\int_\Omega \left| \sum_{k=1}^n \Delta X_k^2 \right|^j d\mathbf{P} \right]^{1/j}.$$

Therefore, we have obtained

$$\| s_n(X) \|_{2j}^2 \leqslant j \, \| S_n(X) \|_{2j}^2 ,$$

and (4.113) follows by the monotone convergence theorem. Combining Burkholder's inequality (4.44) and (4.113), we have that

$$\|s(X)\|_{2j} \leqslant c_{2j}\|X\|_{2j} \tag{4.114}$$

for each positive integer j.

To finish the proof of (i), we apply Marcinkiewicz's interpolation theorem, see [160]. First we remark that (4.114) is true for complex martingales provided ΔX_k^2 is replaced by $|\Delta X_k|^2$ in the definition of $s(X)$, as (4.114) for the real case directly implies an inequality of the same form for the complex case.

Now consider the operator T defined on complex $L^2 = L^2(\Omega, \mathscr{F}, \mathbf{P})$ as follows. If X_∞ denotes the almost everywhere limit of X and $X_\infty \in L^2$, then $T X_\infty = s(X)$ where $X = \{X_k\}$ is the complex martingale defined by

$$X_n = \mathbf{E}[X_\infty \mid \mathscr{F}_n], \quad n \geqslant 1.$$

We use the fact that

$$\|X\|_p \leqslant \|X_\infty\|_p, \quad 1 < p < \infty$$

with equality holding if X_∞ is measurable relative to $\mathscr{F}_\infty$, the smallest σ-field containing $\bigcup_{k=1}^{\infty} \mathscr{F}_k$. By the complex version of (4.114),

$$\|T(X_\infty)\|_{2j} \leqslant c_{2j}\|X_\infty\|_{2j}, \quad j = 1, 2, \ldots$$

Clearly, T satisfies

$$T(X_\infty + Z_\infty) \leqslant T(X_\infty) + T(Z_\infty), \quad X_\infty, Z_\infty \in L^2 \tag{4.115}$$

and the other conditions of Marcinkiewicz's interpolation theorem. Therefore, if $2 \leqslant p \leqslant 2j$, we have that

$$\|T(X_\infty)\|_p \leqslant c_p\|X_\infty\|_p, \quad X_\infty \in L^p, c_p \leqslant \max(c_2, c_{2p}). \tag{4.116}$$

In summary, if $X \in L^p$, $2 \leqslant p < \infty$, and then

$$\|s(X)\|_p = \|T(X_\infty)\|_p \leqslant c_p\|X_\infty\|_p = c_p\|X\|_p,$$

and the function $p \to c_p$ is bounded on each compact subinterval of $[2, \infty)$.

(ii) We may assume in the proof of (ii) that $\mathscr{F}_\infty = \mathscr{F}$. Then T, the operator defined in the proof of (i), is an isometry in L^2, where now it is enough to consider real L^2. Therefore, by (4.115),

$$2\int_\Omega X_\infty \cdot Y_\infty \, d\mathbf{P} = \|X_\infty + Y_\infty\|_2^2 - \|X_\infty\|_2^2 - \|Y_\infty\|_2^2$$

$$= \|T(X_\infty + Y_\infty)\|_2^2 - \|T(X_\infty)\|_2^2 - \|T(Y_\infty)\|_2^2 \leqslant 2\int_\Omega T X_\infty \cdot T Y_\infty \, d\mathbf{P}$$

for all X_∞ and Y_∞ in L^2.

Now, let $1 < p \leqslant 2 \leqslant q$, $p^{-1} + q^{-1} = 1$. If $X_\infty \in L_2$ and $B = \{Y_\infty \in L^2 : \|Y_\infty\|_q \leqslant 1\}$, then

$$
\begin{aligned}
\|X_\infty\|_p = \sup_{Y_\infty \in B} \int_\Omega X_\infty \cdot Y_\infty \, d\mathbf{P} &\leqslant \sup_{Y_\infty \in B} \int_\Omega T(X_\infty) \cdot T(Y_\infty) \, d\mathbf{P} \\
&\leqslant \sup_{Y_\infty \in B} \|T(X_\infty)\|_p \, \|T(Y_\infty)\|_q \leqslant c_q \, \|T(X_\infty)\|_p
\end{aligned}
$$

by (4.116).

If $X \in \mathcal{M}$ and $\|X\|_2 = \|s(X)\|_2 < \infty$, then X converges almost everywhere to X_∞ in L^2, by Corollary 3.18. Therefore, using (3.31), we have that

$$
\|X^*\|_p \leqslant Cp\|X\|_p = C_p \, \|X_\infty\|_p \leqslant C_p' \, \|T(X_\infty)\|_p = C_p' \|s(X)\|_p. \tag{4.117}
$$

Now, we show that (4.117) holds without the assumption that $\|s(X)\|_2 < \infty$. This will complete the proof of (ii) since, by (i), the function $p \to C_p'$ is bounded on compact subintervals of $(1, 2]$ and, by the proof of Lemma 4.7,

$$
\|X^*\|_p^p \leqslant 4\|s(X)\|_p^p, \quad 0 < p \leqslant 4/3.
$$

Let $X \in \mathcal{M}$ and $1 < p \leqslant 2 \leqslant q$, $p^{-1} + q^{-1} = 1$. To prove (4.117) we may assume that $\|s(X)\|_p < \infty$; then $\mathbf{E}\left[\Delta X_k^2 \mid \mathscr{F}_{k-1}\right] < \infty$, $k \geqslant 1$. Let m be a positive integer. If $1 \leqslant k \leqslant m$, let

$$
\delta_{mk} = \left\{
\begin{array}{l}
\Delta X_k \quad \text{on} \quad \left\{ \mathbf{E}\left[\Delta X_k^2 \mid \mathscr{F}_{k-1}\right] \leqslant m \right\} \\[2mm]
\Delta X_k / \left(\mathbf{E}\left[\Delta X_k^2 \mid \mathscr{F}_{k-1}\right] \right)^{\frac{1}{2}} \ \text{elsewhere}
\end{array}
\right.
$$

if $k > m$, let $\delta_{mk} = 0$. Then $\delta_m = (\delta_{m1}, \delta_{m2}, \ldots)$ is a martingale difference sequence relative to $\mathscr{F}_1, \mathscr{F}_2, \ldots$, and the martingale $Y_m = \{Y_{m1}, Y_{m2}, \ldots\}$ with this difference sequence satisfies

$$
\|s(Y_m)\|_2 < \infty.
$$

Therefore,

$$
\|Y_m^*\|_p \leqslant c_p \, \|s(Y_m)\|_p \leqslant c_p \|s((Y \circ X))\|_p,
$$

using the fact that $|\delta_{mk}|$ increases to $|\Delta(Y \circ X)_k|$ as $m \to \infty$. Since $\lim Y_{mn} = (Y \circ X)_n$,

$$
X^* \leqslant \liminf_{m \to \infty} X_m^*
$$

by Fatou's lemma,

$$
\|X^*\|_p \leqslant \liminf_{m \to \infty} \|Y_m^*\|_p \leqslant c_p \|s(X)\|_p.
$$

(iii) This part of the theorem is an immediate corollary of Theorem 4.34 with $\Phi(t) = t^p, 0 < p \leqslant 2$.

(iv) This part follows from (4.111) applied to the operator s and the function $\Phi(t) = t^p, 2 \leqslant p \leqslant \varrho$. Observe that s satisfies condition $\mathfrak{B}$. Condition (R1) is satisfied with $p_1 = 1$ by Lemma 4.7; condition (R2) holds for $p_2 = 2$ by (i) of this theorem. Now let us assume that X satisfies condition **A**, in order to check RΦ with $\Phi(t) = t^p$ for some $p, 2 \leqslant p \leqslant \varrho$ and $T = s$. Using conditions (A3) and then (A2) we have

$$
\begin{aligned}
\left\| \Delta(Y \circ X)^* \right\|_p^p &\leqslant \int_\Omega \sum_{k=1}^\infty |Y_k|^p \, |\Delta X_k|^p \, d\mathbf{P} \leqslant \int_\Omega \sum_{k=1}^\infty |Y_k|^p \, \mathbf{E}\left[|\Delta X_k|^p \mid \mathscr{F}_{k-1} \right] d\mathbf{P} \\
&\leqslant \int_\Omega \sum_{k=1}^\infty |Y_k|^p \left(\mathbf{E}\left[|\Delta X_k|^\varrho \mid \mathscr{F}_{k-1} \right] \right)^{p/\varrho} d\mathbf{P} \\
&\leqslant c^{p/\varrho} \int_\Omega \sum_{k=1}^\infty |Y_k|^p \, d\mathbf{P} \leqslant c \int_\Omega \left(\sum_{k=1}^\infty Y_k^2 \right)^{p/2} d\mathbf{P} \\
&= c \int_\Omega \left(\sum_{k=1}^\infty Y_k^2 \, \mathbf{E}[\Delta X_k^2 \mid \mathscr{F}_{k-1}] \right)^{p/2} d\mathbf{P} \\
&= c \int_\Omega \left(\sum_{k=1}^\infty \mathbf{E}[Y_k^2 \Delta X_k^2 \mid \mathscr{F}_{k-1}] \right)^{p/2} d\mathbf{P} = c \| s((Y \circ X)) \|_p^p.
\end{aligned}
$$

Therefore, RΦ holds for every $p, 2 \leqslant p \leqslant \varrho$. Part (iv) now follows from (4.111). $\square$

As already mentioned, part (iii) of the previous theorem is a special case of the following theorem

Theorem 4.35 *Let Φ be Young's function, $X = \{X_n\}$ be a $\{\mathscr{F}_n\}$-martingale, and $(Y \circ X) \in \mathcal{M}(X)$ be a martingale transform of X by a multiplier sequence $Y = \{Y_k\}$. Suppose that conditions (A1) and (A2) hold, then*

$$
\int_\Omega \Phi(s((Y \circ X)) d\mathbf{P} \leqslant c \int_\Omega \Phi\left((Y \circ X)^* \right) \mathbf{P}.
$$

Condition (A2) alone is not sufficient to imply the conclusions of Theorem 4.35 and parts (iii) and (iv) of Theorem 4.34, see Example 8.2 of [44]).

Proof We assume that conditions (A1) and (A2) hold and apply Theorem 4.32 to the operator s. This operator satisfies condition $\mathfrak{B}$ and the measurability condition (L3), and, as shown in the proof of Theorem 4.34 (iv), condition (L2) holds. Therefore, only (L1) and $L\Phi$ must be checked. We show that (L1) holds with $s_1 = 1$.

Define an operator T on $\mathcal{M}$, the set of all $\{\mathscr{F}_n\}$-martingales X, by

$$T(X) = \left[\sum_{k=1}^{\infty} \mathbf{E}\left(|\Delta X_k| \mid \mathscr{F}_{k-1}\right)^2\right]^{\frac{1}{2}}.$$

Then T satisfies conditions 1) and 3) in the definition of class $\mathscr{B}$, see Definition 4.10,

$$T(X + Y) \leqslant T(X) + T(Y),$$

and

$$\|T(X)\|_2 \leqslant \|s(X)\|_2 = \|X\|_2$$

$$\|T(X)\|_1 \leqslant \left\|\sum_{k=1}^{\infty} \mathbf{E}\left[|\Delta X_k| \mid \mathscr{F}_{k-1}\right]\right\|_1 = \left\|\sum_{k=1}^{\infty} |\Delta X_k|\right\|_1.$$

Since $TX \leqslant s(X)$, T satisfies (B2), from the Definition 4.11, but not the local condition (2) of class $\mathscr{B}$ mappings. However, it can be proved that the results remain valid if condition (2) of class $\mathscr{B}$ mappings is replaced by condition (B2), see [44] p. 280, and [87].

Now consider the operator s. Since (A2) holds

$$s((Y \circ X)) = \left(\sum_{k=1}^{\infty} \mathbf{E}[\Delta(Y \circ X)_k^2 \mid \mathscr{F}_{k-1}]\right)^{\frac{1}{2}} = \left(\sum_{k=1}^{\infty} \mathbf{E}[Y_k^2 \Delta X_k^2 \mid \mathscr{F}_{k-1}]\right)^{\frac{1}{2}}$$

$$= \left(\sum_{k=1}^{\infty} Y_k^2 \mathbf{E}[\Delta X_k^2 \mid \mathscr{F}_{k-1}]\right)^{\frac{1}{2}} = \left(\sum_{k=1}^{\infty} Y_k^2\right)^{\frac{1}{2}}.$$

By (A1),

$$\delta |Y_k| \leqslant |Y_k| \mathbf{E}\left[|\Delta X_k| \mid \mathscr{F}_{k-1}\right] = \mathbf{E}\left[|(Y \circ X)_k| \mid \mathscr{F}_{k-1}\right]$$

so that

$$\delta s((Y \circ X)) \leqslant T(Y \circ X)$$

for all $(Y \circ X) \in \mathcal{M}(X)$. Therefore,

$$\lambda \mathbf{P}\{s((Y \circ X)) > \lambda\} \leqslant \lambda \mathbf{P}\left\{\delta^{-1} T(Y \circ X) > \lambda\right\} \leqslant c\|(Y \circ X)\|_1$$

so that (L1) holds with $s_1 = 1$. Finally, observe that in this case $\Delta^* = Y^*$, see (4.108), then by Lemma 4.10,

$$\mathbf{P}\left\{\Delta^* > \lambda\right\} = \mathbf{P}\left\{Y^* > \lambda\right\} \leqslant c\mathbf{P}\left\{c(Y \circ X)^* > \lambda\right\},$$

which in turn implies LΦ for any Φ:

$$\int_\Omega \Phi\left(\Delta^*\right) d\mathbf{P} \leqslant c \int_\Omega \Phi\left((Y \circ X)^*\right) d\mathbf{P}.$$

Therefore, by Theorem 4.32 and Corollary 4.33, we obtain

$$\int_\Omega \Phi(s((Y \circ X))) d\mathbf{P} \leqslant c \int_\Omega \Phi\left((Y \circ X)^*\right) d\mathbf{P}. \qquad \square$$

4.8 Local Properties of Martingales

In Theorem 1.1 we saw that a Rademacher series $\sum_{k=1}^{\infty} c_k r_k$ converges almost everywhere if and only if $\sum_{k=1}^{\infty} |c_k|^2 < \infty$. If we now consider $\{\xi_k\}_k$ a sequence of independent centered random variables, with variance one,[9]

$$\mathbf{E}\left[\xi_k\right] = 0, \quad \mathbf{E}\left[\xi_k{}^2\right] = 1, \quad k = 1, 2, \dots$$

then, when is the assertion that $\sum_{k=1}^{n} v_k \xi_k$ converges a.e. is equivalent to $\sum_{k=1}^{\infty} v_k^2 < \infty$ for any numerical sequence $\{v_k\}_{k=1}^{\infty}$?

At least three sufficient conditions for this equivalence are known:

(i) Uniform boundedness of the random variables $\{\xi_k\}_{k=1}^{\infty}$, due to Khinchine and Kolmogorov [107];
(ii) Uniform integrability of the sequence $\left\{\xi_k^2\right\}_{k=1}^{\infty}$, due to Kac and Steinhaus [104];
(iii) The Marcinkiewicz-Zygmund condition : $\mathbf{E}\left[|\xi_k|\right] \geqslant \delta > 0$ uniformly in k given in [117]. In other words, if the Marcinkiewicz-Zygmund condition holds for $\{\xi_k\}$ independent random variables with zero means and variance one then for every sequence of constants v_n, $\sum^{\infty} v_n \xi_n$ converges a.e. if and only if $\sum_1^{\infty} v_n^2 < \infty$. The Marcinkiewicz-Zygmund theorem [117, Theorem 4] is far more general than either (ii) or (i) and is the starting point.

Observe that the Marcinkiewicz and Zygmund condition, which was the starting point of this discussion, is in the case of martingales, is called regular martingales, see Definition 4.8. Then we want to prove for the transforms of regular martingales the same equivalence but also add other equivalent conditions. This is what are called *local properties of martingales*, see [44].

We will give alternative proofs of the local behavior for martingale transforms of $X = \{X_n\}_{n \geqslant 1}$ a regular $\{\mathscr{F}_n\}$-martingale, using good λ inequalities. More precisely,

---

[9] Thus $\{\xi_k\}_k$ are then orthonormal.

we will prove that the following sets are a.e. equivalents, i.e., their indicator functions
are equals a.e.,

$$
A = \left\{ \lim_{n \to \infty} (Y \circ X)_n = (Y \circ X) \text{ exists and is finite} \right\},
$$

$$
B = \left\{ s(Y \circ X) = \left(\sum_{n=1}^{+\infty} \mathbf{E}[\Delta(Y \circ X)_n^2 | \mathscr{F}_{n-1}] \right)^{1/2} < \infty \right\},
$$

$$
C = \left\{ S(Y \circ X) = \left(\sum_{n=1}^{\infty} \Delta(Y \circ X)_n^2 \right)^{1/2} < \infty \right\},
$$

$$
D = \left\{ (Y \circ X)^* = \sup_n |(Y \circ X)_n| < \infty \right\},
$$

$$
E = \left\{ \sup_n (Y \circ X)_n < \infty \right\}.
$$

The equivalences among the sets A, B, and C were proved initially by Gundy in
[85,89,90]; the equivalences with D and E are due to Chow [55] and Davis [64],
respectively.

The proofs that we are going to give are a modification of the ones given in [44],
based only on good lambda inequalities, see [124].

Theorem 4.36 *Let $X = \{X_n\}_{n=1}^{\infty}$ be a $\{\mathscr{F}_n\}$-martingale that satisfies condition
A, $Y = \{Y_n\}_{n \geqslant 1}$ a multiplier sequence and $(Y \circ X)$ the martingale transform of
X by Y, then the following sets are a.e. equivalents:*

$$
A = \left\{ \lim_{n \to \infty} (Y \circ X)_n = (Y \circ X) \text{ exists and is finite} \right\},
$$

$$
B = \left\{ s(Y \circ X) = \left(\sum_{n=1}^{+\infty} \mathbf{E}[\Delta(Y \circ X)_n^2 | \mathscr{F}_{n-1}] \right)^{1/2} < \infty \right\},
$$

$$
C = \left\{ S(Y \circ X) = \left(\sum_{n=1}^{\infty} \Delta(Y \circ X)_n^2 \right)^{1/2} < \infty \right\},
$$

$$
D = \left\{ (Y \circ X)^* = \sup_n |(Y \circ X)_n| < \infty \right\},
$$

$$
E = \left\{ \sup_n (Y \circ X)_n < \infty \right\}.
$$

To prove this theorem, we need some technical results, which are good λ inequalities for general operators T (see [30,123]).

First, from Proposition 4.12 and Proposition 4.13, taking $T = S$, i.e., $\gamma = 1$ in
(B1) condition, since $\Delta(Y \circ X)^* \leqslant S(Y \circ X)$, we immediately get

Corollary 4.37 *If $\lambda > 0, 0 < \delta < \beta - 1$, $\{X_n\}_{n \geqslant 1}$ is a $\{\mathscr{F}_n\}$-martingale satisfying
condition A, $Y = \{Y_n\}_{n \geqslant 1}$ is a multiplier sequence and $(Y \circ X)$ is the martingale*

transform of X by Y, then:

$$\mathbf{P}\{(Y \circ X)^* > \beta\lambda, \, S(Y \circ X) \vee Y^* \leqslant \delta\lambda\} \leqslant \frac{(C\delta)^2}{(\beta - \delta - 1)^2} \, \mathbf{P}\{(Y \circ X)^* > \lambda\}$$

(4.118)

and

$$\mathbf{P}\{S(Y \circ X) > \beta\lambda, \, (Y \circ X)^* \vee \Delta(Y \circ X)^* \vee Y^* \leqslant \delta\lambda\} \leqslant \frac{(C\delta)^2}{(\beta - \delta - 1)^2} \, \mathbf{P}\{S(Y \circ X) > \lambda\}.$$

(4.119)

We will also need the following good λ-inequalities :

Proposition 4.14 *If $\beta > 1$, $0 < 4\delta < \beta - 1$ and $\{X_n\}_n$ is a $\{\mathscr{F}_n\}$-martingale satisfying condition* **A**, *then:*

$$\mathbf{P}\{(Y \circ X)^* > \beta\lambda, \, \sup_n (Y \circ X)_n^+ \vee Y^* \leqslant \delta\lambda\} \leqslant \frac{(C\delta)^2}{(\beta - 4\delta - 1)^2} \mathbf{P}\{(Y \circ X)^* > \lambda\},$$

(4.120)

where $(Y \circ X)_n^+$ is the positive part of $(Y \circ X)_n$.

Proof Let $\lambda > 0$ and

$$\mu = \inf\{n \geqslant 0 : |(Y \circ X)_n| > \lambda\},$$
$$\nu = \inf\{n \geqslant 0 : |(Y \circ X)_n| > \beta\lambda\},$$
$$\sigma = \inf\{n \geqslant 0 : (Y \circ X)_n^+ \vee |Y_{n+1}| > \delta\lambda\}$$

be stopping times, and consider the stopped martingale,

$$(W \circ X) = \, {}^{\mu}(Y \circ X)^{\nu \wedge \sigma},$$

therefore

$$(W \circ X)_n = \sum_{k=1}^{n} \chi_{\{\mu < k \leqslant \nu \wedge \sigma\}} Y_k \Delta X_k$$

for all $n \in \mathbb{N}$, and the multiplier sequence $\{W_k = \chi_{\{\mu < k \leqslant \nu \wedge \sigma\}} Y_k\}$ is uniformly bounded by $\delta\lambda$ and then, by $4\delta\lambda$. Let $\tau = \inf\{n \geqslant 0 : |(W \circ X)_n| > 4\delta\lambda\}$, by Corollary 4.31 we have

$$\|((W \circ X)^{\tau})^*\|_2^2 \leqslant (C4\delta\lambda)^2 \mathbf{P}\{s(W \circ X) > 0\} \leqslant (C4\delta\lambda)^2 \mathbf{P}\{W^* > 0\}$$
$$\leqslant (4C\delta\lambda)^2 \mathbf{P}\{\mu < +\infty\} \leqslant (C\delta\lambda)^2 \mathbf{P}\{(Y \circ X)^* > \lambda\}. \quad (4.121)$$

On the other hand, let us observe that

$$(W \circ X)^* = \sup_n |{}^\mu (Y \circ X)_n^{\nu \wedge \sigma}| = \sup_n |(Y \circ X)_n^{\nu \wedge \sigma} - (Y \circ X)_n^{\mu}|$$

$$\leqslant \sup_n |(Y \circ X)_n^{\nu \wedge \sigma}| + |\sup_n |(Y \circ X)_n^{\mu}|$$

$$\leqslant |(Y \circ X)_{\nu \wedge \sigma}| + |(Y \circ X)_\mu| \leqslant 4 \sup_n (Y \circ X)_n^+,$$

and, on the set

$$\{(Y \circ X)^* > \beta \lambda, \ \sup_n (Y \circ X)_n^+ \vee Y^* \leqslant \delta \lambda\},$$

$$|(W \circ X)_n| \leqslant (W \circ X)^* \leqslant 4\delta\lambda \quad \text{and} \quad \mu \leqslant \nu < +\infty, \sigma = +\infty, \tau = +\infty.$$

Thus, on the set $\{\mu \leqslant \nu < +\infty, \sigma = +\infty\}$,

$$(W \circ X)^* = \sup_n |{}^\mu (Y \circ X)_n^{\nu}| \geqslant {}^\mu (Y \circ X)_\nu^{\nu} = |(Y \circ X)_\nu^{\nu} - (Y \circ X)_\nu^{\mu}|$$

$$\geqslant |(Y \circ X)_\nu| - |(Y \circ X)_\mu| > \beta\lambda - \lambda(1 + 4\delta) = (\beta - 4\delta - 1)\lambda.$$

Moreover, let us observe that

$$|(Y \circ X)_\mu| \leqslant |(Y \circ X)_{\mu-1}| + |\Delta(Y \circ X)_\mu| \leqslant \lambda + \Delta(Y \circ X)^*$$
$$\leqslant \lambda + 4 \sup_n (Y \circ X)_n^+ \leqslant \lambda + 4\delta\lambda = (1 + 4\delta)\lambda.$$

By (4.121) and Doob's maximal inequality,

$$\mathbf{P}\{(Y \circ X)^* > \beta\lambda, \sup_n (Y \circ X)_n^+ \vee Y^* \leqslant \delta\lambda\} \leqslant \mathbf{P}\{\mu \leqslant \nu < +\infty, \sigma = +\infty, \tau = +\infty\}$$

$$\leqslant \mathbf{P}\{(W \circ X)^* > (\beta - 4\delta - 1)\lambda, \tau = +\infty\}$$

$$\leqslant \mathbf{P}\{((W \circ X)^\tau)^* > (\beta - 4\delta - 1)\lambda\}$$

$$\leqslant \frac{1}{(\beta - 4\delta - 1)^2 \lambda^2} \|((W \circ X)^\tau)^*\|_2^2$$

$$= \frac{(C\delta)^2}{(\beta - 4\delta - 1)^2} \mathbf{P}\{(Y \circ X)^* > \lambda\}. \qquad \square$$

Also, by Proposition 4.13, we have for the conditional quadratic function s, taking $T = s$,

Corollary 4.38 *If $\lambda > 0$, $0 < \delta < \beta - 1$, and $\{X_n\}_{n \geqslant 1}$ a is $\{\mathscr{F}_n\}$-martingale satisfying condition* **A***, then*

$$\mathbf{P}\{s(Y \circ X) > \beta\lambda, (Y \circ X)^* \vee Y^* \leqslant \delta\lambda\} \leqslant \frac{(C\delta)^2}{(\beta - \delta - 1)^2} \mathbf{P}\{s(Y \circ X) > \lambda\}.$$

$$(4.122)$$

On the other hand,

Proposition 4.15 *If* $\lambda > 0, 0 < \delta < \beta - 1,$ *and* $X = \{X_n\}_{n \geq 1}$ *is a* $\{\mathscr{F}_n\}$*-martingale satisfying condition* **A***, then*

$$\mathbf{P}\{s(Y \circ X) > \beta\lambda, \ \sup_n (Y \circ X)_n^+ \vee Y^* \leq \delta\lambda\} \leq \frac{C\delta^2}{(\beta - \delta - 1)^2} \mathbf{P}\{s(Y \circ X) > \lambda\}.$$

$$(4.123)$$

Proof Again, let $\lambda > 0$ and

$$\mu = \inf\{n : s_{n+1}(Y \circ X) > \lambda\},$$
$$\nu = \inf\{n : s_{n+1}(Y \circ X) > \beta\lambda\},$$
$$\sigma = \inf\{n : (Y \circ X)_n^+ \vee |Y_{n+1}| \geq \delta\lambda\}$$

be stopping times, and consider the stopped martingale

$$(W \circ X) =^\mu (Y \circ X)^{\nu \wedge \sigma},$$

then, the multiplier sequence $\{W_n\}_{n \geq 1}$ is uniformly bounded by $\delta\lambda$ and therefore by $4\delta\lambda$. Set

$$\tau = \inf\{n : |(W \circ X)_n| > 4\delta\lambda\},$$

then, by Corollary 4.31

$$\begin{aligned}
\|((W \circ X)^\tau)^*\|_2^2 &\leq C(\delta\lambda)^2 \mathbf{P}\{s(W \circ X) > 0\} \leq C(\delta\lambda)^2 \mathbf{P}\{W^* > 0\} \\
&\leq C(\delta\lambda)^2 \mathbf{P}\{\mu < +\infty\} \leq C(\delta\lambda)^2 \mathbf{P}\{s(Y \circ X) > \lambda)\}.
\end{aligned}$$

$$(4.124)$$

Thus, by the sublinearity of s and by Lemma 4.8, with $\gamma = 1$, we have

$$\begin{aligned}
\beta\lambda &< s_{\nu+1}(Y \circ X) \leq s(Y \circ X)^{\nu+1} \leq s(Y \circ X)^\mu + s^\mu(Y \circ X)^\nu + s^\nu(Y \circ X)^{\nu+1} \\
&\leq s(Y \circ X)^\mu + s^\mu(Y \circ X)^\nu + |Y_{\nu+1}| \leq s_\mu(Y \circ X) + s(W \circ X) + |Y_{\nu+1}| \\
&\leq \lambda + s(W \circ X) + |Y_{\nu+1}|,
\end{aligned}$$

and therefore

$$s(W \circ X) > \beta\lambda - \lambda - |Y_{\nu+1}|.$$

On the set $\{s(Y \circ X) > \beta\lambda, \sup_n (Y \circ X)_n^+ \vee Y^* \leq \delta\lambda\}$, we have

$$\mu \leq \nu < +\infty, \sigma = +\infty, \tau = +\infty.$$

Therefore, by using Tchebychev's inequality and (4.124), we get

$$\mathbf{P}\{s(Y \circ X) > \beta\lambda, \sup_n (Y \circ X)_n^+ \vee Y^* \leqslant \delta\lambda\} \leqslant \mathbf{P}\{\mu \leqslant \nu < +\infty, \sigma = +\infty, \tau = +\infty\}$$

$$\leqslant \mathbf{P}\{s(W \circ X)^\tau > (\beta - \delta - 1)\lambda, \tau = +\infty\}$$

$$\leqslant \mathbf{P}\{s(W \circ X)^\tau > (\beta - \delta - 1)\lambda\}$$

$$\leqslant \frac{C}{(\beta - \delta - 1)^2\lambda^2}\|((W \circ X)^\tau)^*\|_2^2$$

$$\leqslant \frac{C\delta^2}{(\beta - \delta - 1)^2}\mathbf{P}\{s(Y \circ X) > \lambda\}. \qquad \square$$

Finally, the following technical result is needed, see [89, Lemma 2].

Lemma 4.11 *Let $\{X_n\}_{n \geqslant 1}$ be a regular $\{\mathscr{F}_n\}$- martingale, $Y = \{Y_n\}_{n \geqslant 1}$ be $\{\mathscr{F}_{n-1}\}$ multiplier sequence, and $(Y \circ X)$ is the transform of X by Y. If $\sup_n |Y_n \Delta X_n| < \infty$ a.e. on a set F, then $\sup_n |Y_n| < \infty$ a.e. in F.*

Proof The key of the proof is Lévy's version of Borel-Cantelli's lemma, see [129, Corollary, p. 151].

We have seen that assuming $\mathbf{E}[(\Delta X_k)^2 \mid \mathscr{F}_{k-1}] = 1$, then, thanks to Paley-Zygmund's lemma, Lemma 4.4, (4.70) and (4.71) are equivalents. Let $\lambda > 0, \gamma > 0$ be the constant in (4.71). Since $\sup_n |Y_n \Delta X_n| < \infty$ we can choose M sufficiently large so that if $I_n = \chi_{\{|Y_n \Delta X_n| \geqslant M\}}$ then $\sum_{n=1}^\infty I_n < +\infty$ a.e. on a subset $F' \subseteq F$ such that $\mathbf{P}\left(F'\right) \geqslant (1 - \epsilon)\mathbf{P}(F)$.

Let $J_n = \chi_{\{|Y_n| \geqslant M/\lambda\}}$ and $K_n = \chi_{\{|\Delta X_n| \geqslant \lambda\}}$, then $J_n K_n \leqslant I_n$ and

$$\sum_{n=1}^\infty \gamma J_n \leqslant \sum_{n=1}^\infty J_n \mathbf{P}\{|\Delta X_n| > \lambda \mid \mathscr{F}_{n-1}\} = \sum_{n=1}^\infty \mathbf{E}\left[J_n K_n \mid \mathscr{F}_{n-1}\right]$$

$$\leqslant \sum_{n=1}^\infty \mathbf{E}\left[I_n \mid \mathscr{F}_{n-1}\right] < +\infty,$$

a.e. on F. Then, by Lévy's version of Borel-Cantelli's lemma, this implies that

$$\sum_{n=1}^\infty J_n < +\infty,$$

a.e. on F', and since $\epsilon > 0$ is arbitrary, we conclude $\limsup |Y_n| < +\infty$ a.e. on F. $\qquad \square$

We are ready to consider the proof of Theorem 4.36

Proof First, observe that the inclusions

$$A = \{ \lim_{n \to +\infty} (Y \circ X)_n = (Y \circ X) \text{ exists and is finite}\} \subset \{\sup_n |(Y \circ X)_n| < +\infty\} = D$$

and

$$A = \{ \lim_{n \to +\infty} (Y \circ X)_n = (Y \circ X) \text{ exists and is finite}\} \subset \{\sup_n (Y \circ X)_n < +\infty\} = E$$

are trivially true.

Now, let us prove that the sets $A = \{\lim_{n \to +\infty}(Y \circ X)_n = (Y \circ X)$ exists and is finite$\}$ and $B = \{s(Y \circ X) < +\infty\}$ are a.e. equivalents.

To prove the inclusion $B \subset A$ a.e., let us consider $\lambda > 0$ and define the stopping times:

$$\sigma = \inf\{n : |Y_{n+1}| > \lambda\},$$
$$v = \inf\{n : s_{n+1}(Y \circ X) > \lambda\},$$

and

$$\mu = \sigma \wedge v.$$

Consider the stopped martingales

$$(W \circ X) = (Y \circ X)^{\mu},$$

$$(W' \circ X) = (Y \circ X)^{\sigma},$$

and the stopping time,

$$\tau = \inf\{n : s_{n+1}(W' \circ X) > \lambda\}.$$

By definition

$$\mu \leqslant \sigma \wedge \tau,$$

and $\{W'_n\}_{n \geqslant 1}$ is uniformly bounded by λ. Applying now Corollary 4.31 to the martingale transform $\{(W' \circ X)_n\}_{n \geqslant 1}$, we have

$$||(W \circ X)^*||_2 \leqslant ||(Y \circ X)^{\sigma \wedge \tau})^*||_2 = ||((W' \circ X)^{\tau})^*||_2 \leqslant c\lambda.$$

Thus, by Jensen's inequality,

$$||(W \circ X)^*||_1 \leqslant c\lambda, \tag{4.125}$$

then, by martingale transform convergence theorem, Theorem 4.7, the martingale transform $\{(W \circ X)_n\}_{n \geqslant 1}$, converges a.e.

On the set $\{s(Y \circ X) < \lambda\}$, since X is a regular martingale, we have, for any $n \geqslant 1$

$$|Y_{n+1}| \leqslant \left(\sum_{k=1}^{\infty} Y_k^2 \right)^{1/2} = \left(\sum_{k=1}^{\infty} \mathbf{E}\left[(Y_k \Delta X_k)^2 \mid \mathscr{F}_{k-1} \right] \right)^{1/2} = s(Y \circ X) < \lambda,$$

and therefore

$$\sigma = +\infty, \quad \nu = +\infty, \quad \text{and therefore } \mu = +\infty.$$

Hence

$$(Y \circ X)_n = (W \circ X)_n \text{ on } \{s(Y \circ X) < \lambda\}.$$

Since $\{s(Y \circ X) < +\infty\} = \bigcup_{\lambda \in \mathbb{Q}} \{s(Y \circ X) < \lambda\}$, we have

$$(Y \circ X)_n = (W \circ X)_n \text{ on } \{s(Y \circ X) < +\infty\},$$

thus

$$\{(Y \circ X)_n\}_{n \geqslant 1} \text{ converges a.e.}$$

on the set $\{s(Y \circ X) < +\infty\}$. Therefore $B \subset A$ a.e.

Now let us consider the opposite inclusion $A \subset B$ a.e.

Taking $\beta \to +\infty$ in Proposition 4.15, we get

$$\mathbf{P}\{s(Y \circ X) = +\infty, \ \sup_n (Y \circ X)_n^+ \vee Y^* < \delta\lambda\} = 0$$

for all $\delta > 0$ and $\lambda > 0$, therefore

$$\mathbf{P}\{s(Y \circ X) = +\infty, \ \sup_n (Y \circ X)_n^+ \vee Y^* < +\infty\} = 0,$$

which implies

$$s(Y \circ X) < +\infty \quad \text{a.e.}$$

on the set $\{\sup_n (Y \circ X)_n^+ \vee Y^* < +\infty\}$. Then, by Lemma 4.11

$$Y^* < +\infty \quad \text{a.e.}$$

on the set $\{\Delta(Y \circ X)^* < +\infty\}$, and therefore

$$\{\sup_n (Y \circ X)_n^+ < +\infty\} \subset \{\Delta(Y \circ X)^* < +\infty\} \subset \{Y^* < +\infty\}.$$

Thus,

$$\{\sup_n(Y \circ X)_n^+ \vee Y^* < +\infty\} = \{\sup_n(Y \circ X)_n^+ < +\infty\}, \qquad (4.126)$$

so we have

$$E = \{\sup_n(Y \circ X)_n < +\infty\} \subset \{\sup_n(Y \circ X)_n^+ < +\infty\} \subset \{s(Y \circ X) < +\infty\} = B.$$

Therefore, as we know that trivially $A \subset E$, we have $A \subset B$ a.e..

Let us prove now that the set $C = \{S(Y \circ X) < +\infty\}$ and $D = \{(Y \circ X)^* < +\infty\}$ are a.e. equivalents. Let us consider first the inclusion $C \subset D$ a.e.

Taking $\beta \to +\infty$ in Corollary 4.37, we have

$$\mathbf{P}\{(Y \circ X)^* = +\infty, \ S(Y \circ X) \vee Y^* \leqslant \delta\lambda\} = 0$$

for all $\lambda > 0$, thus

$$\mathbf{P}\{(Y \circ X)^* = +\infty, \ S(Y \circ X) \vee Y^* \leqslant +\infty\} = 0.$$

That is,

$$(Y \circ X)^* < +\infty \text{ a.e.}$$

on the set $\{S(Y \circ X) \vee Y^* < +\infty\}$. Again, by Lemma 4.11, we have

$$\{S(Y \circ X) < +\infty\} \subset \{\Delta(Y \circ X)^* < +\infty\} \subset \{Y^* < +\infty\},$$

and thus

$$\{S(Y \circ X) \vee Y^* < +\infty\} = \{S(Y \circ X) < +\infty\},$$

therefore

$$(Y \circ X)^* < +\infty \text{ a.e.}$$

on the set $\{S(Y \circ X) < +\infty\}$, and this implies

$$C = \{S(Y \circ X) < +\infty\} \subset \{(Y \circ X)^* < +\infty\} = D.$$

Now, let us prove the opposite inclusion $D \subset C$ a.e. Taking $\beta \to +\infty$ in Corollary 4.37, we get

$$\mathbf{P}\{S(Y \circ X) = +\infty, \ (Y \circ X)^* \vee \Delta^*(Y \circ X) \vee Y^* \leqslant +\infty\} = 0,$$

thus

$$S(Y \circ X) < +\infty \text{ a.e.}$$

on the set $\{(Y \circ X)^* \vee \Delta(Y \circ X)^* \vee Y^* < +\infty\}$, so applying again Lemma 4.11, we have $\Delta(Y \circ X)^* \leqslant 2(Y \circ X)^*$ and $\{\Delta^*(Y \circ X) < +\infty\} \subset \{Y^* < +\infty\}$, thus

$$\{(Y \circ X)^* < +\infty\} \subset \{\Delta(Y \circ X)^* < +\infty\} \subset \{Y^* < +\infty\},$$

and therefore

$$\{(Y \circ X)^* \vee \Delta(Y \circ X)^* \vee Y^* < +\infty\} = \{(Y \circ X)^* < +\infty\}.$$

Then

$$S(Y \circ X) < +\infty \quad \text{a.e.}$$

on the set $\{(Y \circ X)^* < +\infty\}$, that is

$$D = \{(Y \circ X)^* < +\infty\} \subset \{S(Y \circ X) < +\infty\} = C \text{ a.e.} \tag{4.127}$$

Let us see now that the sets $D = \{(Y \circ X)^* = \sup_n |(Y \circ X)_n| < +\infty\}$ and $E = \{\sup_n (Y \circ X)_n < +\infty\}$ are a.e. equivalents. It is clear that $E \subset D$. Then we just need to prove the inclusion $D \subset E$.

Taking $\beta \to +\infty$ in Proposition 4.14, we get

$$\mathbf{P}\{(Y \circ X)^* = +\infty, \quad \sup_n (Y \circ X)_n^+ \vee Y^* < \delta\lambda\} = 0$$

for all $\lambda > 0$. This implies

$$\mathbf{P}\{(Y \circ X)^* = +\infty, \quad \sup_n (Y \circ X)_n^+ \vee Y^* < +\infty\} = 0,$$

and therefore, we have

$$(Y \circ X)^* < +\infty \quad \text{a.e.}$$

on the set $\{\sup_n (Y \circ X)_n^+ \vee Y^* < \infty\}$, but then, by (4.126)

$$(Y \circ X)^* < +\infty \quad \text{a.e.}$$

on the set $\{\sup_n (Y \circ X)_n^+ < +\infty\}$. Since

$$\{\sup_n (Y \circ X)_n < +\infty\} \subset \{\sup_n (Y \circ X)_n^+ < +\infty\}$$

we have

$$(Y \circ X)^* < +\infty \text{ a.e.}$$

on the set $\{\sup_n (Y \circ X)_n < +\infty\}$. Thus

$$E = \{\sup_n (Y \circ X)_n < +\infty\} \subset \{(Y \circ X)^* < +\infty\} = D \text{ a.e.}$$

Finally, let us see that the sets $D = \{(Y \circ X)^* < \infty\}$ and $B = \{s(Y \circ X) < \infty\}$ are a.e. equivalents. In order to prove the inclusion $D \subset B$ a.e. taking $\beta \to +\infty$ in Corollary 4.38, we have

$$\mathbf{P}\{s(Y \circ X) = +\infty, (Y \circ X)^* \vee Y^* \leqslant \delta\lambda\} = 0$$

for all $\lambda > 0$; therefore,

$$\mathbf{P}\{s(Y \circ X) = +\infty, (Y \circ X)^* \vee Y^* < +\infty\} = 0.$$

This implies

$$s(Y \circ X) < +\infty \text{ a.e.}$$

on the set $\{(Y \circ X)^* \vee Y^* < +\infty\}$, so, by Lemma 4.11

$$\{\Delta^*(Y \circ X) < +\infty\} \subset \{Y^* < +\infty\} \text{ and } \{(Y \circ X)^* < +\infty\} \subset \{\Delta^*(Y \circ X) < +\infty\},$$

and then,

$$\{(Y \circ X)^* \vee Y^* < +\infty\} = \{(Y \circ X)^* < +\infty\}.$$

Therefore

$$s(Y \circ X) < +\infty \text{ a.e.}$$

on the set $\{(Y \circ X)^* < +\infty\}$. Thus

$$D = \{(Y \circ X)^* < +\infty\} \subset \{s(Y \circ X) < +\infty\} = B \text{ a.e.}$$

Let us prove the other inclusion $B \subset D$ a.e. Let $\lambda > 0$, consider σ, ν and $\mu = \sigma \wedge \nu$ the stopping times, defined at the beginning of the proof, and let

$$(W \circ X) = (Y \circ X)^\mu$$

be the stopped martingale; by (4.125), we have

$$(W \circ X)^* < +\infty \text{ a.e.}$$

Since $\mu = +\infty$ on the set $\{s(Y \circ X) < +\infty\}$, this implies

$$(Y \circ X)_n = (W \circ X)_n \text{ on the set } \{s(Y \circ X) < +\infty\}$$

for all $n \geqslant 1$, thus, we have

$$\{s(Y \circ X) < +\infty\} \subset \{\mu = +\infty\} \subset \{(Y \circ X)^* = (W \circ X)^* < +\infty\}.$$

Therefore we obtain

$$B = \{s(Y \circ X) < +\infty\} \subset \{(Y \circ X)^* < +\infty\} = D, \text{ a.e.} \qquad \square$$

Observe that if the (A1) condition does not hold, one can obtain a sequence $\{Y_n\}_{n \geqslant 1}$ of independent random variables such that the series

$$\sum_{n=1}^{\infty} (Y_n)^2 = +\infty,$$

even though the series

$$\sum_{n=1}^{\infty} Y_k \Delta X_k < \infty$$

a.e., see [116].

4.9   Martingale Hardy Spaces

We have considered $\mathcal{M}$ the set of all $\{\mathcal{F}_n\}$-martingales defined on $(\Omega, \mathcal{F}, \mathbf{P})$, see Definition 3.2. Now, we will consider subsets of $\mathcal{M}$ with more mathematical structure, the martingale Hardy spaces. Martingale theory and the theory of the martingale Hardy spaces can excellently be applied in the theory of complex functions and the theory of the classical Hardy spaces, see the book of Durret [73] for the case of continuous parameter. The subject emerged from the work of C. Fefferman and E. Stein on classical Hardy spaces and functions of bounded mean oscillation, see [77], and the related work of D. L. Burkholder, B. Davis, R. Gundy, and Silverstein on martingale theory. In this section, we will only consider discrete martingale Hardy spaces.

Definition 4.14 The martingale Hardy spaces for $0 < p \leqslant \infty$; denote by H_p^s, H_p^S and H_p^* respectively are the spaces of $\{\mathcal{F}_n\}$-martingales defined as follows

$$H_p^s = \{X \in \mathcal{M} : \|X\|_{H_p^s} < \infty\}, \text{ where } \|X\|_{H_p^s} := \|s(X)\|_p, \qquad (4.128)$$

$$H_p^S = \{X \in \mathcal{M} : \|X\|_{H_p^S} < \infty\}, \text{ where } \|X\|_{H_p^S} := \|S(X)\|_p, \qquad (4.129)$$

and

$$H_p^* = \{X \in \mathcal{M} : \|X\|_{H_p^*} < \infty\}, \text{ where } \|X\|_{H_p^*} := \|X^*\|_p. \qquad (4.130)$$

When it does not lead to confusion we will substitute H_p for H_p^S. This terminology is due to the fact that, for an appropriate choice of $(\Omega, \mathcal{F}, \mathbf{P})$ these spaces can be identified with the classical H_p spaces (see [108]), i.e., the sets of functions $F(z)$ analytic in $|z| < 1$ and such that

$$\sup_{0 \leqslant r < 1} \int_{-\pi}^{\pi} \left| F\left(re^{i\theta}\right) \right|^p d\theta < \infty.$$

Now, let us also consider the BMO and Lipschitz spaces for measurable functions defined on the probability space $(\Omega, \mathscr{F}, \mathbf{P})$.

Definition 4.15 The space of bounded mean oscillation BMO_q, $1 \leqslant q < \infty$, is the space of functions $f \in L^q(\Omega)$ for which

$$\|f\|_{BMO_q} := \sup_{n \in \mathbb{N}} \left\| \left(\mathbf{E}[|f - \mathbf{E}[f|\mathscr{F}_{n-1}]|^q \, |\mathscr{F}_n] \right)^{1/q} \right\|_\infty < \infty.$$

and the space of bounded mean oscillation BMO_q^+, $1 \leqslant q < \infty$, is the space of functions $f \in L^q(\Omega)$ for which

$$\|f\|_{BMO_q^+} := \sup_{n \in \mathbb{N}} \left\| \left(\mathbf{E}[|f - \mathbf{E}[f|\mathscr{F}_n]|^q \, |\mathscr{F}_n] \right)^{1/q} \right\|_\infty < \infty.$$

Definition 4.16 The Lipschitz space $\Lambda_q(\alpha)$, $1 \leqslant q < \infty, \alpha \geqslant 0$, is defined as the space of space of functions $f \in L^q(\Omega)$ for which

$$\|f\|_{\Lambda_{\bar{q}}(\alpha)} := \sup_{n \in \mathbb{N}} \sup_{A \in \mathscr{F}_n} P(A)^{-1/q-\alpha} \left(\int_A |f - \mathbf{E}[f|\mathscr{F}_{n-1}]|^q \, dP \right)^{1/q} < \infty$$

and the Lipschitz space $\Lambda_q^+(\alpha)$, $1 \leqslant q < \infty, \alpha \geqslant, 0$ is defined as the space functions $f \in L^q(\Omega)$ for which

$$\|f\|_{\Lambda_{q+}(\alpha)} := \sup_{n \in \mathbb{N}} \sup_{A \in \mathscr{F}_n} P(A)^{-1/q-\alpha} \left(\int_A |f - \mathbf{E}[f|\mathscr{F}_n]|^q \, dP \right)^{1/q} < \infty.$$

If $\alpha = 0$, $\Lambda_q(0) = BMO_q$.

Observe that from Doob's L^p inequality, Corollary 3.9 inequality (3.31) implies that

$$H_p^* \sim L_p \quad \text{if} \quad p > 1.$$

Burkholder and Gundy [29, 30, 44] proved that the martingale Hardy spaces H_p^S and H_p^* are equivalent whenever $1 < p < \infty$. A few years later Davis [64] extended this result to the case $p = 1$, using Burkholder's and Davis' inequalities (4.46) and (4.48), since, they can be written as

$$c_p \|f\|_{H_p^S} \leqslant \|f\|_{H_p^*} \leqslant C_p \|f\|_{H_p^S}, \quad 1 \leqslant p < \infty.$$

Theorem 4.39 *For the martingale Hardy spaces H_p^s, H_p^S and H_p^* we have the following inclusion relations.*

(i) For X a $\{\mathscr{F}_n\}$-martingale

$$\|X\|_{H_p^s} \leqslant C_p \|X\|_{H_p^S}, \quad 2 \leqslant p < \infty, \tag{4.131}$$

and therefore

$$H_p^s \subset H_p^S, \quad 2 \leqslant p < \infty. \tag{4.132}$$

(ii) For X a $\{\mathscr{F}_n\}$-martingale

$$\|X\|_{H_p^*} \leqslant C_p \|X\|_{H_p^s}, \quad 0 < p \leqslant 2, \tag{4.133}$$

and therefore

$$H_p^* \subset H_p^s, \quad 0 < p \leqslant 2. \tag{4.134}$$

(iii) For X a $\{\mathscr{F}_n\}$-martingale

$$\|X\|_{H_p^S} \leqslant C_p \|X\|_{H_p^s}, \quad 0 < p \leqslant 2, \tag{4.135}$$

and therefore

$$H_p^S \subset H_p^s, \quad 0 < p \leqslant 2. \tag{4.136}$$

(iv) For X a $\{\mathscr{F}_n\}$-martingale

$$\|X\|_{H_p^s} \leqslant C_p \|X\|_{H_p^*}, \quad 2 \leqslant p < \infty, \tag{4.137}$$

and therefore

$$H_p^s \subset H_p^*, \quad 2 \leqslant p < \infty. \tag{4.138}$$

The proof of Theorem 4.39 can be obtained from Theorem 4.28, (i) and (ii) of Theorem 4.34, Doob's L^p inequality (3.31) and Burkholder's inequality (4.46).

F. Weiss in his book [173] gives alternative proofs of these inclusion results using the atomic decomposition method. For completeness, we will limit ourselves to introducing the major concepts, which provide the building blocks for the atomic decomposition argument. For more details see [173], Chap. 2.

Definition 4.17 A measurable function a is a (p, ∞) atom of the first, second, or third category (or simply $(1, p, \infty)$ atom, $(2, p, \infty)$ atom, and $(3, p, \infty)$ atom respectively) if there exists a stopping time v such that

$$a_n := \mathbf{E}[a|\mathscr{F}_n] = 0 \text{ if } v \geqslant n,$$

and the respective inequalities hold

$$\|s(a)\|_\infty \geqslant \mathbf{P}(\{v \neq \infty\})^{-1/p}$$

$$\|S(a)\|_\infty \geqslant \mathbf{P}(\{v \neq \infty\})^{-1/p}$$

$$\|a^*\|_\infty \geqslant \mathbf{P}(\{v \neq \infty\})^{-1/p}$$

Let us mention one atomic decomposition result.

Theorem 4.40 (Atomic decomposition) *If the $\{\mathcal{F}_n\}$-martingale $X = \{X_n\}_n$ is in H_p^s, $0 < p \leqslant \infty$ then there exists a sequence $\{a^k,\, k \in \mathbb{Z}\}$ of $(1, p, \infty)$ atoms and a sequence $\{\mu = \{\mu_k, k \in \mathbb{Z}\} \in l_p\}$ of real numbers such that for all $n \in \mathbb{N}$*

$$\sum_{k=-\infty}^{\infty} \mu_k \mathbf{E}[a^k | \mathcal{F}_n] = X_n$$

and

$$\left(\sum_{k=-\infty}^{\infty} |\mu_k|^p \right)^{1/p} \leqslant C_p \|X\|_{H_p^*}.$$

Also, $\sum_{k=-\infty}^{\infty} \mu_k a^k$ converges to X in H_p^s norm.

Conversely, if $0 < p \leqslant 1$ and the $\{\mathcal{F}_n\}$-martingale X has a decomposition of the above type then $X \in H_p^s$ and

$$\|X\|_{H_p^s} \sim \left(\sum_{k=-\infty}^{\infty} |\mu_k|^p \right)^{1/p},$$

where the infimum is taken over all such decompositions.

Finally, let us mention some results on duality. There is a close analogy between certain martingales and analytic functions. In particular, it is to be expected that any of the results about H_p which carry over to the Fefferman-Stein theory. The above results on martingale inequalities, as well as the atomic decomposition argument in general, are of much help in establishing duality relations between BMO and Hardy spaces.

However, the proof of some basic duality results does not require these techniques. An example is Fefferman's proof see [76] of the duality of H_1 and BMO, consisting of the statement that

$$|\mathbf{E}[f\phi]| \leqslant c\|f\|_{H_1} \|\phi\|_{BMO}, \, f \in H_1, \phi \in BMO$$

frequently referred to as *Fefferman's inequality*, and a theorem asserting that every functional $L(f)$ such that

$$|L(f)| \leqslant C\|f\|_{H_1}$$

is of the form $L(f) = \mathbf{E}[f\phi]$ with $\|\phi\|_{BMO} \leqslant cB$, where the expectation does not have to make sense in the sense of the Lebesgue integral, but can be defined as $\mathbf{E}[f\phi] = \lim_{n\to\infty} \mathbf{E}[f_n\phi_n]$.

Shortly after Fefferman's proved that the dual Banach space to the Hardy space H_1 was equivalent to the space BMO of functions of bounded mean oscillation, a martingale analogue was proved by Fefferman and Stein [77], Garsia [83], and Herz [96], independently. In this Richard Gundy played a role was of the highest value.

Theorem 4.41 *The dual space of H_1^s is BMO_2.*

As a corollary, it can be obtained an inequality due to Rosenthal [145] and Burkholder [30] in which the H_q^* norm is estimated by the sum of the H_q^s norm and the L_q norm of the supremum of the martingale differences.

Furthermore, in [97] Herz gave a description of the dual of H_p^s in case $0 < p < 1$, too, and considering a sequence of atomic σ-algebras he proved that its dual space is equivalent to $\Lambda_2(\alpha)$ ($\alpha = 1/p - 1$).

Theorem 4.42 *The dual space of H_p^s is $\Lambda_2(\alpha)(0 < p \leqslant 1, \alpha = 1/p - 1)$.*

Also with the atomic decomposition, it is proved that $\Lambda_1(\alpha)$ is equivalent to $\Lambda_2(\alpha)$ ($\alpha \geqslant 0$) in the regular case.

Observe, the spaces H_1^s and H_1^* are non-reflexive. It is interesting to ask whether it can be found a subspace of BMO, as in the classical case, the dual of which is one of the Hardy spaces.

Finally, we have, a fundamental duality result for the H_p^s spaces

Theorem 4.43 *If $1 < p < \infty$ then the dual of H_p^s is H_q^s where $1/p + 1/q = 1$.*

For more details on martingales Hardy spaces we refer to Weisz's book [173] or Garsia [82].

4.10   Appendix

Let us give the proof of Lemma 4.3 in full detail.

Proof

- If $p = 2$, $p^* = 2$, then, $u(x, y) = x \cdot y$, in both regions

$$A_1 = \left\{ (x, y) \in \mathbb{R}^2 : x > 0,\ 0 < y < x \right\}$$

$$A_2 = \left\{ (x, y) \in \mathbb{R}^2 : x > 0,\ -x < y < 0 \right\}$$

and therefore in the whole plane.

Thus, we have

(i) $v(x, y) = u(x, y) = x \cdot y$ in the whole plane.

(ii) If $h \cdot k \leqslant 0$

$$u(x + h, y + k) = (x + h) \cdot (y + k) = xy + xk + yh + h \cdot k \leqslant xy + xk + yh$$
$$= u(x, y) + u_x(x, y)h + u_y(x, y)k.$$

(iii) $u(x, y) = x \cdot y \leqslant 0$, if $x \cdot y \leqslant 0$.

- If $p > 2$ then as $\frac{1}{p} + \frac{1}{q} = 1$ and therefore $p = \frac{q}{q-1} > 2$ then $q > 2q - 2$ and $q < 2$, thus $p^* = p$. Hence,

$$
A_1 = \left\{ (x, y) \in \mathbb{R}^2 : x > 0, \ (1 - 2/p)\, x < y < x \right\}
$$
$$
A_2 = \left\{ (x, y) \in \mathbb{R}^2 : x > 0, \ -x < y < (1 - 2/p)\, x \right\}
$$

and

$$
u(x, y) = \begin{cases} \alpha_p x^p \left[1 - \frac{p(x-y)}{2x} \right] & \text{if } (x, y) \in A_1 \\ \left(\frac{x+y}{2} \right)^p - (p-1)^p \left(\frac{x-y}{2} \right)^p & \text{if } (x, y) \in A_2. \end{cases}
$$

Observe that in A_2, $x > y$ and that is why we have eliminated the absolute value in the definition of the function v. Thus u is infinitely differentiable out of the lines

$$
x = y, \ x = -y, \ y = (1 - 2/p)x, \ x = (1 - 2/p)y.
$$

Let $x > 0$ be fixed,

$$
u_x(x, y) = \begin{cases} p\alpha_p x^{p-1} \left[1 - \frac{p(x-y)}{2x} \right] - \alpha_p x^p \left(\frac{p}{2} \frac{y}{x^2} \right) & \text{if } (x, y) \in A_1 \\ \frac{p}{2} \left(\frac{x+y}{2} \right)^{p-1} - (p-1)^p \frac{p}{2} \left(\frac{x-y}{2} \right)^{p-1} & \text{if } (x, y) \in A_2. \end{cases}
$$

Evaluating u_x at $\left(x^-, \left(1 - \frac{2}{p} \right) x \right) \in A_1$, we get

$$
u_x \left(x^-, \left(1 - \frac{2}{p} \right) x \right) = \alpha_p x^{p-1} \left(1 - \frac{p}{2} \right).
$$

Analogously, evaluating u_x at $\left(x^+, \left(1 - \frac{2}{p} \right) x \right) \in A_2$ we get

$$
u_x \left(x^+, \left(1 - \frac{2}{p} \right) x \right) = \alpha_p x^{p-1} \left(1 - \frac{p}{2} \right)
$$

thus

$$
u_x \left(x^-, \left(1 - \frac{2}{p} \right) x \right) = u_x \left(x^+, \left(1 - \frac{2}{p} \right) x \right) = \alpha_p x^{p-1} \left(1 - \frac{p}{2} \right),
$$

therefore u_x exists and it is continuous at those points. Analogously, it can be proved u_x exists and it is continuous at (x, x), $(x, -x)$ and $\left(x, \left(1 - \frac{2}{p} \right)^{-1} x \right)$. Also

$$
u_y(x, y) = \begin{cases} \frac{p}{2} \alpha_p x^{p-1} & \text{if } (x, y) \in A_1 \\ \frac{p}{2} \left(\frac{x+y}{2} \right)^{p-1} + (p-1)^p \frac{p}{2} \left(\frac{x-y}{2} \right)^{p-1} & \text{if } (x, y) \in A_2. \end{cases}
$$

Again, evaluating u_y at $\left(x^-, \left(1 - \frac{2}{p}\right)x\right) \in A_1$, we get

$$u_y\left(x^-, \left(1 - \frac{2}{p}\right)x\right) = \frac{p}{2}\alpha_p x^{p-1}$$

and evaluating u_y at $\left(x^+, \left(1 - \frac{2}{p}\right)x\right) \in A_2$ we get

$$u_y\left(x^+, \left(1 - \frac{2}{p}\right)x\right) = \frac{p}{2}\left[\left(\frac{p-1}{p}\right)^{p-1} x^{p-1} + (p-1)^p \left(\frac{x}{p}\right)^{p-1}\right] = \frac{p}{2}\alpha_p x^{p-1}$$

thus

$$u_y\left(x^-, \left(1 - \frac{2}{p}\right)x\right) = u_y\left(x^+, \left(1 - \frac{2}{p}\right)x\right) = \frac{p}{2}\alpha_p x^{p-1}.$$

Therefore u_x and u_y exist and are continuous at those points. By the symmetry conditions u_x and u_y exist and are continuous on the whole $\mathbb{R}^2$.
Additionally, out of those lines, the second partial derivative u_{xx} satisfies

$$u_{xx} \leqslant 0, \ u_{yy} \leqslant 0 \ \text{and} \ u_{xy} \geqslant 0.$$

In fact,

$$u_{xx}(x, y) = \begin{cases} p(p-1)\alpha_p x^{p-2}[1 - \frac{p(x-y)}{2x}] - p^2\alpha_p x^{p-1}\frac{y}{x^2} + p\alpha_p x^p \frac{y}{x^3} & \text{if } (x, y) \in A_1 \\ \frac{p}{4}(p-1)\left(\frac{x+y}{2}\right)^{p-2} - (p-1)^{p+1}\frac{p}{4}\left(\frac{x-y}{2}\right)^{p-2} & \text{if } (x, y) \in A_2 \end{cases}$$

$$= \begin{cases} p(p-1)\alpha_p x^{p-2}[1 - \frac{p(x-y)}{2x} - \frac{y}{x}] & \text{if } (x, y) \in A_1 \\ \frac{p}{4}(p-1)[\left(\frac{x+y}{2}\right)^{p-2} - (p-1)^p \left(\frac{x-y}{2}\right)^{p-2}] & \text{if } (x, y) \in A_2 \end{cases}$$

$$= \begin{cases} \frac{p}{2}(p-1)(p-2)\alpha_p x^{p-2}[\frac{y}{x} - 1] & \text{if } (x, y) \in A_1 \\ \frac{x^{p-2}(p-1)^{p-2}p(2-p)}{p^{p-2}} & \text{if } (x, y) \in A_2 \end{cases}$$

as

$$\left[\left(\frac{x+y}{2}\right)^{p-2} - (p-1)^p \left(\frac{x-y}{2}\right)^{p-2}\right] < \frac{x^{p-2}(p-1)^{p-2}p(2-p)}{p^{p-2}} \tag{4.139}$$

on A_2. Hence, as $y < x$ on A_1 and $p > 2$, we get $u_{xx}(x, y) \leqslant 0$. Also

$$u_{yy}(x, y) = \begin{cases} 0 & (x, y) \in A_1 \\ \frac{p}{4}(p-1)\left[\left(\frac{x+y}{2}\right)^{p-2} - (p-1)^p \left(\frac{x-y}{2}\right)^{p-2}\right] & (x, y) \in A_2 \end{cases}$$

and therefore, by (4.139), $u_{yy}(x, y) \leqslant 0$. Finally,

$$u_{xy}(x, y) = \begin{cases} \frac{p}{2}(p-1)\alpha_p x^{p-2} & (x, y) \in A_1 \\ \frac{p}{4}(p-1)\left[\left(\frac{x+y}{2}\right)^{p-2} + (p-1)^p \left(\frac{x-y}{2}\right)^{p-2}\right] & (x, y) \in A_2 \end{cases}.$$

Thus, clearly, $u_{xy}(x, y) \geqslant 0$.

Let $h \cdot k \leqslant 0$ with $h + k \neq 0$ and define $g : \mathbb{R} \to \mathbb{R}$ such that

$$g(t) = u(x + ht, y + kt).$$

If (h, k) and (x, y) are fixed, g takes values only on the line $(x, y) + (h, k)t$. Moreover, if $h + k = 0$, the line $y = -x$ would be on the domain of g and therefore of u, but this cannot happen. Thus, except for a finite number of values of t the function g is twice continuously differentiable on a neighborhood of $(x + ht, y + kt)$, since u is a C^∞ function. Hence by the chain rule, we get, the inequalities above

$$g''(t) = u_{xx}(x + ht, y + kt)h^2 + u_{yy}(x + ht, y + kt)k^2 + 2u_{xy}(x + ht, y + kt)h \cdot k \leqslant 0.$$

By continuity, g' is decreasing on $\mathbb{R}$, and g is a concave function, and then

$$g\left(\lambda t_0 + (1 - \lambda)t\right) \geqslant \lambda g\left(t_0\right) + (1 - \lambda)g(t), \quad \lambda \in [0, 1];$$

that can be rewritten as

$$g(t) \leqslant g\left(\lambda t_0 + (1 - \lambda)t\right) - \lambda g\left(t_0\right) + \lambda g(t).$$

Now, if $t_0 = 0$, $t = 1$ and $\lambda = 1$, then as $g(1) = g(0) + g(1) - g(0)$, by the mean value theorem

$$g(1) = g(0) + g'(\xi)$$

for $\xi \in (0, 1)$.

Given that g' is decreasing $g(1) \leqslant g(0) + g'(0)$ then

$$u(x + h, y + k) \leqslant u(x, y) + u_x(x, y)h + u_y(x, y)k.$$

Thus, we have obtained (4.56).

On the other hand, observe that $A_2 \subset \{(x, y) : v(x, y) \leqslant 0\}$ since

$$-x < y < \left(1 - \frac{2}{p}\right)x,$$

which implies

$$0 < \left(\frac{x + y}{2}\right)^p < x^p \left(1 - \frac{1}{p}\right)^p$$

and

$$-(p - 1)^p x^p < -(p - 1)^p \left(\frac{x - y}{2}\right)^p < -(p - 1)^p \frac{x^p}{p^p}.$$

Adding the last two inequalities, we get

$$-(p-1)^p x^p < \left(\frac{x+y}{2}\right)^p - (p-1)^p \left(\frac{x-y}{2}\right)^p < x^p \left(1-\frac{1}{p}\right)^p - (p-1)^p \frac{x^p}{p^p}$$

but $x^p \left(1-\frac{1}{p}\right)^p - (p-1)^p \frac{x^p}{p^p} = x^p \left[\left(\frac{p-1}{p}\right)^p - \frac{(p-1)^p}{p^p}\right] = 0$.

Hence, we conclude $v(x,y) \leqslant 0$ on A_2, and by hypothesis $u = v$ on A_2. By the symmetry conditions of u we conclude that $\{(x,y) : v(x,y) \leqslant 0\}$ contains the second and fourth quadrants.

Therefore, in order to see that $v \leqslant u$, it is enough to prove that $v(x,y) \leqslant u(x,y)$ on A_1. Let $x > 0$ fixed, and let $a = \left(1-\frac{2}{p}\right)x, b = x$, since $p > 2$ then $a < b$. Define $f : \mathbb{R} \to \mathbb{R}$, $f(y) = u(x,y) - v(x,y)$, we want to prove that $f(y) \geqslant 0$ on $[a,b]$. Let us observe that:

(i) $f'(y) = u_y(x,y) - v_y(x,y)$, both partial derivatives are continuous on $[a,b]$, then $f'(y)$ is continuous on $[a,b]$.
(ii) $f''(y) = u_{yy}(x,y) - v_{yy}(x,y)$, since $u_{yy}(x,y) = 0$, $f''(y) = -v_{yy}(x,y)$ for any $y \in (a,b)$.
(iii) Let us see that $f(a) = 0$.

$$f(a) = f\left(\left(1-\frac{2}{p}\right)x\right) = u\left(x, \left(1-\frac{2}{p}\right)x\right) - v\left(x, \left(1-\frac{2}{p}\right)x\right)$$

$$= \alpha_p x^p \left(1 - \frac{p\left(x-\left(1-\frac{2}{p}\right)x\right)}{2x}\right) - \left(\frac{x+\left(1-\frac{2}{p}\right)x}{2}\right)^p$$

$$+ (p-1)^p \left(\frac{x-\left(1-\frac{2}{p}\right)x}{2}\right)^p$$

$$= \alpha_p x^p \left(1 - \frac{p\left(\frac{2x}{p}\right)}{2x}\right) - \left(1-\frac{1}{p}\right)^p x^p + (p-1)^p \left(\frac{x}{p}\right)^p.$$

$$= \alpha_p x^p (1-1) - \left(\frac{p-1}{p}\right)^p x^p + \left(\frac{p-1}{p}\right)^p x^p = 0.$$

Moreover, $f'(a) = 0$, since

$$f'(a) = f'\left(\left(1-\frac{2}{p}\right)x\right) = u_y\left(x, \left(1-\frac{2}{p}\right)x\right) - v_y\left(x, \left(1-\frac{2}{p}\right)x\right)$$

$$= \frac{p}{2}\alpha_p x^{p-1} - \frac{p}{2}\left(1-\frac{1}{p}\right)^{p-1} x^{p-1} - (p-1)^p \frac{p}{2} \left(\frac{x}{p}\right)^{p-1}$$

$$= \frac{p}{2}\left(\frac{p-1}{p}\right)^{p-1} x^{p-1} \left[p - 1 - p\left(\frac{p-1}{p}\right)\right] = 0.$$

(iv) Let us see that $f(b) \geqslant 0$.

$$f(b) = f(x) = u(x, x) - v(x, x)$$
$$= \alpha_p x^p \left(1 - \frac{p(x-x)}{2x}\right) - \left(\frac{x+x}{2}\right)^p + (p-1)^p \left(\frac{x-x}{2}\right)^p$$
$$= \alpha_p x^p - x^p = x^p \left(\alpha_p - 1\right).$$

Observe that if $p > 2$ then $p - 1 > 1$ and therefore $\frac{p-1}{p} > \frac{1}{p}$ which implies $\left(\frac{p-1}{p}\right)^{p-1} > \frac{1}{p}$ which in turn is equivalent to $p \left(\frac{p-1}{p}\right)^{p-1} > 1$ but as $\alpha_p = p \left(\frac{p-1}{p}\right)^{p-1}$, then we have that $f(b) \geqslant 0$.

Now, due to the fact that $f''(a) = -v_{yy}\left(x, \left(1 - \frac{2}{p}\right)x\right) \geqslant 0$ and $f''(b) = -v_{yy}(x, x) \leqslant 0$, then there exists $c \in (a, b)$, such that $f'(y) \geqslant 0 = f'(a)$ if $y \in (a, c]$ and $f'(y) \leqslant 0$ if $y \in [c, b)$. Hence, $f'(y) \geqslant 0$ if $y \in [a, c]$, which implies that f is non-negative on $[a, c]$. Since f is concave on the interval $[c, b]$, f is non-negative on $[c, b]$. Thus f is non-negative on A_1 and then $v \leqslant u$ on A_1, therefore $v \leqslant u$ on $\mathbb{R}^2$. This ends the proof for the case $p > 2$.

- The proof of the case $p < 2$ is essentially analogous. $\qquad\qquad\square$

4.11 Exercises

(1) Given $X = \{X_n\}$ a $\{\mathscr{F}_n\}$-martingale, where $\{\Delta X_n\}$ are its martingale differences, and $p \in (1, 2]$ then

$$\mathbf{E}\left[\sup_{n \geqslant 1} |X_n|^p\right] = \mathbf{E}\left[\sup_{n \geqslant 1} \left|\sum_1^n \Delta X_j\right|^p\right] \leqslant A_p \sum_1^\infty \mathbf{E}\,|\Delta X_n|^p$$

for some finite constant A_p.

(2) Suppose that a gambler with initial capital $X_0 \in (0, 1)$ wins or loses (at each play) whatever amount he stakes with probabilities p and q respectively, where $0 < p \leqslant \frac{1}{2}$, the successive plays being independent. What is the best gambling scheme for boosting his capital to one dollar? Prove that a bold strategy is optimal, namely, whenever his capital $X_n \leqslant \frac{1}{2}$, the entire capital should be staked, whereas whenever $X_n > \frac{1}{2}$, only the amount $1 - X_n$ (needed to obtain his goal) should be bet.

(3) Paley's Theorem, Theorem 1.7, establish that the Walsh system $\{\psi_k^{(n)}\}$ is complete on $L^2([0, 1), m)$. Moreover,

(i) Let $f \in L^1([0, 1), m)$ be given, and set

$$X_n(f) = \sum_{m=1}^{n} \sum_{k=1}^{2^{m-1}} c_k^{(m)} \psi_k^{(m)}(t) \quad \text{for } n \in \mathbb{N}.$$

Show that $X_n(f) = \mathbf{E}[f \mid \mathscr{F}_n]$, where $\mathscr{F}_n = \sigma\left(\left\{[\frac{k}{2^n}, \frac{k+1}{2^n}] : 0 \leqslant k \leqslant 2^{n-1}\right\}\right)$, and therefore that $\{X_n(f)\}$ is a $\{\mathscr{F}_n\}$-martingale. In particular, $\{X_n(f)\}$ converges to f both a.e., as well as in $L^1([0, 1), m)$.

(ii) Show that for each $1 \leqslant p \leqslant \infty$ there exists constants c_p, C_p such that

$$c_p \|f\|_p \leqslant \left[\int_{[0,1)} \left(\sum_{n=1}^{\infty} \left[\sum_{k=1}^{2^{n-1}} \left(\int_{[0,1)} f(s)\psi_k^{(n)}(s)ds\right) \psi_k^{(n)}(t)\right]^2\right)^{\frac{p}{2}} dt\right]^{\frac{1}{p}} \leqslant C_p \|f\|_p$$

$$(4.140)$$

for any $f \in L^1([0, 1), m)$.

(iii) Additionally, R. Gundy in [86] observed that for any numerical sequence $\left\{c_k^{(n)}\right\}$, the dyadic partial sums

$$\left\{W_{2^n} = \sum_{m=0}^{2^n} \sum_{k=1}^{2^{m-1}} c_k^{(m)} \psi_k^{(m)}(t)\right\}_n \qquad (4.141)$$

is a martingale, prove Gundy's observation in detail. Besides, Gundy observed that (4.141) can be considered as martingale transform of the martingale $\left\{\sum_{k=0}^{n} a_k r_k(t)\right\}_n$, the partial sums of Rademacher series $\sum_{k=0}^{\infty} a_k r_k(t)$. Prove that for $1 < p < \infty$ there exists a constant C_p such that

$$\left[\int_{[0,1)} \left(\sum_{n=1}^{\infty} \sum_{k=1}^{2^{n-1}} \left(\int_{[0,1)} f(s)\psi_k^{(n)}(s)ds\right) \psi_k^{(n)}(t)\right)^p dt\right]^{\frac{1}{p}} \leqslant C_p \|f\|_p$$

$$(4.142)$$

for any $f \in L^1([0, 1), m)$.

(4) *The switching principle.* Suppose $\{X_n^1\}$ and $\{X_n^2\}$ are $\{\mathscr{F}_n\}$-supermartingales, and τ is a stopping time so that $X_\tau^1 \geqslant X_\tau^2$. Then

$$Y_n = X_n^1 \chi_{\{\tau > n\}} + X_n^2 \chi_{\{\tau \leqslant n\}} \text{ is a supermartingale.}$$

$$Z_n = X_n^1 \chi_{\{\tau \geqslant n\}} + X_n^2 \chi_{\{\tau < n\}} \text{ is a supermartingale.}$$

(5) *Dubins' inequality.* For every positive $\{\mathscr{F}_n\}$-supermartingale $\{X_n\}$, the number of upcrossings U of $[a, b]$ satisfies

$$\mathbf{P}\{U \geqslant k\} \leqslant \left(\frac{a}{b}\right)^k \mathbf{E}[\min(X_0/a, 1)].$$

To prove this, we let $\tau_0 = -1$ and for $j \geqslant 1$ let

$$\tau_{2j-1} = \inf\left\{ n > \tau_{2j-2} : X_n \leqslant a \right\},$$
$$\tau_{2j} = \inf\left\{ \tau_{2j-1} : X_n \geqslant b \right\}.$$

Let $Y_n = 1$ for $0 \leqslant n < \tau_1$ and for $j \geq 1$

$$Y_n = \begin{cases} (b/a)^{j-1}\,(X_n/a) & \text{for } \tau_{2j-1} \leqslant n < \tau_{2j}, \\ (b/a)^j & \text{for } \tau_{2j} \leqslant n < \tau_{2j+1}. \end{cases}$$

 (i) Use the switching principle of the previous exercise and induction to show that $Z_n^j = Y_{n \wedge \tau_j}$ is a supermartingale.
(ii) Use $\mathbf{E}[Y_{n \wedge \tau_{2k}}] \leqslant \mathbf{E}[Y_0]$ and let $n \to \infty$ to get *Dubins' inequality*.

(6) (Blackwell and Dubins) Let $\{X_n\}$ be random variables on $(\Omega, \mathscr{F}, \mathbf{P})$ such that $\mathbf{E}\left[\sup_{n \geq 1} |X_n|\right] < \infty$ and $X_n \to X$ almost everywhere. Let $\{\mathscr{G}_n\}$ be sub- σ-algebras which are either non-increasing or non-decreasing. Show that in either case there exists σ-algebra $\mathscr{G}$ (identify $\mathscr{G}$ in the two cases) such that

$$\lim_{\substack{j \to \infty \\ n \to \infty}} \mathbf{E}\left[X_n \mid \mathscr{G}_j\right] = \mathbf{E}[X \mid \mathscr{G}] \text{ a.e.}$$

(7) Given the remainder formula in the discrete Itô's formula (DIF), (4.36)

$$R_n(X, f(X)) = \sum_{k=1}^{n} \int_{X_{k-1}}^{X_k} \left[f(x) - \frac{f(X_{k-1}) + f(X_k)}{2} \right] dx$$
$$= -\frac{1}{12} \sum_{k=1}^{n} f''(\eta_k)(\Delta X_k)^3 .$$

Find conditions on the martingale $\{X_n\}_n$ such that $R_n(X, f(X)) \to 0$ in probability, see (4.37).

Applications of Martingale Theory

5

Martingale theory is a very important branch of probability. As we have seen in previous chapters it is at the root of several concepts both in probability theory and in analysis. Thus, it is not by chance that there is a large number of applications in both areas. In this chapter, without the pretension of being exhaustive, we are going to discuss some of the applications of martingale theory to both areas.

5.1 Applications to Probabilities

5.1.1 Sum of Independent Random Variables

As we have mentioned before, one of the basic examples of martingales is the sum of centered independent random variables. Thus, we can say that martingale theory is a generalization of the theory of sums of centered independent random variables. Thus, it is not a surprise that various results of the theory of sums of independent centered random variables can be obtained from the results of the martingale theory. For instance, *Kolmogorov's inequality*:

Theorem 5.1 (Kolmogorov) *Let $\{\xi_k\}$ be a sequence of independent random variables such that for any $k \geqslant 0 \, \mathbf{E}[\xi_k] = 0$ and $\mathbf{E}[\xi_k] < \infty$, and let $S_n = \sum_{k=1}^{n} \xi_k$, $n \geqslant 1$; then for each $\lambda > 0$*

$$\mathbf{P}\left(\left\{\max_{1 \leqslant k \leqslant n} |S_k| \geqslant \lambda\right\}\right) \leqslant \frac{1}{\lambda^2} \mathbf{E}\left[(S_n)^2\right] = \frac{1}{\lambda^2} \sum_{k=1}^{n} \mathbf{E}\left[\xi_k^2\right].$$

W. Urbina-Romero and R. Rios, *An Introduction to the Modern Martingale Theory and Applications*, Texts in Applied Mathematics 81,
https://doi.org/10.1007/978-3-031-88903-5_5

Proof Since $\{S_n\}$ is a $\{\mathscr{F}_n^\xi\}$-martingale where $\mathscr{F}_n^\xi = \sigma(\xi_1, \xi_2, \ldots, \xi_n)$, then $\{S_n^2\}$ is a $\{\mathscr{F}_n^\xi\}$-submartingale, hence this an immediate consequence Doob's L^p inequality (3.30), Corollary 3.9, for $p = 2$. $\qquad\square$

Also, as already mentioned, one of the direct motivations of Burkholder's and Davis inequalities, (4.45) and (4.48), was the Marcinkiewicz-Zygmund inequality (4.39). But now, thanks to the generalizations of Theorems 4.17 and 4.18 made in Theorem 4.23, we are able to obtain a more general inequality for independent random variables.

Theorem 5.2 *Let $\xi_1, \xi_2, \ldots$ be a sequence of independent random variables with $\mathbf{E}[\xi_k] = 0$ and Φ a convex Young's function such that $\mathbf{E}[\Phi(\xi_k)] < \infty$, for all $k \geqslant 1$, then there exist universal constants c and C such that*

$$c\mathbf{E}\left[\Phi\left(\left(\sum_{k=1}^{n}\xi_k^2\right)^{1/2}\right)\right] \leqslant \mathbf{E}\left(\Phi\left(\max_{1\leqslant k\leqslant n}|S_k|\right)\right) \leqslant C\mathbf{E}\left[\Phi\left(\left(\sum_{k=1}^{n}\xi_k^2\right)^{1/2}\right)\right] \qquad (5.1)$$

for $n \geqslant 1$, where $S_k = \xi_1 + \cdots + \xi_k$, $k \geqslant 1$. The choice of c and C depends only on Φ.

On the other hand, let us consider now applications of the martingale theory to other results on independent random variables. First, notice that the result obtained in Theorem 1.2 is a particular case of the following result.

Theorem 5.3 (0–1 Law) *Let $\{\xi_n\}$ be a sequence of independent random variables and let ξ be a measurable random variable with respect to the σ-algebra $\mathscr{G}_\infty = \bigcap_{n=1}^{\infty} \sigma(\xi_n, \xi_{n+1}, \ldots)$ then ξ is constant with probability one.*

Proof First, let us assume that $\mathbf{E}[|\xi|] < \infty$ then for each n, ξ is independent of the variables $\xi_1, \ldots \xi_n$, and therefore

$$\mathbf{E}[\xi \mid \xi_1, \ldots \xi_n] = \mathbf{E}[\xi] \text{ a.e.}$$

By the martingale convergence theorem, Theorem 3.14, we get

$$\mathbf{E}[\xi \mid \xi_1, \ldots \xi_n] \to \mathbf{E}[\xi \mid \xi_\infty] = \xi,$$

thus

$$\xi = \mathbf{E}[\xi] \text{ a.e.}$$

Now, if $\mathbf{E}[|\xi|] = \infty$. Let $\zeta_n = \xi \chi_{\{|\xi|\leqslant n\}}$ then, by the previous argument, we have that $\zeta_n = \mathbf{E}[\zeta_n]$ a.e. for any n which is impossible unless ξ is constant with probability 1. $\qquad\square$

Remember that if $\{\xi_n\}$ is a sequence of independent centered random variables (in other words $\mathbf{E}[\xi_n] = 0$) then the sequence of partial sums $\{S_n\}$, where $S_n = \sum_{k=1}^{n} \xi_k$, is a martingale and therefore $\{|S_n|\}$ is a submartingale and then $\{\mathbf{E}[|S_n|]\}_n$ is a non-decreasing sequence. Then, by the martingale convergence theorem, Theorem 3.14, we get that $\sum_{k=1}^{n} \xi_n$ converges with probability one if $\lim_{n\to\infty} \mathbf{E}[|S_n|] < \infty$.

Martingale theory allows us to prove the general case of the law of large numbers for sequences of independent random variables, which implies the result that was discussed for the particular case of Rademacher functions in Theorem 1.4.

Theorem 5.4 (Strong law of the large numbers) *Let $\{\xi_n\}$ be a sequence of independent random variables, identically distributed and such that $\mathbf{E}\,|\xi_1| < \infty$. Then,*

$$\lim_{n\to\infty} \frac{1}{n} \sum_{k=1}^{n} \xi_k = \mathbf{E}[\xi_1] \ \ a.e. \ and \ in \ L^1.$$

Proof Let $\zeta_n = \sum_{k=1}^{n} \xi_k$ and for each $n \geqslant 1$ let $\mathscr{F}_{-n} = \sigma\left(\{\xi_n, \xi_{n+1}, \ldots\}\right)$ the reverse filtration

$$\mathscr{F}_{-1} \supset \mathscr{F}_{-2} \supset \cdots \supset \mathscr{F}_{-n} \supset \cdots \supset \mathscr{F}_{-\infty},$$

where $\mathscr{F}_{-\infty} = \bigcap_n \mathscr{F}_{-n}$.

Define $X_{-n} = \mathbf{E}\left[\xi_1 \mid \xi_n, \xi_{n+1,\ldots}\right] = \mathbf{E}\left[\xi_1 \mid \mathscr{F}_{-n}\right]$, then $\{X_{-n}\}$ is a reverse $\{\mathscr{F}_{-n}\}$-martingale and therefore by the theorem of convergence for reverse martingales, Theorems 3.23 and 3.24,

$$\lim_{n\to\infty} X_{-n} = X_{-\infty} \ \text{a.e. and in } L^1.$$

On the other hand, it is clear that

$$X_{-n} = \mathbf{E}\left[\xi_1 \mid \mathscr{F}_n\right] = \mathbf{E}\left[\xi_1 \mid \xi_n, \xi_{n+1}, \xi_{n+2} \cdots\right] = \mathbf{E}\left[\xi_1 \mid \xi_n, \xi_{n+1}, \xi_{n+2}, \ldots\right].$$

Now, as $\sigma(\xi_1, \ldots, \xi_n)$ and $\sigma(\xi_{n+1}, \xi_{n+2}\cdots)$ are independent and $\{\xi_1, \xi_n\} \subset \sigma(\xi_1, \ldots, \xi_n)$ then $\{\xi_1, \xi_n\}$ and $\{\xi_{n+1}, \xi_{n+2}\cdots\}$ are independent, hence $\mathbf{E}\left[\xi_1 \mid \xi_n, \xi_{n+1,\ldots}\right] = \mathbf{E}\left[\xi_1 \mid \xi_n\right]$ a.e. and therefore

$$\lim_{n\to\infty} \mathbf{E}\left[\xi_1 \mid \xi_1 + \cdots + \xi_n\right] = \lim_{n\to\infty} \mathbf{E}\left[\xi_1 \mid \xi_n\right] = X_{-\infty} \ \text{a.e. and in } L^1.$$

Finally, by symmetry, for $j \leqslant n$

$$\mathbf{E}\left[\xi_1 \mid \xi_1 + \cdots + \xi_n\right] = \mathbf{E}\left[\xi_j \mid \xi_1 + \cdots + \xi_n\right]$$

$$= \frac{1}{n} \sum_{i=1}^{n} \mathbf{E}\left[\xi_i \mid \xi_1 + \cdots + \xi_n\right]$$

$$= \frac{1}{n} \mathbf{E}\left[\sum_{i=1}^{n} \xi_i \,\middle|\, \xi_1 + \cdots + \xi_n\right] = \frac{1}{n} \sum_{j=1}^{n} \xi_j.$$

Now, we need to identify the limit. It is clear that $X_{-\infty}$ belongs to the tail σ-algebra (it is not affected by any change of a finite number of terms) and therefore by the 0–1 law, Theorem 5.3,

$$X_{-\infty} = \mathbf{E}\left[X_{-\infty}\right] \text{ a.e. },$$

i.e., $X_{-\infty}$ is constant. As $\{X_{-\infty}, \ldots, X_{-n}, \ldots X_{-2}, X_{-1}, \xi_1\}$ is a martingale, then

$$\lim_{n\to\infty} \frac{1}{n} \sum_{j=1}^{n} \xi_j = \mathbf{E}\left[X_{-\infty}\right] = \mathbf{E}\left[\xi_1\right].$$

$\square$

For the general statement for martingales, see Theorem 4.11, from which this result follows immediately.

An important consequence of Doob's optional stopping theorem, Theorem 3.1, is *Wald's identities* [167,168], developed in the context of sequential analysis.

Theorem 5.5 (Wald's first identity) *Let $\{\xi_k\}_{k\geq 1}$ be a sequence of identically distributed real-valued random variables. Let $S_n := \sum_{k=1}^{n} \xi_k$, $\mathscr{F}_n := \sigma\left(\xi_1, \ldots, \xi_n\right)$ for each $n \geqslant 0$. Let τ be a $\{\mathscr{F}_n\}$-stopping time. If $\mathbf{E}\left[|\xi_1|\right] < \infty$ and $\mathbf{E}[\tau] < \infty$, then*

$$\mathbf{E}\left[|S_\tau|\right] < \infty,$$

and

$$\mathbf{E}\left[S_\tau\right] = \mathbf{E}[\tau]\mathbf{E}\left[\xi_1\right]. \tag{5.2}$$

Proof First assume ξ_i are non-negative. For each n, $\tau \wedge n$ is a bounded stop time, and $\{S_n - n\mathbf{E}[\xi_1]\}$ is a $\{\mathscr{F}_n\}$-martingale.

So, by Doob's optional stopping, Theorem 3.1,

$$\mathbf{E}\left[S_{\tau \wedge n}\right] = \mathbf{E}[\tau \wedge n]\mathbf{E}\left[\xi_1\right]. \tag{5.3}$$

Since $\tau \wedge n \uparrow \tau$ and $S_{\tau \wedge n} \uparrow S_\tau$, by monotone convergence theorem, applied to (5.3), we get (5.2). Now, $\mathbf{E}[\tau] < \infty$ implies that $\mathbf{E}\left[S_\tau\right] < \infty$. This proves the result for non-negative random variables.

Now for the general case $\mathbf{E}\left[|\xi_1|\right] < \infty$, the above arguments imply that $\mathbf{E}\left[\sum_{k=1}^{\tau} |\xi_k|\right] < \infty$. Also, $|S_{\tau \wedge n}| \leq \sum_{k=1}^{\tau} |X_k|$ and $S_{\tau \wedge n} \to S_\tau$.

Hence, by dominated convergence theorem, we have

$$\mathbf{E}\left[S_\tau\right] = \lim_{n\to\infty} \mathbf{E}\left[S_{\tau \wedge n}\right] = \lim_{n\to\infty} \mathbf{E}[\tau \wedge n]\mathbf{E}\left[\xi_1\right].$$

By monotone convergence theorem it follows that $\mathbf{E}[\tau \wedge n] \uparrow \mathbf{E}[\tau]$, which completes the proof.

$\square$

Theorem 5.6 (Wald's second identity) *Let $\{\xi_k\}_{k\geq 1}$ be a sequence of identically distributed real-valued random variables. Let $S_n := \sum_{k=1}^{n} \xi_k$, $\mathscr{F}_n := \sigma(\xi_1, \ldots, \xi_n)$ for each $n \geq 0$. Let τ be a $\{\mathscr{F}_n\}$-stopping time. Suppose $\mathbf{E}[\xi_1] = 0$ and $\sigma^2 = \mathbf{E}[\xi_1^2] < \infty$. Then*

$$\mathbf{E}[S_\tau^2] = \sigma^2 \mathbf{E}[\tau]. \tag{5.4}$$

Proof It is easy to see that the stochastic process $\{M_n := S_n^2 - n\sigma^2\}$ is a $\{\mathscr{F}_n\}$-martingale. By Doob's optional stopping, Theorem 3.1, we have

$$\mathbf{E}[S_{\tau \wedge n}^2] = \mathbf{E}[\tau \wedge n]\sigma^2.$$

The right side converges to $\mathbf{E}[\tau]\sigma^2$ by monotone convergence theorem. Now, by Fatou's lemma,

$$\mathbf{E}[S_\tau^2] \leq \liminf_{n \to \infty} \mathbf{E}[S_{\tau \wedge n}^2] = \mathbf{E}[\tau]\sigma^2 < \infty.$$

The finite variance of $\{\xi_i\}$ implies $\mathbf{E}[\xi_i \xi_j]$ is finite. Note that for each i, ξ_i is independent of $\mathscr{F}_{i-1}$ and $\chi_{\{\tau \geq i\}} \in \mathscr{F}_{i-1}$. Hence we have

$$\mathbf{E}[\xi_i \chi_{\{\tau \geq i\}} \xi_j \chi_{\{\tau \geq j\}}] = 0 \text{ for } i \neq j.$$

This implies that

$$\sup_{m \geq n} \mathbf{E}\left[(S_{\tau \wedge m} - S_{\tau \wedge n})^2\right] = \sup_{m \geq n} \sum_{i=n+1}^{m} \mathbf{E}\left[\xi_i \chi_{\{\tau \geq i\}}\right]^2 \leq \sup_{m \geq n} \sum_{i>n} \mathbf{E}\left[\xi_i^2 \chi_{\{\tau \geq i\}}\right]$$

$$\leq \sum_{i>n} \mathbf{E}[\xi_i^2] \mathbf{E}[\chi_{\{\tau \geq i\}}] = \mathbf{E}[\xi_1^2] \sum_{i>n} \mathbf{E}[\chi_{\{\tau \geq i\}}] \to 0$$

as $n \to \infty$, since $\mathbf{E}[\tau] < \infty$. Thus $\{S_{\tau \wedge n}\}$ is Cauchy in L^2, and hence converges in L^2. But then

$$\mathbf{E}[S_{\tau \wedge n}^2] \to \mathbf{E}[S_\tau^2].$$

$\square$

Theorem 5.7 (Wald's third identity) *Let $\{\xi_k\}_{k\geq 1}$ be a sequence of identically distributed real-valued random variables. Let $S_n := \sum_{k=1}^{n} \xi_k$, $\mathscr{F}_n := \sigma(\xi_1, \ldots, \xi_n)$ for each $n \geq 0$. Assume that $\mathbf{E}[\exp(\theta \xi_1)] = \exp(\psi(\theta)) < \infty$ for all $\theta > 0$, and for some function ψ. Then, for every bounded stopping time τ,*

$$\mathbf{E}\left[\exp(\theta S_\tau - \tau \psi(\theta))\right] = 1.$$

Proof Since τ is bounded, the result follows directly from Doob's optional stopping, Theorem 3.1, applied to the non-negative martingale $\{M_n = \exp(\theta S_n - n\psi(\theta))\}$. $\square$

The last result of this section is the proof, using martingale theory, of one of the classical results on the convergence of sums of independent random variables with finite variance, which is the generalization of the result of Rademacher, Kolmogorov, and Khintchine, for Rademacher functions discussed in Theorem 1.1.

Theorem 5.8 *Let $\{\xi_n\}_n$ be a sequence of independent random variables such that for any $n \in \mathbb{N}$, $\mathbf{E}[\xi_n] = 0$ and $\sigma_n^2 = \mathrm{Var}\,(\xi_n) < \infty$. Then*

(i) If $\sum_{k=1}^{\infty} \sigma_k^2 < \infty$ then $\sum_{k=1}^{\infty} \xi_k < \infty$ a.e.
(ii) If the random variables $\{\xi_n\}$ are uniformly bounded, i.e., $|\xi_n| \leqslant M < \infty$, for any $n \in \mathbb{N}$, the converse is true, if

$$\sum_{k=1}^{\infty} \xi_k < \infty \text{ a.e. then } \sum_{k=1}^{\infty} \sigma_k^2 < \infty.$$

Proof Let $\mathscr{F}_n^{\xi} = \sigma\,(\xi_1, \ldots, \xi_n)$, the σ-algebra generated by $\{\xi_1, \ldots, \xi_n\}$, $S_n = \sum_{k=1}^{n} \xi_k$, $\Sigma_n = \sum_{k=1}^{n} \sigma_k^2$, and $\zeta_n = S_n^2 - \Sigma_n$. Trivially, $\xi_1, \xi_2, \ldots, \xi_n$ are $\mathscr{F}_n^{\xi}$-measurable and ξ_{n+1} is independent of $\mathscr{F}_n^{\xi}$.

(i) By the conditional expectation properties i), ii), and v) of Theorem 2.1 we get

$$\mathbf{E}\left[S_{n+1} \mid \mathscr{F}_n^{\xi}\right] = \mathbf{E}\left[S_n + \xi_{n+1} \mid \mathscr{F}_n^{\xi}\right] = S_n + \mathbf{E}\left[\xi_{n+1}\right] = S_n,$$

and therefore $\{S_n\}_n$ is a $\{\mathscr{F}_n^{\xi}\}$-martingale with $\mathbf{E}[S_n] = 0$ and $\mathbf{E}\left[S_n^2\right] = \Sigma_n$. Now, if $\sum_{k=1}^{\infty} \sigma_k^2 < \infty$, then

$$\sup_{n \in \mathbb{N}} \mathbf{E}\left[|S_n|\right] \leqslant \sup_{n \in \mathbb{N}} \mathbf{E}\left[S_n^2\right]^{1/2} < \infty.$$

Thus, by Propositions 3.9 and 3.11, we get $\sum_{k=1}^{\infty} \xi_k = \lim_{n \to \infty} S_n$ exists a.e.
(ii) Again, by the conditional expectation properties i), ii), and v) we get

$$\begin{aligned}
\sigma_{n+1}^2 &= \mathbf{E}\left[\xi_{n+1}^2\right] = \mathbf{E}\left[\xi_{n+1}^2 \mid \mathscr{F}_n^{\xi}\right] = \mathbf{E}\left[(S_{n+1} - S_n)^2 \mid \mathscr{F}_n^{\xi}\right] \\
&= \mathbf{E}\left[S_{n+1}^2 \mid \mathscr{F}_n^{\xi}\right] - 2 S_n \mathbf{E}\left[S_{n+1} \mid \mathscr{F}_n^{\xi}\right] + S_n^2 \\
&= \mathbf{E}\left[S_{n+1}^2 \mid \mathscr{F}_n^{\xi}\right] - S_n^2,
\end{aligned}$$

i.e.,

$$\mathbf{E}\left[S_{n+1}^2 \mid \mathscr{F}_n^{\xi}\right] = S_n^2 + \sigma_{n+1}^2.$$

Therefore

$$\mathbf{E}\left[S_{n+1}^2 - \Sigma_{n+1} \mid \mathscr{F}_n^\xi\right] = S_n^2 - \Sigma_n,$$

and hence $\{\zeta_n\}_n$ is a $\{\mathscr{F}_n^\xi\}$-martingale.

Now, for each $j \in \mathbb{N}$, let us consider a $\mathscr{F}_n^\xi$ stopping time, defined

$$\tau_j = \begin{cases} \inf\{n \in \mathbb{N} : |S_n| > j\} & \text{if } \{n \in \mathbb{N} : |S_n| > j\} \neq \varnothing, \\ \infty & \text{if } \{n \in \mathbb{N} : |S_n| > j\} = \varnothing. \end{cases}$$

By Doob's optional stopping time theorem, Theorem 3.1, $\left\{\zeta_{\tau_j \wedge n}\right\}_n$ is a $\left\{\mathscr{F}_n^\xi\right\}$-martingale such that $n \in \mathbb{N}$,

$$0 = \mathbf{E}\left[\zeta_1\right] = \mathbf{E}\left[\xi_{\tau_j \wedge n}\right] = \mathbf{E}\left[S_{\tau_j \wedge n}^2 - \Sigma_{\tau_j \wedge n}\right].$$

By hypothesis $\{\xi_n\}$ are uniformly bounded, i.e., for any $n \in \mathbb{N}$ we have $|\xi_n| \leqslant M$, then $\left|S_{\tau_j} - S_{\tau_j - 1}\right| = \left|\xi_{\tau_j}\right| \leqslant M$ if τ_j is finite. Therefore $\left|S_{\tau_j \wedge n}\right| \leqslant M + j$, thus

$$\mathbf{E}\left[\Sigma_{\tau_j \wedge n}\right] = \mathbf{E}\left[S_{\tau_j \wedge n}^2\right] \leqslant (M + j)^2,$$

for all $n \in \mathbb{N}$. Given that $\sum_{k=1}^\infty \xi_k$ converges a.e., its partial sums S_n are bounded a.e. Hence, there exists $j \in \mathbb{N}$ such that $\mathbf{P}\{\tau_j = \infty\} > 0$, and as for any $n \in \mathbb{N}$

$$\Sigma_n \mathbf{P}\{\tau_j \geqslant n\} \leqslant \mathbf{E}\left[\Sigma_{\tau_j \wedge n}\right] \leqslant (M + j)^2,$$

then

$$\Sigma_\infty = \sum_{k=1}^\infty \sigma_k^2 < \infty.$$

$\square$

5.2 Applications to Analysis

5.2.1 Applications to Measure Theory: The Radon-Nikodým Theorem

The Radon-Nikodým theorem is used in several instances in martingale theory, starting with the basic notion of conditional expectation; we will prove it using martingale theory. Nevertheless, observe that there is no circularity since the existence of conditional expectation can be proved without the need to use the Radon-Nikodým theorem.

Given $(\Omega, \mathscr{F}, \mu)$ a measure space and f an integrable function, i.e., $\int_\Omega f d\mu$ exists, then the set function

$$v(A) = \int_A f d\mu = \int_\Omega f \chi_A d\mu, \quad A \in \mathscr{F}$$

defines a signed measure. Moreover v is absolutely continuous with respect to μ, i.e., for $\epsilon > 0$, there exists $\delta > 0$ such that if $A \in \mathscr{F}$ and $\mu(A) < \delta$, then $v(A) < \epsilon$. The Radon-Nikodým establishes the converse statement when μ is σ-finite.

Theorem 5.9 *Let μ be a σ-finite measure and v be a signed measure on a σ-field $\mathscr{F}$ of Ω. If v is absolutely continuous with respect to μ, then there exists a measurable function $f : \Omega \to \overline{\mathbb{R}}$ such that*

$$v(A) = \int_A f d\mu, \quad A \in \mathscr{F}.$$

Moreover, if g is another such function, $g = f$ a.e.

In measure theory the proof of this theorem is divided into cases, starting with the one where μ and v are finite measures.

Here we use martingale theory to present the proof in the case where $(\Omega, \mathscr{F}, \mathbf{P})$ is a probability space, $\mathscr{F}$ is a separable σ-algebra, (i.e., $\mathscr{F} = \sigma(\{F_n : n \in \mathbb{N}\})$ for some sequence $\{F_n\}_n$ of subsets of Ω), and $\mathbf{P}$ is a probability measure.

Theorem 5.10 *Let $(\Omega, \mathscr{F}, \mathbf{P})$ be a probability space, $\mathscr{F}$ is a separable σ-algebra, and v is a finite measure on $(\Omega, \mathscr{F})$ absolutely continuous with respect to $\mathbf{P}$. Then, there exists $f \in L^1(\Omega, \mathscr{F}, P)$ such that*

$$v(E) = \int_E f d\mathbf{P} \quad \textit{for all} \quad E \in \mathscr{F}.$$

Here f, is unique a.e, is called a version of the Radon-Nikodým of v with respect to $\mathbf{P}$ on $(\Omega, \mathscr{F})$, and is denoted as $\frac{dv}{d\mathbf{P}} = f$.

Proof Since $\mathscr{F}$ is separable, and therefore $\mathscr{F} = \sigma(\{F_n : n \in \mathbb{N}\})$ for some sequence $\{F_n\}_n$ of subsets of Ω, let $\mathscr{F}_n = \sigma(\{F_1, \ldots, F_n\})$, $n = 1, 2, 3, \ldots$. Then, for each $n \in \mathbb{N}$, $\mathscr{F}_n$ is the collection of all finite unions of sets of the form $B_1 \cap B_2 \cap \cdots \cap B_n$, where $B_i = F_i$ or $B_i = F_i^c$. Since there are at most 2^n disjoint sets of this form, then $\mathscr{F}_n$ has at most 2^{2^n} elements. Thus, $\mathscr{F}_n$ consists of $2^{r(n)}$ possible unions of atoms $A_1^n, A_2^n, \ldots, A_{r(n)}^n$ of $\mathscr{F}_n$, where an atom A of $\mathscr{F}_n$ is a set of $\mathscr{F}_n$ such that $\varnothing$ is the unique proper subset of A which is an element of $\mathscr{F}_n$. Let us define the function $f_n : \Omega \to \mathbb{R}$ as

$$f_n(\omega) = \sum_{k=1}^{r(n)} \beta_k^n \chi_{A_k^n}(\omega),$$

where

$$\beta_k^n = \begin{cases} 0 & \text{if } \mathbf{P}\left(A_k^n\right) = 0, \\ \frac{\nu(A_k^n)}{\mathbf{P}(A_k^n)} & \text{if } \mathbf{P}\left(A_k^n\right) > 0. \end{cases}$$

Then, f_n is $\mathscr{F}_n$-measurable, $\mathbf{E}\left(|f_n|\right) < \infty$ and for $E = \bigcup_{j=1}^{s} A_{i_j}^n \in \mathscr{F}_n$, a disjoint union

$$\int_E f_n d\mathbf{P} = \int_E \sum_{k=1}^{r(n)} \beta_k^n \chi_{A_k^n} d\mathbf{P} = \sum_{k=1}^{r(n)} \int_\Omega \beta_k^n \chi_{A_k^n \cap E} d\mathbf{P}$$

$$= \sum_{j=1}^{s} \beta_{i_j}^n \mathbf{P}\left(A_{i_j}^n\right) = \sum_{j=1}^{s} \nu\left(A_{i_j}^n\right) = \nu(E)$$

and since $E \in \mathscr{F}_n \subset \mathscr{F}_{n+1}$

$$\int_E f_{n+1} d\mathbf{P} = \nu(E) = \int_E f_n d\mathbf{P}.$$

Then, $\mathbf{E}\left[f_{n+1} \mid \mathscr{F}_n\right] = f_n$ and therefore $\{f_n\}_n$ is a non-negative $\{\mathscr{F}_n\}$-martingale. Then, using Corollary 3.15 we get that $f_\infty = \lim_{n\to\infty} f_n$ exists.

Let us prove that $\{f_n\}_n$ is uniformly integrable. As ν is absolutely continuous with respect to $\mathbf{P}$, given $\epsilon > 0$, there exists $\delta > 0$ such that $\mathbf{P}(E) < \delta$ implies $\nu(E) < \epsilon$. As ν is finite, there exists $c > 0$ such that $\nu(\Omega) < c\delta$. Now, by Tchebyshev's inequality, for $n = 1, 2, 3, \ldots$,

$$\mathbf{P}\{f_n > c\} < \frac{\mathbf{E}\left[f_n\right]}{c} = \frac{\nu(\Omega)}{c} < \frac{c\delta}{c} = \delta.$$

Therefore, for $E = \{f_n > c\} \in \mathscr{F}_n$, we get

$$\int_{\{f_n > c\}} |f_n| \, d\mathbf{P} = \int_{\{f_n > c\}} f_n d\mathbf{P} = \nu\left(E\right) < \epsilon.$$

Thus, $\{f_n\}_n$ is uniformly integrable. Then, by Proposition 3.12 $\{f_n\}_n$ converges in L^1 and $f_n = \mathbf{E}\left[f_\infty \mid \mathscr{F}_n\right]$. Hence, for $E \in \mathscr{F}_n$

$$\nu(E) = \int_E f_n = \int_E f_\infty d\mathbf{P}.$$

Therefore, for any $E \in \bigcup_{n=1}^{\infty} \mathscr{F}_n$, and then for any $E \in \sigma\left(\bigcup_{n=1}^{\infty} \mathscr{F}_n\right) = \mathscr{F}_\infty = \mathscr{F}$, we have

$$\nu(F) = \int_F X_n d\mathbf{P} = \int_F X_\infty d\mathbf{P}.$$

$\square$

5.2.2   Applications to Lipschitz Functions

First of all, remember the following definition:

Definition 5.1 A function $f : [a, b] \to \mathbb{R}$ is Lipschitz continuous on $[a, b]$ if there exists $M \in (0, \infty)$ such that for any $x, y \in [a, b]$

$$|f(x) - f(y)| \leqslant M|x - y|.$$

Moreover, it is also known that if a function f is Lipschitz continuous on $[a, b]$ then it is absolutely continuous on $[a, b]$.[1] Moreover, it is known that a function $f : [a, b] \to \mathbb{R}$ is absolutely continuous if and only if there exists a function $h \in L^1[a, b]$ such that

$$f(x) - f(a) = \int_a^x h(t)dt, \quad x \in [a, b].$$

We will prove a similar result for Lipschitz functions directly using martingale theory.

Theorem 5.11 *If* $f : [0, 1] \to \mathbb{R}$ *is Lipschitz continuous, then there exists* $h \in L^1[a, b]$ *such that*

$$f(x) - f(0) = \int_0^x h(t)dt, \quad x \in [0, 1].$$

Proof Let us consider the probability space $(\Omega, \mathscr{F}, P) = ([0, 1), \mathscr{B}([0, 1)), m)$ and let $\{X_n\}_n$ be a sequence of random variables defined as

$$X_n(t) = \sum_{k=0}^{2^n - 1} \frac{k}{2^n} \chi_{\left[\frac{k}{2^n}, \frac{k+1}{2^n}\right)}(t).$$

Let $\mathscr{F}_n^X = \sigma\left(\{X_1, \ldots, X_n\}\right) = \sigma\left(\left\{[0, \frac{1}{2^n}), [\frac{1}{2^n}, \frac{2}{2^n}) \cdots [\frac{2^n-1}{2^n}, 1)\right\}\right)$ the sigma-algebra generated by $\{X_1, \ldots, X_n\}$ and define

$$Y_n = \frac{f\left(X_n + 2^{-n}\right) - f\left(X_n\right)}{2^{-n}}.$$

---

[1] A function $f : [a, b] \to \mathbb{R}$ is absolutely continuous on $[a, b]$ if for any $\epsilon > 0$ there exists $\delta > 0$ such that for any finite family of disjoint open intervals $(a_1, b_1), \ldots, (a_k, b_k)$ of $[a, b]$ such that $\sum_{k=1}^n (b_k - a_k) < \delta$, then

$$\sum_{k=1}^n |f(b_k) - f(a_k)| < \epsilon.$$

Then $\{Y_n\}_n$ is a $\{\mathscr{F}_n^X\}$-martingale, since clearly Y_n is $\mathscr{F}_n^X$-measurable and integrable. Now, for any $n \in \mathbb{N}$ we have

$$
X_{n+1}(t) = \begin{cases} X_n(t) & \text{for } t \in \left[\frac{k}{2^n}, \frac{2k+1}{2^{n+1}}\right), k = 0, \ldots, 2^n - 1, \\ X_n(t) + \frac{1}{2^{n+1}} & \text{for } t \in \left[\frac{2k+1}{2^{n+1}}, \frac{k+1}{2^n}\right), k = 0, \ldots, 2^n - 1. \end{cases}
$$

Therefore,

$$
f(X_{n+1}) = \sum_{k=0}^{2^n-1} \left(f(X_n) \chi_{\left[\frac{k}{2^n}, \frac{2k+1}{2^{n+1}}\right)} + f\left(X_n + \frac{1}{2^{n+1}}\right) \chi_{\left[\frac{2k+1}{2^{n+1}}, \frac{k+1}{2^n}\right)} \right).
$$

As $f(X_n)$ and $f\left(X_n + \frac{1}{2^n}\right)$ are $\mathscr{F}_n^X$-measurable and bounded, then by property iii) of Theorem 2.1

$$
\mathbf{E}\left[f(X_{n+1}) \mid \mathscr{F}_n^X\right] = f(X_n) \sum_{k=0}^{2^n-1} \mathbf{E}\left[\chi_{\left[\frac{k}{2^n}, \frac{2k+1}{2^{n+1}}\right)} \mid \mathscr{F}_n^X\right]
$$
$$
+ f\left(X_n + \frac{1}{2^{n+1}}\right) \sum_{k=0}^{2^n-1} \mathbf{E}\left[\chi_{\left[\frac{2k+1}{2^{n+1}}, \frac{k+1}{2^n}\right)} \mid \mathscr{F}_n^X\right],
$$

as

$$
\mathbf{E}\left[\chi_{\left[\frac{k}{2^n}, \frac{2k+1}{2^{n+1}}\right)} \mid \mathscr{F}_n^X\right] = \mathbf{E}\left[\chi_{\left[\frac{2k+1}{2^{n+1}}, \frac{k+1}{2^n}\right)} \mid \mathscr{F}_n^X\right] = \frac{1}{2} \chi_{\left[\frac{k}{2^n}, \frac{k+1}{2^n}\right]},
$$

then

$$
\mathbf{E}\left[f(X_{n+1}) \mid \mathscr{F}_n^X\right] = \frac{1}{2} f(X_n) \sum_{k=0}^{2^n-1} \chi_{\left[\frac{k}{2^n}, \frac{k+1}{2^n}\right)} + \frac{1}{2} f\left(X_n + \frac{1}{2^{n+1}}\right) \sum_{k=0}^{2^n-1} \chi_{\left[\frac{k}{2^n}, \frac{k+1}{2^n}\right)}
$$
$$
= \frac{1}{2} f(X_n) + \frac{1}{2} f\left(X_n + \frac{1}{2^{n+1}}\right).
$$

Analogously

$$
\mathbf{E}\left[f\left(X_{n+1} + \frac{1}{2^{n+1}}\right) \mid \mathscr{F}_n^X\right] = \frac{1}{2} f\left(X_n + \frac{1}{2^{n+1}}\right) + \frac{1}{2} f\left(X_n + \frac{1}{2^n}\right).
$$

Therefore

$$
\mathbf{E}\left[Y_{n+1} \mid \mathscr{F}_n^X\right] = \mathbf{E}\left[Y_{n+1} \mid X_n\right] = 2^{n+1} \mathbf{E}\left[f\left(X_{n+1} + 2^{-(n+1)}\right) - f(X_{n+1}) \mid X_n\right]
$$
$$
= 2^n \left[f\left(X_n + 2^{-n}\right) - f(X_n)\right] = Y_n.
$$

Moreover, $\{Y_n\}_{n \in \mathbb{N}}$ is uniformly integrable, since

$$|Y_n| = \left| \frac{f\left(X_n + 2^{-n}\right) - f\left(X_n\right)}{2^{-n}} \right| \leqslant M \frac{1}{2^{-n}} \left|2^{-n}\right| = M,$$

where M is the Lipschitz constant of f.

Observe that $\mathscr{F} = \mathscr{B}([0,1]) = \sigma \left(\bigcup_{n \geqslant 1} \mathscr{F}_n^X \right)$. Hence, by Propositions 3.12 and 3.13 there exists a random variable $h = h(t)$ such that $Y_n \to h$ a.e. and $Y_n = \mathbf{E}\left[h \mid \mathscr{F}_n^X\right]$, i.e.:

$$\int_E Y_n d\mathbf{P} = \int_E h d\mathbf{P}, \quad \text{for any } E \in \mathscr{F}_n^X.$$

Taking $E = \left[0, \frac{k}{2^n}\right) \in \mathscr{F}_n^X$, we get

$$f\left(\frac{k}{2^n}\right) - f(0) = \int_0^{\frac{k}{2^n}} Y_n d\mathbf{P} = \int_h^{\frac{k}{2^n}} h(t) d\mathbf{P}.$$

As the set $D = \left\{ \frac{k}{2^n} : n \in \mathbb{N}, k = 0, \ldots, 2^n \right\}$ is dense in [0,1] and the continuous functions $f(x) - f(0)$ and $\int_0^x h(t)dt$ are equal on D, then they are equal on $[0,1]$, thus

$$f(x) - f(0) = \int_0^x h(t)dt \quad \text{for any } x \in [0,1].$$

$\square$

Observe that if $g : [a,b] \to \mathbb{R}$ is Lipschitz continuous, considering the function $f : [0,1] \to \mathbb{R}$ defined[2] as $f(t) := g(a + (b-a)t), t \in [0,1]$ is also Lipschitz continuous and therefore by Theorem 5.11,

$$g(x) - g(a) = f\left(\frac{x-a}{b-a}\right) - f(0) = \int_0^{\frac{x-a}{b-a}} h(t)dt = \int_a^x h\left(\frac{u-a}{b-a}\right) \frac{du}{b-a}, \quad x \in [a,b].$$

5.2.3 Applications to Harmonic Analysis

Let us now look very briefly at some notions of harmonic analysis where martingales appear naturally. Recall that if we assume an integrable and periodic function of period 1, then the Hardy-Littlewood maximal function of f, denoted as Mf, is defined as

$$Mf(x) = \sup_I \frac{1}{|I|} \int_I |f(u)| du,$$

---

[2] i.e., $g(x) = f(\frac{x-a}{b-a}), x \in [a,b]$.

where the supremum is taken over all open intervals I contained in $[0, 1]$, which contain x.

We can also define a dyadic maximal function $M_d f$ taking the supremum only over dyadic intervals. It is trivial that, for any x,

$$M_d f(x) \leqslant M f(x),$$

and, moreover, it can be proved that (see, for instance, [71])

$$M f(x) \leqslant M_d f(x),$$

and therefore $M_d f$ and $M f$ are equivalent.

Now, consider the interval $[0, 1]$, the σ-algebra generated by the dyadic intervals of length 2^{-n} on $[0, 1]$, that is, $\mathscr{F}_n^d = \sigma \left\{ \left(\frac{k}{2^n}, \frac{k+1}{2^n} \right) : k = 0, 1, 2, \ldots, 2^n - 1 \right\}$ then by (2.5),

$$\mathbf{E}\left[f \mid \xi_n \right](x) = \sum_{k=1}^{2^n - 1} 2^n \left(\int_{k/2^n}^{(k+1)/2^n} f(u) du \right) \chi_{\left[\frac{k}{2^n}, \frac{k+1}{2^n} \right]}(x),$$

and therefore the dyadic maximal function of f, $M_d f$ is simply the supremum in n, i.e.:

$$M_d f(x) = \sup_n \mathbf{E}\left[f \mid \mathscr{F}_n^d \right](x).$$

On the other hand, we know that given a filtration $\{\mathscr{F}_n\}$, defining $X_n = \mathbf{E}[f \mid \mathscr{F}_n]$ then $\{X_n\}$ is a $\{\mathscr{F}_n\}$-martingale. Now, the Hardy-Littlewood inequalities for $M_d f$

$$m \{t \mid M_d f(t) > \lambda\} \leqslant \frac{C}{\lambda} \|f\|_1 \text{ and } \|M_d f\|_p \leqslant A_p \|f\|_p, \, 1 < p \leqslant \infty$$

are consequences of Doob's maximal inequalities for martingales, Theorem 3.7 and Corollary 3.9.

Now, a basic result for Walsh series is that for the 2^N partial sums $W_{2^N} f(t) = \sum_{n=0}^{2^N - 1} c_n \psi_n(t)$, the following equality holds

$$W_{2^N} f = \frac{1}{|J_{N,k}|} \int_{J_{N,k}} f(u) du,$$

where $J_{N,k} = [0, 1]$ if $N = 1$ and for $N = 2^n + j, 1 \leqslant j \leqslant 2^n, J_{N,k} = I_{n+1}^k = \left[\frac{k-1}{2^{n+1}}, \frac{k}{2^{n+1}} \right]$ if $k = 1, \ldots, 2j$ and $J_{N,k} = I_n^{k-1} = \left[\frac{k-1}{2^n}, \frac{k}{2^n} \right]$ if $k = 2j + 1, \ldots, n$. Therefore,

$$M_d f = \sup_N W_{2^N} f.$$

Another application in harmonic analysis is the use of *good-λ inequalities* to study the boundedness of classical integral operators. As we have seen in Theorem 4.22

of Chap. 4, good-λ inequalities are inequalities relating distribution functions which are enough to get integral inequalities and they are the main tool to obtain the generalization of Burkholder-Gundy inequalities and several other integral inequalities; see Theorems 4.23, 4.24, 4.25, for instance.

As has been mentioned before the results of [44] include the rudiments of the good-λ-inequalities that later were formulated in a more elegant but more cryptic form in [30], and fundamental integral inequalities, comparing the maximal function and the square function, or quadratic variation, of martingales having controlled jumps or continuous paths. A very large share of the extensive applications of these kinds of martingale inequalities, both in probability and other areas of mathematics, involves continuous path martingales. Good λ inequalities are a very powerful technique to obtain integral inequalities and even local or pointwise estimates of an operator, and they have been used not only in probability as we saw in the last part of Chap. 4, but also extensively in analysis, see, for instance, A. Torchinsky [160, Chap. 13] or García-Cuerva and Rubio de Francia [80]. For instance, they are very efficient in establishing the continuity of Calderón-Zygmund operators in $L^p(\omega)$ spaces for any Muckenhoupt A_p-weight ω that was proved by Coifman and Fefferman [58]. In the particular case of the Hilbert transform, this result is due to Hunt, et al. [99].

The L^p-boundedness of very general singular integrals and Fourier multiplier operators can be proved using martingale transforms, see, for example, Burkholder [35] and McConnell [120]. On the other hand, in the groundbreaking paper [91], Gundy and Varopoulos gave a representation for Riesz transforms in terms of stochastic integrals arising from composing the harmonic extension of the given function with Brownian motion in the upper half-space.

The Beurling-Ahlfors transform S is defined on $L^p(\mathbb{C})$, $1 < p < \infty$ as

$$Sf(x) = -\frac{1}{\pi} p.v. \int_{\mathbb{C}} \frac{f(\omega)}{(z-\omega)^2} d\omega,$$

where $d\omega$ is the two-dimensional Lebesgue measure and the integral is understood as a Cauchy principal value. By the Calderón-Zygmund theory S is a bounded operator on $L^p(\mathbb{C})$, $1 < p < \infty$. Moreover, its Fourier multiplier is $\xi/\bar{\xi}$, and therefore it is an isometry on $L^2(\mathbb{C})$ by Plancherel's theorem. Its norm on the other L^p spaces is unknown, the well-known Iwaniec's conjecture asserts that

$$\|S\|_p = p^* - 1 := \max\{p, p/(p-1)\} - 1.$$

The best current estimate is due to R. Bañuelos and P. Janakiraman.

$$\|S\|_p = 1.575(p^* - 1).$$

Additionally, D. Burkholder, R. Gundy, and M. Silverstein in an exceptionally important paper [46] completed the work of Hardy and Littlewood on the characterization of the Hardy H^p-spaces via the integrability of certain maximal functions.

5.2.4   Bellman Functions

In essence, the Bellman function method was introduced by D. Burkholder to find sharp estimates for the L^p- norm inequality of martingale transforms in [36,40,42], as we discussed in Chap. 4. Later it became clear that the scope of the method is quite wide in harmonic analysis.

Bellman functions method can be credited for helping to solve several old harmonic analysis problems and for unifying the approach to many others. In the first case, one would name the (sharp weighted) estimates of such classical operators as the Ahlfors-Beurling transform and the Hilbert and Riesz transform as mentioned earlier. In the second case, one can name all kinds of dimension-free estimates of weighted and unweighted Riesz transforms.

The simplest way to understand the sharp L^p-estimates for martingale transform obtained by D. Burkholder is to operate with one of the so-called Burkholder's function:

$$u_p(x, y) = p \left(1 - \frac{1}{p^*}\right)^{p-1} \left(|y| - (p^* - 1)\, |x|\right) \left(|x| + |y|\right)^{p-1},$$

where $p^* := \max\left(p, \frac{p}{p-1}\right)$, $1 < p < \infty$. However, the main question is of course how to get this function? Where did it come from? Burkholder explains in many details the way this function (and several of its relatives) is obtained. It is almost (but not quite) the least bi-concave majorant of the function

$$|y|^{p-1} - \left(p^* - 1\right)^p |x|^p.$$

It is obtained by solving a certain PDE and performing certain manipulations with the solution after that.

Thus, roughly speaking the Bellman function method makes apparent the hidden scaling properties of a given harmonic analysis problem. Conversely, given a harmonic analysis problem with certain scaling properties, one can (formally) associate it with a non-linear PDE, the so-called Bellman equation of the problem. For more details, see, for instance, [163, 164].

For $1 < p < \infty$, the best constant for inequality (1.27), β_p is given by

$$\beta_p = p^* - 1,$$

with $p^* = \max\{p, p/(p - 1)\}$. This result was obtained by D. Burkholder in [36], and it remains the same if we allow the coefficients to take values in a Hilbert space $\mathbf{H}$, see [40]. This result can be further generalized to sequences $\left\{d_k^{(n)}\right\}_n$, $\left\{e_k^{(n)}\right\}_n$ with $\mathcal{H}$-valued terms such that $\left|d_k^{(n)}\right| \leqslant \left|e_k^{(n)}\right|$ for each n, i.e., they are differentially subordinate, see Definition 4.1, then

$$\left\|\sum_{n=0}^{N-1}\sum_{k=1}^{2^n-1} d_k^{(n)} \chi_k^{(n)}\right\|_p \leqslant (p^* - 1) \left\|\sum_{n=0}^{N-1}\sum_{k=1}^{2^n-1} e_k^{(n)} \chi_k^{(n)}\right\|_p, \tag{5.5}$$

$n = 0, \ldots N - 1; k = 1, \ldots, 2^{n-1}, \; 1 < p < \infty$, and the constant $p^* - 1$ is optimal (i.e., cannot be replaced by a smaller one). The original proof of this fact exploits the properties of certain special functions, the associated Bellman function.

As we have already mentioned Burkholder's sharp martingale inequalities have been widely used to obtain tight bounds for a large class of operators, including many classical Fourier multipliers. For more detail, see the paper by R. Bañuelos and A. Osękowski [10] and references contained therein.

In the late 90s, T. L. Nazarov, S. R. Trail, and A. Volberg, see [125, 128], proposed a different, dual approach to the above $p^* - 1$ problems. Namely, they proved that (1.28), (5.5) can be deduced from the existence of a Bellman function B_p defined as follows.

Definition 5.2 A Bellman function B_p is a function defined on the set

$$\mathcal{D} = \left\{ (\zeta, \eta, Z, H) \in \mathcal{H} \times \mathcal{H} \times [0, \infty) \times [0, \infty) : Z \geqslant |\zeta|^p, H \geqslant |\eta|^q \right\}$$

satisfying the following conditions:

(i) $0 \leqslant B_p(\zeta, \eta, Z, H) \leqslant (p^* - 1) Z^{1/p} H^{1/q}$ on $\mathcal{D}$.
(ii) For any $a_{\pm} = (\zeta_{\pm}, \eta_{\pm}, Z_{\pm}, H_{\pm}) \in \mathcal{D}$, we have the concavity-type condition:

$$B_p \left(\frac{a_- + a_+}{2} \right) - \frac{B_p(a_-) + B_p(a_+)}{2} \geqslant \left| \frac{\zeta_+ - \zeta_-}{2} \right| \left| \frac{\eta_+ - \eta_-}{2} \right|.$$

The existence of such a function can be extracted from Burkholder's works [36, 40] via a dual formulation. As shown later by Nazarov and Volberg [126] and Dragičević and Volberg [70], this special object can be further exploited to yield interesting tight L^p bounds for classical Riesz transforms in the Lebesgue setting and for the Gaussian Riesz transforms in the Gaussian setting, see [162].

Finding explicit formulas for Bellman functions is in general a rather non-trivial matter, and there is an intriguing question about an explicit formula for B_p. What is even more surprising in this case is that this problem has been solved thus far only in the particular case $p = 2$ where the explicit expression is very easy to obtain. Indeed, for this value of the parameter p, Nazarov, Treil, and Volberg, see [125, 126].

$$\mathbb{B}_2(\zeta, \eta, Z, H) = \sqrt{\left(Z - |\zeta|^2 \right) \left(H - |\eta|^2 \right)}$$

works just fine. In [9] can be found some attempts to find B_p explicitly for other values of p, but with no success. Nevertheless, the authors managed to construct, for each $1 < p < \infty$, a function that satisfies (ii) and a version of (i), in which $p^* - 1$ is replaced by a slightly larger constant. The purpose of that paper is to fill this gap and give an explicit formula for B_p satisfying (i) and (ii), for all $1 < p < \infty$. For further details, we refer to [8].

The Bellman function techniques have become a powerful tool to study sharp inequalities for many operators of great importance in harmonic analysis and its applications to PDEs, quasiconformal mappings, and also for weighted martingale inequalities that will be discussed in the next section.

5.2.5 Weighted Martingales Inequalities

Weighted inequalities for martingales extend classical martingales inequalities, studied in Chaps. 3 and 4, like, for instance, (3.33), (4.44), (4.54), and (4.46), have weighted versions. They first appeared in the 1970s, see, for instance, [102], but they have been developing slowly. One reason is that some decomposition theorems and covering theorems, which depend on algebraic structure and topological structure, are invalid in probability space. It is important to observe that Davis decomposition, Theorem 4.6, and other decompositions can be adapted to weighted spaces, facilitating proof in this context.

Definition 5.3 A positive random variable W defined on $(\Omega, \mathscr{F}, \mathbf{P})$ is an A_p weight,[3] $p > 1$ if

$$\sup_n \left(\operatorname{ess\,sup} \mathbf{E}\left[W \mid \mathscr{F}_n\right] \mathbf{E}\left[\left(\frac{1}{W} \right)^{1/(p-1)} \Big| \mathscr{F}_n \right]^{p-1} \right) < \infty. \tag{5.6}$$

In the case $p = 1$, W is an A_1 weight if there exists a positive constant C such that

$$\sup_n \mathbf{E}[W \mid \mathscr{F}_n] \leqslant CW \quad \text{a.e.} \tag{5.7}$$

Given a weight W we defined the weighted norm of a random variable ξ

$$\|\xi\|_{p,W} = \left(\mathbf{E}[|\xi|^p W] \right)^{1/p} = \left(\int_\Omega |\xi|^p W d\mathbf{P} \right)^{1/p},$$

and the weighted L^p-space, L^p_W, as

$$L^p_W = \left\{ \xi(\Omega, \mathscr{F}, \mathbf{P}) \to \mathbb{R} : \xi \text{ random variable such that } \|\xi\|^p_{p,W} < \infty \right\}.$$

---

[3] Following Muckenhoupt [121] definitions in the harmonic analysis context, $\omega \in A_p \; p > 1$

$$\sup_Q \left(\frac{1}{|Q|} \int_Q \omega(x) dx \right) \left(\frac{1}{|Q|} \int_Q \omega(x)^{-1/(p-1)} dx \right)^{p-1} < \infty,$$

where the supremum is taking over all Q cubes with sides parallel to the axes, and for $p = 1$ there is a $C_1 > 0$ such that

$$M\omega(x) \leq C\omega(x) \quad \text{a.e.},$$

where Mf is the Hardy-Littlewood maximal function, see [71,121,122].

Then several results can be proved, for instance, Izumisawa and Kazamaki [102], proved an analog of inequality Doob's maximal inequality (3.33).

Theorem 5.12 *If W is an A_{p_0} weight for some $p_0 > 1$, then there exists a constant C_p such that for every **P**-uniformly integrable $\{\mathscr{F}_n\}$-martingale X*

$$\|X_n^*\|_{p,W} \leqq C_p\|X_n\|_{p,W} \quad p > p_0, \tag{5.8}$$

for all $n \geqslant 1$.

On the other hand, we have a weighted version of (4.54) due to A. Osękowsk [132].

Theorem 5.13 *If W is an A_1 weight $\{X_n\}$ is a $\{\mathscr{F}_n\}$-martingale and $Y = \{Y_n\}_n$ be a multiplier sequence, then the transform of X by Y, $(Y \circ X) = \{(Y \circ X)_n\}_n$ then there exists a constant C_p, independent of X and Y such that for $1 < p < \infty$*

$$\|(Y \circ X)_n\|_{L_{p,W}} \leqslant C_p\|X_n\|_{L_{p,W}}, \tag{5.9}$$

and, for some universal constant c_1,[4]

$$\lambda\mathbf{P}\{(Y \circ X) \geqslant \lambda\} \leqslant c_1\|X\|_{L_{1,W}}. \tag{5.10}$$

One possible approach to weighted theory in martingale spaces is very closely related to Burkholder's method (see [31]). This is the so-called Bellman's method, which also rests on the construction of an appropriate special function, a Bellman function. The technique has been used very intensively mostly in analysis, in the study of Carleson embedding theorems, BMO estimates, square function inequalities, bounds for maximal operators, estimates for weights, and many other related results. In martingale spaces, this theory was further developed in a series of papers by R. Bañuelos and A. Osękowski, see, for instance, [7,9,10] and the Osękowski's monograph [131].

As an application of these types of inequalities, it can be obtained by a corresponding weak-type estimate for the Haar system, and dyadic paraproducts, which have had an important impact on harmonic analysis, for more details we refer to [17,127,128,137,138].

5.2.6 Applications to the Semigroup Theory

Let us now establish a relationship between martingales and certain types of semigroups. This relation allows a general form for semigroups of the Littlewood Paley

---

[4] $(\Omega, \mathscr{F}, \mathbf{P})$ is a probability space with some additional condition (equipped with a β-regular tree $\mathscr{T}$), and the constant C_p, c_1 depends of several parameters, see [132, Theorem 5.3].

inequality; for more details, see Stein [153] Chap. 4. This link was established by G. C. Rota in [146].

Definition 5.4 Let $(M, \mathcal{M}, \mu)$ be a σ-finite measurable space, $\{T_t\}_{0 \leqslant t < \infty}$ is a family of operators, such that each of them sends functions of M into functions of M, $\{T_t\}_{0 \leqslant t < \infty}$, satisfies the semigroup property

$$T_{t_1 + t_2} = T_{t_1} T_{t_2} \text{ and } T_0 = \text{ identity,}$$

for any $f \in L^2(M, \mu)$ $T_t f \in L^2(M, \mu)$ and $\lim_{t \to \infty} T_t f$ exists in $L^2(M, \mu)$. Additionally, assume that

(i) Contraction property: for any $t \geqslant 0$, $\|T_t f\|_p \leqslant \|f\|_p$, $1 \leqslant p \leqslant +\infty$.
(ii) Symmetry property: for any $t \geqslant 0$, T_t is a self-adjoint operator in $L^2(M, \mu)$,

$$\int_M T_t f(x) g(x) \mu(dx) = \int_M f(x) T_t g(x) \mu(dx). \tag{5.11}$$

(iii) Positivity property: for any $t \geqslant 0$, $T_t f \geqslant 0$ if $f \geqslant 0$.
(iv) Conservative property: for any $t \geqslant 0$, $T_t 1 = 1$.

A family of operators $\{T_t\}_{0 \leqslant t < \infty}$ satisfying i)-iv) is called a *symmetric semigroup of diffusions* or *Markov semigroup* (sometimes (iii) and (iv) are not included but (i) and (ii) are absolutely essential).

Assuming only properties (i) and (ii) it can be proved, using the spectral theorem, that the map $t \to T_t$ has an analytic continuation, see Theorem 1, Chap. 3 of [153], in particular, $t \to T_t$ is continuous.

The heat and the Poisson semigroups in $(\mathbb{R}^d, m)$ are probably the most well-known examples of semigroups, see, for instance, [152, 170]. Semigroups associated with classical orthogonal polynomial expansions can be found, for instance, at [161, 162].

Given that we have worked with discrete parameter martingales but the semigroups depend on a continuous parameter (time), in order to establish a relation between the two we need to discretize the parameter of the semigroups. Hence, we will consider the powers of an operator Q in $L^p(M, \mu)$ satisfying

(i') $\|Qf\|_p \leqslant \|f\|_p$, $1 \leqslant p < \infty$.
(ii') Q is self-adjoint in L^2.
(iii') $Qf \geqslant 0$ if $f \geqslant 0$.
(iv') $Q1 = 1$.

Let us now consider the Rota's result [146] which establishes the relation between martingales and semigroups. We will use the formulation and the proof given by E. Stein in [153], see Theorem 9, Sect. 4, Chap. 4. Here we will use the notion of conditional expectation on $(M, \mathcal{M}, \mu)$ a σ-finite measure space, see Problem 6 of Chap. 2.

Theorem 5.14 (Alternierende Verfahren) *Given Q an operator in $L^p(M, \mu)$ satisfying the conditions (i'), (ii'), (iii'), and (iv'), above then there exists a measure space $(\Omega, \mathscr{F}, \mathbf{P})$, a decreasing collection $\{\mathscr{F}_n\}$ of sub-σ-algebras of $\mathscr{F}$ and another sub-σ-algebra $\mathscr{G}$ contained also in $\mathscr{F}$ such that*

(i) *The spaces of measure $(M, \mathscr{M}, \mu)$ and $(\Omega, \mathscr{G}, \mathbf{P})$ are isometric under the natural injection $i : \Omega \to M$. The isomorphism induced by i between $L^p(\Omega, \mathscr{G}, \mathbf{P})$ and $L^p(M, \mathscr{M}, \mu)$ is going to be denoted by $\hat{i}$.*

(ii) *Let $Y \in L^p(\Omega, \mathscr{G}, \mathbf{P})$ then*

$$Q^{2n}(\hat{i}(Y)) = \hat{i}\left(\mathbf{E}\left[\mathbf{E}\left[Y \mid \mathscr{G}\right] \mid \mathscr{F}_n\right]\right). \tag{5.12}$$

Proof First, we construct the measure space $(\Omega, \mathscr{F}, \mathbf{P})$ using the classical Markov process construction.

Let us consider the product space $\Omega = M \times M \times M \times \cdots = M^{\mathbb{N}}$ and let $\mathscr{F}$ be the σ-algebra generated by the cylinder sets, i.e., all the sets of the form

$$A_0 \times A_1 \times \cdots \times A_n \times M \times M \times \cdots$$

with $n \in \mathbb{N}$ and $A_i \in \mathscr{M}$. In order to define $\mathbf{P}$, by the classical Caratheodory's argument, see [155], it is enough to define it on $\mathscr{C}$ the Boolean algebra of cylinder sets.

Let $C = A_0 \times A_1 \times \cdots \times A_n \times M \times M \times \cdots \in \mathscr{C}$. Let us start with the indicator function of the set A_n, $\chi_{A_n} : M \to \mathbb{R}$, then take $Q(\chi_{A_n})$; then multiplying by $\chi_{A_{n-1}}$, we obtain $\chi_{A_{n-1}} Q\left(\chi_{A_n}\right)$; iterating this process, we finally get the function

$$\chi_{A_0} Q\left(\chi_{A_1} Q\left(\chi_{A_2} \cdots Q\left(\chi_{A_{n-1}} Q\left(\chi_{A_n}\right)\right)\right) \cdots\right).$$

Now, define

$$\mathbf{P}(C) = \int_M \chi_{A_0}(t) Q\left(\chi_{A_1} Q\left(\chi_{A_2} \cdots Q\left(\chi_{A_{n-1}} Q\left(\chi_{A_n}\right)\right)\right) \cdots\right)(t)\mu(dt).$$

It can be proved that $\mathbf{P}$ is well-defined, non-negative, and finitely additive on the family of cylinder sets $\mathscr{C}$. Observe that (iv'), $Q1 = 1$, is needed to show that $\mathbf{P}$ is well defined, since $(A_1 \times A_2 \times \cdots \times A_n) \times M \times M \times M \times \cdots$ and $(A_1 \times A_2 \times \cdots \times A_n \times M) \times M \times M \cdots$ are representations of the same cylinder set and

$$\int_M \chi_{A_0}(t) Q\left(\chi_{A_1} Q\left(\chi_{A_2} \cdots Q\left(\chi_{A_{n-1}} Q\left(\chi_{A_n}\right)\right)\right) \cdots\right)(t)\mu(dt)$$

$$= \int_M \chi_{A_0}(t) Q\left(\chi_{A_1} Q\left(\chi_{A_2} \cdots Q\left(\chi_{A_{n-1}} Q\left(\chi_{A_n} 1\right)\right)\right) \cdots\right)(t)\mu(dt)$$

$$= \int_M \chi_{A_0}(t) Q\left(\chi_{A_1} Q\left(\chi_{A_2} \cdots Q\left(\chi_{A_{n-1}} Q\left(\chi_{A_n} Q(1)\right)\right)\right) \cdots\right)(t)\mu(dt)$$

$$= \int_M \chi_{A_0}(t) Q\left(\chi_{A_1} Q\left(\chi_{A_2} \cdots Q\left(\chi_{A_{n-1}} Q\left(\chi_{A_n} Q(\chi_M)\right)\right)\right) \cdots\right)(t)d\mu(dt).$$

Then, by Caratheodory's theorem $\mathbf{P}$ can be extended to a countably additive measure on $\mathscr{F}$.

Now define $\mathscr{G} = \{A_0 \times M \times M \times M \times \cdots \mid A_0 \in \mathscr{M}\}$, and set

$$\mathscr{F}_n = \{\overbrace{M \times M \times M \times \cdots \times M}^{n+1} \times S \mid S \in \mathscr{F}, \text{ so that } S \subseteq M \times M \times M \times \cdots\}.$$

Obviously

$$\cdots \subseteq \mathscr{F}_{n+1} \subseteq \mathscr{F}_n \subseteq \cdots \subseteq \mathscr{F}_1 \subseteq \mathscr{F}_0 = \mathscr{F}.$$

The mapping $\mathrm{i} : \Omega \to M$ defined as the projection in the first component, i.e., $\mathrm{i}(x) = \mathrm{i}(x_0, x_1, x_2, \ldots) = x_0$, where $x = (x_0, x_1, x_2, \ldots) \in \Omega$, sets up an isomorphism of measure spaces between $(\Omega, \mathscr{F}, \mathbf{P})$ and $(M, \mathscr{M}, d\mu)$, since if $C = A_0 \times M \times M \times \ldots$, $\mathrm{i}(C) = A_0 \in \mathscr{M}$, and

$$\mathbf{P}(C) = \int_\Omega \chi_C(x)d\mathbf{P}(x) = \int_M \chi_{A_0}(t)d\mu(t) = \int_M \chi_{\mathrm{i}(C)}(t)d\mu(t) = \mu(\mathrm{i}(C)).$$

Now, let $\hat{\mathrm{i}}$ be the induced isomorphism from $L^p(\Omega, \mathscr{G}, \mathbf{P})$ to $L^p(M, \mathscr{M}, \mu)$ by i, which is defined as the map that makes the following diagram commutative:

$$\begin{array}{ccc}
\Omega & \xrightarrow{\ Y\ } & \mathbb{R} \\
{\scriptstyle i}\Big\downarrow & \nearrow{\scriptstyle f} & \\
M & &
\end{array}$$

thus $\hat{\mathrm{i}}(Y) = f$, i.e., $Y = f \circ \mathrm{i}$. For instance, if $Y = \chi_{A_0 \times M \times M \cdots}$, i.e.:

$$Y(x_0, x_1, \ldots) = \begin{cases} 1, & \text{if } x_0 \in A_0 \\ 0, & \text{if } x_0 \notin A_0, \end{cases}$$

then $\hat{\mathrm{i}}(Y) = \chi_{A_0}$. Analogously, if $Y = \chi_{M \times M \times M \cdots \times A_n \times M \times M \cdots}$ then $\hat{\mathrm{i}}(Y) = \chi_{A_n}$.

In order to prove (ii), we make two claims:

(a) If $Y : \Omega \to \mathbb{R}$ is such that $Y(x)$ depends only on x_n, i.e., $Y(x) = Y(x_n)$ then

$$\mathbf{E}[Y \mid \mathscr{G}](x) = \hat{\mathrm{i}}^{-1} Q^n \hat{\mathrm{i}}(Y)(x_0).$$

(b) If $Y : \Omega \to \mathbb{R}$ is such that $Y(x)$ depends only on x_0, i.e., $Y(x) = Y(x_0)$, thus Y is $\mathscr{G}$-measurable, then

$$\mathbf{E}[Y \mid \mathscr{F}_n](x) = \hat{\mathrm{i}}^{-1} Q^n \hat{\mathrm{i}}(Y)(x_n).$$

From these two claims, it is obvious that if $f \in L^p(M, \mathcal{M}, \mu)$, $Y = \hat{i}^{-1}(f) \in L^p(\Omega, \mathcal{F}, \mathbf{P})$ (i.e., $f = \hat{i}(Y)$) and depends only on x_0 then, by (b)

$$\mathbf{E}\left[\hat{i}^{-1}(f) \mid \mathcal{F}_n\right](x) = \hat{i}^{-1} Q^n \hat{i}(\hat{i}^{-1} f)(x_n) = \hat{i}^{-1} Q^n f(x_n),$$

hence, this conditional expectation depends only on x_n and by (a)

$$\mathbf{E}\left[\mathbf{E}\left[\hat{i}^{-1}(f) \mid \mathcal{F}_n\right] \mid \mathcal{G}\right](x) = \mathbf{E}\left[\hat{i}^{-1} Q^n f \mid \mathcal{G}\right](x_n)$$
$$= \hat{i}^{-1} Q^n \left(\hat{i}\left(\hat{i}^{-1} Q^n f\right)\right)(x_0) = \hat{i}^{-1}\left(Q^{2n} f\right)(x_0).$$

Thus

$$\mathbf{E}\left[\mathbf{E}\left[\hat{i}^{-1}(f) \mid \mathcal{F}_n\right] \mid \mathcal{G}\right](x) = \hat{i}^{-1}\left(Q^{2n} f\right)(x_0),$$

and hence

$$\mathbf{E}\left[\mathbf{E}\left[Y \mid \mathcal{F}_n\right] \mid \mathcal{G}\right](x) = \hat{i}^{-1} Q^{2n}(i(Y))(i(x)).$$

which is (ii).

Thus it is enough to prove (a) and (b), which are, basically, conditional expectation computations

- Proof of (a): We are given the candidate, $\hat{i}^{-1} Q^n(i(Y))(x_0)$ for the conditional expectation of Y with respect to $\mathcal{G}$. Since $Q^n Y(x_0)$ is obviously $\mathcal{G}$ measurable, it is enough to check that

$$\int_B Y(x_n)\, d\mathbf{P}(x) = \int_B \hat{i}^{-1} Q^n(\hat{i}(Y))(x_0)\, d\mathbf{P}(x) \qquad (5.13)$$

for $B \in \mathcal{G}$, i.e., $B = A_0 \times M \times M \times M \times \cdots$. Both sides of (5.13) are equal to $\int_{A_0} Q^n Y(x)dx$ if Y is the indicator function of a subset of M, as follows from the definition of $\mathbf{P}$ since if $Y = \chi_C$ the indicator function of $C = M \times M \times \cdots \times B_n \times M \times \ldots$, then

$$\int_B Y(x_n)\, d\mathbf{P}(x) = \int_B \chi_C(x)\, d\mathbf{P}(x) = \int_\Omega \chi_B(x)\, \chi_C(x)\, d\mathbf{P}(x)$$
$$= \int_\Omega \chi_{B \cap C}(x)\, d\mathbf{P}(x)$$
$$= \int_\Omega \chi_{A_0 \times M \times \cdots \times B_n \times M \times \cdots}(x)\, d\mathbf{P}(x)$$
$$= \mathbf{P}\{A_0 \times M \times M \times \cdots \times B_n \times M \times \cdots\}$$
$$= \int_M \chi_{A_0}(t) Q\left(\chi_M Q\left(\chi_M \cdots \left(Q(\chi_M Q\left(\chi_{B_n}\right))\right) \cdots \right)\right)(t) d\mu(t)$$
$$= \int_{A_0} Q^n\left(\chi_{B_n}\right)(t) d\mu(t) = \int_{A_0} Q^n(\hat{i}(Y))(t) d\mu(t)$$
$$= \int_B \hat{i}^{-1} Q^n(\hat{i}(Y))(x_0)\, d\mathbf{P}(x),$$

by using the change of variable $x_0 = \mathrm{i}(x)$.

On the other hand, both sides of (5.13) are linear in Y, and well behaved under the limit processes. So (5.13) is valid for all Y.

Observe that so far we have not used the self-adjointness of Q.

- Proof of (b): As in (a), the problem reduces to showing that

$$\int_B Y(x_0)\,d\mathbf{P}(x) = \int_B \hat{\mathrm{i}}^{-1}\,Q^n\,(\hat{\mathrm{i}}(Y))\,(x_n)\,d\mathbf{P}(x) \tag{5.14}$$

for any $B \in \mathscr{F}_n$ and therefore

$$B = \overbrace{M \times M \times \cdots \times M}^{n} \times A_n \times A_{n+1} \times \cdots \times A_N \times M \times \cdots$$

Again, let $Y = \chi_C$ with $C = A_0 \times M \times M \times \ldots$, then

$$\int_B Y(x_0)\,d\mathbf{P}(x) = \int_\Omega \chi_C(x)\chi_B(x)\,d\mathbf{P}(dx)$$

$$= \int_\Omega \chi_{A_0 \times M \times \cdots \times M \times A_n \times A_{n+1} \times \cdots \times A_N \times M \times \cdots}(x)\,d\mathbf{P}(x)$$

$$= \mathbf{P}\left(A_0 \times M \times \cdots \times M \times A_n \times A_{n+1} \times \cdots \times A_N \times M \times \cdots\right)$$

$$= \int_M \chi_{A_0}(t)Q\left(\chi_M \cdots Q\left(\chi_{A_n}Q\left(\chi_{A_{n+1}}Q\left(\chi_{A_{n+2}}\cdots Q\left(\chi_{A_N}\right)\right)\right)\right)\cdots\right)(t)\,d\mu(t)$$

$$= \int_M \chi_{A_0}(t)Q^n\left(\chi_{A_n}Q\left(\chi_{A_{n+1}}Q\left(\chi_{A_{n+n}}\cdots Q\left(\chi_{A_N}\right)\right)\right)\cdots\right)(t)\,d\mu(t)$$

$$= \int_M \hat{\mathrm{i}}(Y)(t)Q^n\left(\chi_{A_n}Q\left(\chi_{A_{n+1}}Q\left(\chi_{A_{n+2}}\cdots Q\left(\chi_{A_N}\right)\right)\right)\cdots\right)(t)\,d\mu(t).$$

Observe that from the fact that Q is self-adjoint and $Q\chi_M = Q1 = 1$, we have

$$\int_M Qh(t)\,d\mu(t) = \int_M Qh(t)\cdot 1\,d\mu(t) = \int_M h(t)\cdot Q1\,d\mu(t) = \int_M h(t)\cdot Q\chi_M(t)\,d\mu(t) = \int_M h(t)\,d\mu(t),$$

therefore, by the definition of $\mathbf{P}$ we obtain

$$\int_B \hat{\mathrm{i}}^{-1}Q^n(\hat{\mathrm{i}}(Y))\,(x_n)\,d\mathbf{P}(x) = \int_\Omega \hat{\mathrm{i}}^{-1}Q^n\left(\chi_{A_0}\right)(x_n)\chi_B(x)\,d\mathbf{P}(x)$$

$$= \int_\Omega \hat{\mathrm{i}}^{-1}Q^n\left(\chi_{A_0}\right)(x_n)\,\chi_{M \times M \times \cdots \times M \times A_n \times A_{n+1} \times \cdots \times A_N \times M \times \cdots}(x)\,d\mathbf{P}(x)$$

$$= \int_M Q^n\left(\chi_{A_0}\right)(t)Q\left(\chi_M Q\left(\chi_M Q\cdots\left(\chi_M Q\left(\chi_{A_n}Q\left(\chi_{A_{n+1}}\cdots Q\left(\chi_{A_N}\right)\right)\right)\right)\right)\cdots\right)(t)\,d\mu(t)$$

$$= \int_M Q^n\left(\chi_{A_0}\right)(t)Q^n\left(\chi_{A_n}Q\left(\chi_{A_{n+1}}Q\left(\chi_{A_{n+2}}\cdots Q\left(\chi_{A_N}\right)\right)\right)\cdots\right)(t)\,d\mu(t)$$

$$= \int_M Q^n\left(\chi_{A_0}\right)(t)\chi_{A_n}(t)Q\left(\chi_{A_{n+1}}Q\left(\chi_{A_{n+2}}\cdots Q\left(\chi_{A_m}\right)\right)\cdots\right)(t)\,d\mu(t).$$

Thus,

$$\int_\Omega \hat{\mathrm{i}}^{-1}Q^n(\hat{\mathrm{i}}(Y))\,(x_n)\,d\mathbf{P}(x) = \int_M Q^n(\hat{\mathrm{i}}(Y))(t)\chi_{A_n}(t)Q\left(\chi_{A_{n+1}}Q\left(\chi_{A_{n+2}}\cdots Q\left(\chi_{A_N}\right)\right)\cdots\right)(t)\,d\mu(t)$$

$$= \int_M \hat{\mathrm{i}}(Y)Q^n\chi_{A_n}(t)Q\left(\chi_{A_{n+1}}Q\left(\chi_{A_{n+2}}\cdots Q\left(\chi_{A_N}\right)\right)\cdots\right)(t)\,d\mu(t),$$

using again the fact that Q is self-adjoint. $\square$

Corollary 5.1 *If* $f \in L^p(M, \mathcal{M}, \mu)$, $\lim_{n \to \infty} Q^{2n}(f)$, *exists a.e.*

Proof Set $Y = \hat{i}^{-1}(f)$, then as $Y_n = \mathbf{E}[Y \mid \mathcal{F}_n]$ is a $\{\mathcal{F}_n\}$-reverse martingale, by the convergence theorem, Theorem 3.23 $\lim_{n \to \infty} X_n$ converges a.e. but $\{X_n\}$ is clearly uniformly integrable and therefore the convergence is also in L^p-norm. By the dominated convergence theorem for conditional expectation, Theorem 2.1, property vii), we have that $\lim_{n \to \infty} \mathbf{E}[X_n \mid \mathcal{G}]$ also exists a.e. and by ii) of the previous theorem this is $\lim_{n \to \infty} Q^{2n}(f)$ a.e. $\square$

Corollary 5.2 *Let* $\{T_t\}$ *be a symmetric semigroup of diffusions and let*

$$T^* f(x) := \sup_t |T_t f(x)|$$

the maximal function of the semigroup. Then T^* *is* $L^p(M, \mathcal{M}, \mu)$*-bounded,* $1 < p < \infty$ *and, moreover,*

$$\left\| T^* \right\|_p \leqslant A_p \| f \|_p.$$

Proof This result follows from the theorem above and Doob's maximal inequality Corollary 3.9. Let $Q = T_{1/2^{k+1}}$ for some fixed $k \in \mathbb{N}$. For ii) of the previous theorem and Doob's maximal inequality we have that if $Y \in L^p(\Omega, \mathcal{F}, \mathbf{P})$

$$\left\| \sup_n Q^{2n}(\hat{i}(Y)) \right\|_p = \left\| \sup_n \mathbf{E}[\mathbf{E}[Y \mid \mathcal{F}_n] \mid \mathcal{G}] \right\|_p \leqslant \left\| \mathbf{E}\left[\sup_n \mathbf{E}[Y \mid \mathcal{F}_n] \mid \mathcal{G} \right] \right\|_p$$

$$\leqslant \left\| \sup_n \mathbf{E}[Y \mid \mathcal{F}_n] \right\|_p \leqslant A_p \| Y \|_p.$$

Then, given $f \in L^p(M, \mathcal{M}, \mu)$ can be rewritten as

$$\left\| \sup_n Q^{2n}(f) \right\|_p \leqslant A_p \left\| \hat{i}^{-1} f \right\|_p = A_p \| f \|_p,$$

hence

$$\left\| \sup_n T_{n/2^k}(f) \right\|_p \leqslant A_p \| f \|_p,$$

for any k, with A_p independent of k. Taking $k \to \infty$ by the monotone convergence theorem, we get

$$\| T^* \|_p \leqslant A_p \| f \|_p,$$

remember that T_t is a continuous function on t. $\square$

5.3 Applications to Functional Analysis: Geometry of Banach Spaces

The study of martingales with values in a Banach space goes back to Chatterji [51,52]. A detailed treatment is given in Pisier [141]. The application of martingale theory in the study of the geometry of Banach spaces is very important and consequential. When we deal with martingales with Hilbert space values, the passage from the real case is typically quite easy and does not require much additional effort. However, the extension of for martingales with Banach space values is a much more difficult problem in general. In this section, the basic martingale theory for Bochner integrable functions will be discussed, and then we will see that martingales theory provides effective ways of studying the internal structure of Banach spaces.

5.3.1 Martingales in Banach Spaces and the Radon-Nikodým Property

To consider martingales in Banach spaces we need to use the notion of Bochner integral instead of the notion of Lebesgue integral. We will do a brief review of Bochner integration. For more details, we recommend the book of J. Diestel and J. J. Uhl [68]. As they mention "some have said that the Bochner integral is only the Lebesgue integral with absolute value signs replaced by norm signs." We shall see that often this is the case, but sometimes it is a totally ignorant appraisal of the Bochner integral. In fact, as we shall see later, the failure of the Radon-Nikodým theorem for the Bochner integral lies at the base of some of the most intriguing results in the theory of vector measures and the structure theory of Banach spaces.

The Bochner integral can be traced back to two origins. It first appeared in 1932 due to S. Bochner and later it appeared in 1935 due to N. Dunford. It has also been called Dunford's first integral.

For this section, $(\Omega, \mathscr{F}, \mu)$ will be a finite measure space and $(\mathbf{B}, \| \cdot \|_{\mathbf{B}})$ a real Banach space.

Definition 5.5 A function $f : \Omega \to \mathbf{B}$ is called simple if there exist $x_1, x_2, \ldots, x_n \in \mathbf{B}$ and $E_1, E_2, \ldots, E_n \in \mathscr{F}$ such that

$$f = \sum_{i=1}^{n} x_i \chi_{E_i},$$

where, as usual, χ_{E_i} is the indicator function of the set E_i.

A function $f : \Omega \to \mathbf{B}$ is called μ-measurable if there exists a sequence of simple functions $\{f_n\}$ with $\lim_n \| f_n - f \|_{\mathbf{B}} = 0$ μ- a.e.

As a consequence of Pettis' measurability theorem, it can be proved that a μ-measurable function is almost everywhere the uniform limit of a sequence of countably valued μ-measurable function, see Corollary 3, Chap. 2 [68].

There are other notions of measurability, weak measurability, and weak* measurability but we are not going to consider them here, see [68].

Definition 5.6 A μ-measurable function $f : \Omega \to \mathbf{B}$ is called Bochner integrable if there exists a sequence of simple functions $\{f_n\}$ such that

$$\lim_n \int_\Omega \| f_n - f \|_\mathbf{B} \, d\mu = 0,$$

where $\int_E f d\mu$ is defined for each $E \in \mathscr{F}$ by

$$\int_E f d\mu = \lim_n \int_E f_n d\mu,$$

where $\int_E f_n d\mu$ is defined as

$$\int_E f_n d\mu = \sum_{i=1}^n x_{n,i} \mu(E_{n,i} \cap E),$$

and the convergence is in the norm.

Observe that $\int_E f d\mu \in \mathbf{B}$ for each $E \in \mathscr{F}$.

A concise characterization of Bochner's integrable functions is given next.

Theorem 5.15 *A μ-measurable function $f : \Omega \to \mathbf{B}$ is Bochner integrable if and only if*

$$\int_\Omega \| f \|_\mathbf{B} d\mu < \infty.$$

Proof If f is Bochner integrable, let $\{f_n\}$ be a defining sequence of simple functions for the integral of f. Then

$$\int_\Omega \| f \|_\mathbf{B} d\mu \leqslant \int_\Omega \| f - f_n \|_\mathbf{B} \, d\mu + \int_\Omega \| f_n \|_\mathbf{B} \, d\mu < \infty,$$

for sufficiently large n.

Conversely, suppose f (and consequently $\| f \|$) is μ-measurable and $\int_\Omega \| f \| d\mu < \infty$. It can be proved, see Corollary 1.3 of Chap. 2 of [68], that one can choose a sequence of countably valued measurable functions $\{f_n\}$ such that $\| f - f_n \|_\mathbf{B} \leqslant 1/n$ for each positive integer n. Since $\| f_n \|_\mathbf{B} \leqslant \| f \|_\mathbf{B} + 1/n$ μ-a.e. and μ is finite, then $\int_\Omega \| f_n \|_\mathbf{B} \, d\mu < \infty$. For each positive integer n, write

$$f_n = \sum_{m=1}^\infty x_{n,m} \chi_{E_{n,m}},$$

where $E_{n,i} \cap E_{n,j} = \varnothing$ for $i \neq j$, $E_{n,m} \in \mathscr{F}$, $x_{n,m} \in \mathbf{B}$. For each n, choose p_n so large that

$$\int_{\bigcup_{m=p_n+1}^{\infty} E_{n,m}} \|f_n\|_{\mathbf{B}}\, d\mu < \mu(\Omega)/n$$

and set $g_n = \sum_{m=1}^{p_n} x_{n,m} \chi_{E_{n,m}}$. Then each g_n is a simple function and

$$\int_{\Omega} \|f - g_n\|_{\mathbf{B}}\, d\mu \leqslant \int_{\Omega} \|f - f_n\|_{\mathbf{B}}\, d\mu + \int_{\Omega} \|f_n - g_n\|_{\mathbf{B}}\, d\mu$$

$$\leqslant \mu(\Omega)/n + \mu(\Omega)/n = 2\mu(\Omega)/n.$$

Therefore f is Bochner integrable. $\qquad\qquad\qquad\qquad\qquad\qquad\qquad\qquad\qquad\square$

The Lebesgue dominated convergence theorem for Bochner integral is immediate by applying the scalar version to $\|f_n - f\|_{\mathbf{B}}$.

Further elementary facts about the Bochner integral are collected next.

Theorem 5.16 *If f is a μ-Bochner integrable function, then*

(i) $\lim_{\mu(E)\to 0} \int_E f\, d\mu = 0.$
(ii) For all $E \in \mathscr{F}$

$$\left\| \int_E f\, d\mu \right\|_{\mathbf{B}} \leqslant \int_E \|f\|_{\mathbf{B}}\, d\mu.$$

(iii) If $\{E_n\}$ is a sequence of pairwise disjoint members of $\mathscr{F}$ and $E = \bigcup_{n=1}^{\infty} E_n$, then

$$\int_E f\, d\mu = \sum_{n=1}^{\infty} \int_{E_n} f\, d\mu,$$

where the sum on the right is absolutely convergent.
(iv) If $v(E) = \int_E f\, d\mu$, then v is a vector measure of bounded variation and

$$|v|(E) = \int_E \|f\|_{\mathbf{B}}\, d\mu \quad \text{for all } E \in \mathscr{F}.$$

Proof

- Since $\lim_{\mu(E)\to 0} \int_E \|f\|_{\mathbf{B}}\, d\mu = 0$ for $f \in L^1(\mu)$, statement (i) follows from (ii).
- To prove (ii), note that the triangle inequality establishes (ii) for simple functions. For the general case, pass to the appropriate limit.

- First note that the series $\sum_{n=1}^{\infty} \int_{E_n} f \, d\mu$ is dominated term by term by the convergent series of non-negative numbers $\sum_{n=1}^{\infty} \int_{E_n} \|f\|_{\mathbf{B}} d\mu \left(\leqslant \int_{\Omega} \|f\|_{\mathbf{B}} d\mu < \infty\right)$. Therefore $\sum_{n=1}^{\infty} \int_{E_n} f \, d\mu$ is absolutely convergent. To check its limit note that

$$\left\| \int_{\bigcup_{n=1}^{\infty} E_n} f \, d\mu - \sum_{n=1}^{m} \int_{E_n} f \, d\mu \right\|_{\mathbf{B}} = \left\| \int_{\bigcup_{n=m+1}^{\infty} E_n} f \, d\mu \right\|_{\mathbf{B}},$$

by the finite additivity property of Bochner integral.

Moreover $\lim_m \mu \left(\bigcup_{n=m+1}^{\infty} E_n \right) = 0$, and this implies that

$$\lim_m \left\| \int_{\bigcup_{n=m+1}^{\infty} E_n}^{\infty} f \, d\mu \right\|_{\mathbf{B}} = 0$$

and, consequently,

$$\int_{\bigcup_{n=1}^{\infty} E_n} f \, d\mu = \sum_{n=1}^{\infty} \int_{E_n} f \, d\mu$$

as required.

- To prove (iv), note that if π is a partition of a set $E \in \mathscr{F}$, then

$$\sum_{A \in \pi} \|\nu(A)\|_{\mathbf{B}} = \sum_{A \in \pi} \left\| \int_A f \, d\mu \right\|_{\mathbf{B}}$$
$$\leqslant \sum_{A \in \pi} \int_A \|f\|_{\mathbf{B}} d\mu = \int_E \|f\|_{\mathbf{B}} d\mu.$$

Hence $|\nu|(E) \leqslant \int_E \|f\|_{\mathbf{B}} d\mu$ and ν is of bounded variation. To prove the reverse inequality, let $\varepsilon > 0$ and select a sequence $\{f_n\}$ of simple functions such that

$$\lim_n \int_{\Omega} \|f - f_n\|_{\mathbf{B}} \, d\mu = 0.$$

Fix n_0 such that $\int_{\Omega} \|f - f_{n_0}\|_{\mathbf{B}} \, d\mu < \varepsilon$ and choose a partition π' of E such that

$$\sum_{A \in \pi'} \left\| \int_A f_{n_0} d\mu \right\|_{\mathbf{B}} = \int_E \|f_{n_0}\|_{\mathbf{B}} \, d\mu.$$

Finally, choose a partition π of E refining π' such that

$$|\nu|(E) - \sum_{B \in \pi} \left\| \int_B f \, d\mu \right\|_{\mathbf{B}} < \varepsilon.$$

$\square$

The following theorem exhibits a strong property of Bochner integration that has no analog in the theory of Lebesgue integration.

Theorem 5.17 (Hille) *Let T be a continuous linear operator defined on a Banach space $\mathbf{B}_1$ and having values in a Banach space $\mathbf{B}_2$. If f and Tf are Bochner integrable with respect to μ, then*

$$T\left(\int_E f\, d\mu\right) = \int_E Tf\, d\mu$$

for all $E \in \mathscr{F}$.

Proof Let $\varepsilon > 0$ and select a function $h_\varepsilon = \sum_{n=1}^{\infty} x_n \chi_{E_n}$, where $\{E_n\}$ is a sequence of pairwise disjoint members of $\mathscr{F}$, $E_n \subseteq E$ and $x_n \in \mathbf{B}_1$, such that

$$\sup\left\{\|f(x) - h_\varepsilon(x)\|_{\mathbf{B}_2} : x \in E\backslash N_1\right\} < \varepsilon/2$$

for some μ-null set N_1. Also we can find a function g_ε of the form $g_\varepsilon = \sum_{n,m=1}^{\infty} y_{n,m} \chi_{E_{n,m}}$, where $\left\{E_{n,m}\right\}$ is a sequence of pairwise disjoint members of $\mathscr{F}$, $\bigcup_{m=1}^{\infty} E_{n,m} = E_n$, $y_{n,m} \in \mathbf{B}_2$, such that

$$\sup\left\{\|Tf(x) - g_\varepsilon(x)\|_{\mathbf{B}_2} : x \in E\backslash N_2\right\} < \varepsilon/2$$

for some μ-null set N_2. For each pair (n, m) of positive integers, pick $x_{n,m} \in E_{n,m}$ arbitrarily. Set $\phi = \sum_{n,m} f\left(x_{n,m}\right) \chi_{E_{n,m}}$. Then it follows that

$$\|f(x) - \phi(x)\|_{\mathbf{B}_2} < \varepsilon \quad \text{for } x \notin N_1.$$

and

$$\|Tf(x) - T\phi(x)\|_{\mathbf{B}_2} < \varepsilon \quad \text{for } x \notin N_2.$$

Moreover,

$$\int_\Omega \|f - \phi\|_{\mathbf{B}_2} d\mu < \varepsilon \mu(\Omega) \text{ and } \int_\Omega \|Tf - T\phi\|_{\mathbf{B}_2} d\mu < \varepsilon \mu(\Omega).$$

Also

$$\int_E \phi\, d\mu = \lim_{k,j} \sum_{n=1}^{k} \sum_{m=1}^{j} f\left(x_{n,m}\right) \mu\left(E_{n,m}\right),$$

and

$$\int_E T\phi\, d\mu = \lim_{k,j} \sum_{n=1}^{k} \sum_{m=1}^{j} Tf\left(x_{n,m}\right) \mu\left(E_{n,m}\right).$$

Since T is continuous, it follows that $\int_E \phi d\mu$ belongs to the domain of T and that

$$T\left(\int_E \phi d\mu\right) = \int_E T\phi d\mu.$$

Now replace ε with a sequence $\varepsilon_i \to 0$ and replace ϕ with the corresponding sequence ϕ_i. Then one has

$$\int_E \phi_i d\mu \to \int_E f d\mu, \quad \text{and} \quad \int_E T\phi_i d\mu \to \int_E Tf d\mu.$$

Since $T\left(\int_E \phi_i d\mu\right) = \int_E T\phi_i d\mu$ and T is continuous, it follows that $T\left(\int_E f d\mu\right) = \int_F Tf du$. $\qquad\square$

The indefinite Bochner integrals share one of the most important property with the indefinite Lebesgue integral, Lebesgue's differentiation theorem, see Theorem 9 Chap. 2 of [68].

Next, let us consider the Lebesgue-Bochner spaces. If $1 \leqslant p < \infty$, the space $L^p(\Omega, \mathscr{F}, \mu : \mathbf{B})$ $(= L^p(\mu : \mathbf{B}))$ is the set of all (equivalence classes of) μ-Bochner integrable functions $f : \Omega \to \mathbf{B}$ such that

$$\|f\|_p = \left(\int_\Omega \|f\|_\mathbf{B}^p d\mu\right)^{1/p} < \infty.$$

Normed by the functional $\|\cdot\|_p$ defined above $L^p(\mu, \mathbf{B})$ becomes a Banach space, a fact whose proof is the same as the real-valued case. The space $L^\infty(\Omega, \mathscr{F}, \mu, \mathbf{B})$ $(= L^\infty(\mu, \mathbf{B}))$ is the set of all (equivalence classes of) essentially bounded μ Bochner integrable functions $f : \Omega \to \mathbf{B}$. Normed by the functional $\|\cdot\|_\infty$ defined for $f \in L^\infty(\mu, \mathbf{B})$ by

$$\|f\|_\infty = \text{ess sup} \|f\|_\mathbf{B},$$

$L^\infty(\mu, \mathbf{B})$ becomes a Banach space.

If $\nu : \mathscr{F} \to \mathbf{B}$ is any countably additive continuous measure of bounded variation with finite-dimensional range, then the classical Radon-Nikodým theorem produces a Bochner integrable function f for which $\nu(E) = \int_E f du$. For the general Bochner integral, this is no longer necessarily true.

Example 5.1 A countably additive c_0-valued[5] vector measure of bounded variation that has no Radon-Nikodým derivative. Let μ be Lebesgue measure on $[0, 1]$. For a measurable set $E \subseteq [0, 1]$, write

$$\lambda_n(E) = \int_E \sin\left(2^n \pi t\right) dt$$

---

[5] c_0 is the vector space of real value sequences $\{x_n\}$ such that $\lim_{n \to \infty} x_n = 0$. c_0 is a Banach space with the infinity norm.

and let

$$\nu(E) = (\lambda_1(E), \lambda_2(E), \ldots, \lambda_n(E), \ldots).$$

The Riemann-Lebesgue lemma guarantees that the finitely additive measure F defined above is c_0-valued. Moreover,

$$\|\nu(E)\|_{c_0} \leqslant \sup_n \int_E \left|\sin\left(2^n \pi t\right)\right| dt \leqslant \mu(E)$$

for each measurable set $E \subseteq [0, 1]$. It follows that ν is countably additive, μ-continuous, and of bounded variation.

Now suppose there is a Bochner integrable $f : [0, 1] \to c_0$ such that $\nu(E) = \int_E f \, d\mu$ for each measurable $E \subseteq [0, 1]$. Write $f = (f_1, f_2, \ldots, f_n, \ldots)$. Since the coordinate functionals on c_0 are all continuous linear functionals, each f_n is measurable and the equality $\lambda_n(E) = \int_E f_n d\mu$ holds for each E and each n. Hence $f_n(t) = \sin\left(2^n \pi t\right)$ for almost all $t \in [0, 1]$. Consider

$$E_n = \left\{ t \in [0, 1] : f_n(t) \geqslant 1/\sqrt{2} \right\}.$$

Then $\mu\left(E_n\right) = \frac{1}{4}$ for each n. Moreover

$$\mu\left(\overline{\lim}_j\left(E_j\right)\right) \geqslant \overline{\lim}_j \mu\left(E_j\right) \geqslant \frac{1}{4}.$$

Hence,

$$\mu\left(\{t \in [0, 1] : f(t) \in c_0\}\right) \leqslant 3/4,$$

which is a contradiction with the definition of f.

The failure of the Radon-Nikodým theorem for the Bochner integral is not to be interpreted as a negative aspect of the Bochner integral. Indeed, the failure of a general Radon-Nikodým theorem for the Bochner integral in special cases has powerful repercussions in operator theory, the geometry of Banach spaces, duality theory for Banach spaces, vector-valued probability theory, and integration theory itself.

Definition 5.7 A Banach space $\mathbf{B}$ has the Radon-Nikodým property with respect to $(\Omega, \mathscr{F}, \mu)$ if for each μ-continuous vector measure $\nu : \mathscr{F} \to \mathbf{B}$ of bounded variation there exists $g \in L^1(\mu, \mathbf{B})$ such that

$$\nu(E) = \int_E g \, d\mu$$

for all $E \in \mathscr{F}$.

A Banach space $\mathbf{B}$ has the Radon-Nikodým property if $\mathbf{B}$ has the Radon-Nikodým property with respect to every finite measure space.

Now, in order to define martingales on a Banach space $\mathbf{B}$ observe first that the notion of conditional expectation on $\mathbf{B}$ is completely similar to the scalar case studied in Chap. 2.

Definition 5.8 Let $\mathscr{G}$ be a sub-σ-algebra of $\mathscr{F}$ and $f \in L^1(\mu, \mathbf{B})$. An element g of $L^1(\mu, \mathbf{B})$ is called the conditional expectation of f relative to $\mathscr{G}$ if g is $\mathscr{G}$-measurable and

$$\int_E g d\mu = \int_E f d\mu \text{ for all } E \in \mathscr{G}.$$

In this case, g is denoted by $\mathbf{E}[f \mid \mathscr{G}]$.

The properties of conditional expectation on $\mathbf{B}$ are totally analogous to the ones studied in Chap. 2 for the scalar case, see Theorem 2.1.

Now, considering Definition 3.1, but using the analytic version (3.2), and Bochner integral instead of Lebesgue integral we can define martingales with values in a Banach space $\mathbf{B}$ as follows.

Definition 5.9 A stochastic process $X = \{X_n\}_n$ with values in a Banach space $\mathbf{B}$ is a $\{\mathscr{F}_n\}$-martingale if:

(i) For each n, $X_n \in L^1(\mu, \mathbf{B})$.
(ii) $\{X_n\}$ is $\{\mathscr{F}_n\}$-adapted, i.e., for each n X_n is $\mathscr{F}_n$-measurable.
(iii) For each $n, m \in \mathbb{N}$, if $m > n$, then for any $A \in \mathscr{F}_n$

$$\int_A X_m d\mu = \int_A X_n d\mu. \tag{5.15}$$

Example 5.2

(i) From an infinite tree in B can be obtained a martingale in $L^1([0, 1), \mathbf{B})$.
Remember that an infinite tree $\mathscr{J}$ in a Banach space $\mathbf{B}$ is a sequence $\{x_n\}$ in $\mathbf{B}$ with the property that

$$x_n = \frac{(x_{2n} + x_{2n+1})}{2},$$

for all $n \geqslant 1$. The element x_1 is the trunk of the tree and the element x_j, for $j = 2^k, 2^k + 1, \ldots, 2^{k+1} - 1$ and it is the kth year's growth of the tree.
To realize a tree $\mathscr{J} = \{x_n\}$ as a martingale in $L^1([0, 1), \mathbf{B})$, let μ be Lebesgue measure and consider

$$f_1 = x_1 \chi_{[0,1)}, \quad f_2 = x_2 \chi_{[0,1/2)} + x_3 \chi_{[1/2,1)}$$

and for $k \geqslant 3$

$$f_k = \sum_{i=2^{k-1}}^{2^k-1} x_i \chi_{I_{k,i}},$$

where

$$I_{k,i} = \left[\left(i - 2^{k-1} \right) / 2^{k-1}, \left(i - 2^{k-1} + 1 \right) / 2^{k-1} \right),$$

for $i = 2^{k-1}, 2^{k-1} + 1, \ldots, 2^k - 1$ and $k \geqslant 1$.
Immediately note that

$$\int_{[0,1)} f_1 d\mu = \int_{[0,1)} f_2 d\mu$$

since $(x_2 + x_3)/2 = x_1$. For the same reason, it follows that

$$\int_{I_{k,i}} f_{k+1} d\mu = \int_{I_{k,i}} f_k d\mu$$

for each $i = 2^{k-1}, \ldots, 2^k - 1$ and $k \geqslant 1$.
Thus if $\mathscr{F}_0$ is the trivial σ-algebra consisting of $\varnothing$ and $[0, 1)$ and $\mathscr{F}_k$ is the finite σ-algebra generated by

$$\left\{ I_{k,i}, i = 2^{k-1}, 2^{k-1} + 1, \ldots, 2^k - 1 \right\}, k \geqslant 1,$$

then $\{f_k\}$ is a $\{\mathscr{F}_k\}$-martingale in $L^1([0, 1), \mathbf{B})$.

(ii) An example of a non-trivial infinite tree can be found in $L^1[0, 1)$ which is a non-convergent bounded martingale in $L^1([0, 1), \mathbf{B})$:
Let $x_1 = \chi_{[0,1)}, x_2 = 2\chi_{[0,1/2)}, x_3 = 2\chi_{[1/2,1)}$ and $x_i = 2^{k-1}\chi_{I_{k,i}}$ for $i = 2^{k-1}, \ldots, 2^k - 1$ and $k \geqslant 1$ where the intervals $I_{k,i}$ are the same as above. Furthermore, note that

$$\|x_1 - x_2\|_1 = \int_{[0,1)} \left| \chi_{[0,1)} - 2\chi_{[0,1/2)} \right| d\mu = \int_{[0,1)} 1 d\mu = 1.$$

For the same reason, one has

$$\|x_n - x_{2n}\|_1 = \|x_n - x_{2n+1}\|_1 = 1 \text{ for all } n \in \mathbb{N}.$$

Transferring to the martingale in $L^1([0, 1), \mathbf{B})$ with $\mathbf{B} = L^1[0, 1)$ constructed above, we find that $\|f_n(t)\|_1 = 1$ for all $t \in [0, 1)$ since $\|x_k\|_1 = 1$ for all k and

$$\|f_n(t) - f_{n+1}(t)\|_1 = 1$$

for all $t \in [0, 1)$ (since if $t \in [0, 1)$ and $I_{n,i}$ is selected such that $t \in I_{n,i}$, then $f_n(t) = x_i$ and $f_{n+1}(t)$ is either x_{2i} or x_{2i+1}.

Therefore, $\{f_n\}$ is a $\{\mathscr{F}_k\}$-martingale of uniformly bounded functions in $L^1\big([0, 1) : L^1[0, 1)\big)$ with the property $\|f_n(t) - f_{n+1}(t)\|_1 \equiv 1$ for all n.

It follows immediately that $\{f_n\}$ is not convergent in $L^1\big([0, 1) : L^1[0, 1)\big)$-norm. Observe that this example is impossible in the scalar value case since the difference equation diverges.

Definition 5.10 An infinite tree $\{x_n\}$ in the Banach **B** is called an infinite δ-tree if there is a $\delta > 0$ such that

$$\|x_n - x_{2n}\|_{\mathbf{B}} \geqslant \delta \quad and \quad \|x_n - x_{2n+1}\|_{\mathbf{B}} \geqslant \delta$$

for all n.

Observe that no finite-dimensional Banach space contains a bounded infinite δ-tree and this property is shared by many infinite-dimensional Banach spaces. This fact is a consequence of the convergence theorems for martingales.

The convergence theorems of martingales on Banach spaces with Radon-Nikodým property are essentially similar to those in the scalar case studied in Chap. 3, see Corollaries 3.18 and 3.19.

Theorem 5.18 (Martingale mean convergence theorem) *Let **B** be a Banach space with the Radon-Nikodým property, let* $1 \leqslant p < \infty$, *and let* $\{f_n\}$ *be a* $\{\mathscr{F}_n\}$-*martingale in* $L^p(\mu, B)$. *Then* $\lim_n f_n$ *exists in* $L^p(\mu, \mathbf{B})$-*norm if and only if*

(i) For $p = 1$: $\sup_n \|f_n\|_1 < \infty$ *and* $\{f_n\}$ *is uniformly integrable or*
(ii) For $1 < p < \infty$: $\sup_n \|f_n\|_p < \infty$.

Now let us see what the martingale mean convergence theorem says in terms of infinite δ-trees.

Corollary 5.3 *No Banach space with the Radon-Nikodým property contains a bounded infinite δ-tree.*

The martingale mean convergence theorem has a converse that ties mean martingale convergence to the Radon-Nikodým property.

Theorem 5.19 *Let **B** be a Banach space such that for every finite measure space* $(\Omega, \mathscr{F}, \mu)$ *every bounded uniformly integrable martingale in* $L^1(\mu, \mathbf{B})$ *converges in* $L^1(\mu, \mathbf{B})$-*norm. Then **B** has the Radon-Nikodým property.*

Proof Let $(\Omega, \mathscr{F}, \mu)$ be a fixed finite measure space. Let $\mathscr{P}$ be the class of all partitions of Ω into $\mathscr{F}$ sets and direct $\mathscr{P}$ by refinement. If $\nu : \mathscr{F} \to \mathbf{B}$ is a μ-continuous vector measure of bounded variation, define for any $\pi \in \mathscr{P}$,

$$f_\pi = \sum_{E \in \pi} \frac{\nu(E)}{\mu(E)} \chi_E$$

observing the convention that $0/0 = 0$. Let $\mathscr{B}_\pi$ be the (trivial) σ-field generated by π. Evidently, $\{f_\pi\}$ is a $\{\mathscr{B}_\pi\}$-martingale in $L^1(\mu, \mathbf{B})$ with $\nu(E) = \lim_\pi \int_E f_\pi d\mu$ for every $E \in \mathscr{F}$. Then,

$$\int_\Omega \| f_\pi \|_1 \, d\mu \leqslant |\nu|(\Omega) < \infty$$

for every $\pi \in \mathscr{P}$. Also, since $\nu \ll \mu$, we have $|\nu| \ll \mu$. Hence for each $\varepsilon > 0$, there is $\delta > 0$ such that $|\nu|(E) < \varepsilon$ whenever $\mu(E) < \delta$. Now if $E \in \mathscr{B}_\pi$ and $\mu(E) < \delta$, then

$$\int_E \| f_\pi \|_1 \, d\mu \leqslant |\nu|(E) < \varepsilon.$$

Thus $\{f_\pi\}$ is uniformly integrable. By hypothesis $\lim_\pi \| f_\pi - f \|_1 = 0$ for some $f \in L^1(\mu, B)$. Thus we have

$$\nu(E) = \lim_\pi \int_E f_\pi d\mu = \int_E f d\mu$$

for all $E \in \mathscr{F}$. $\qquad\qquad\square$

Doob's maximal inequality also holds in Banach spaces, see Lemma 7, Chap. 5 of [68] and therefore it can also be proved the a.e. convergence of any L^1-convergent martingale, similarly as Proposition 3.13; see also Theorem 8, Chap. 5 of [68].

Corollary 5.3 preludes the geometric concept of σ-dentability which will be discussed now. First let us construct a martingale which is uniformly integrable and non-convergent. Let $\{f_n\}$ a $\{\mathscr{F}_n\}$-martingale of countably valued functions in $L_1([0, 1), \mathbf{B})$ with the properties that

(i) $\sup_n \| f_n \|_\infty < \infty$ and
(ii) $\| f_n(t) - f_{n+1}(t) \|_\mathbf{B} \geqslant \varepsilon$, for some $\varepsilon > 0$ and all $t \in [0, 1)$.

Let us now see how to construct such a martingale. First, since each f_n is countably valued, it must has the form

$$f_n = \sum_{E \in \Delta_n} x_E \chi_E,$$

where $x_E \in \mathbf{B}\,(f_n(E) = x_E)$ and each $\{\Delta_n\}$ is a partition of Ω of sets in $\mathscr{F}$ of positive μ-measure, i.e., $\Omega = \bigcup_{E \in \Delta_n} E$ and $\{\Delta_n\}$ are disjoint sets. In addition, we may and do assume that $\mathscr{F}_n$ is the σ-field generated by Δ_n. In this case, the inclusion $\mathscr{F}_n \subseteq \mathscr{F}_{n+1}$ means that each $E \in \Delta_n$ can be written as

$$E = \bigcup_{A \subseteq E;\, A \in \Delta_{n+1}} A.$$

Note that

(i) means that the set $D = \{x_A : A \in \Delta_n, n = 1, 2, \ldots\}$ is bounded.
(ii) means that $\|x_A - x_E\|_{\mathbf{B}} \geqslant \varepsilon$ whenever $E \in \Delta_n$, $A \in \Delta_{n+1}$ and $A \subseteq E$.

In addition, the martingale property means that if $E \in \Delta_n$, then

$$x_E \mu(E) = \int_E f_n d\mu = \int_E f_{n+1} d\mu = \sum_{A \subseteq E;\, A \in \Delta_{n+1}} \int_A f_{n+1} d\mu = \sum_{A \subseteq E;\, A \subset \Delta_{n+1}} x_A \mu(A).$$

Thus we have

$$x_E = \sum_{A \subseteq E;\, A \in \Delta_{n+1}} \frac{\mu(A)}{\mu(E)} x_A$$

for each $E \in \Delta_n$. Note that the sum on the right-hand side is a (infinite) convex sum.

Definition 5.11 A subset D of a Banach space $\mathbf{B}$ is a non-σ-dentable set if there exists an $\varepsilon > 0$ such that each $x \in D$ has the form $x = \sum_{i=1}^{\infty} \alpha_i x_i$ where $\sum_{i=1}^{\infty} \alpha_i = 1, \alpha_i > 0, x_i \in D$ and $\|x - x_i\|_{\mathbf{B}} \geqslant \varepsilon$ for all i.

Observe that any bounded infinite δ-tree gives an example of a non-σ-dentable set.

Theorem 5.20 (Maynard) *Suppose $\mathbf{B}$ is a Banach space containing a bounded non-σ-dentable set D. Then there exists $\delta > 0$ and a martingale $\{f_n\}$ in $L^1([0, 1), \mathbf{B})$ such that $f_n([0, 1)) \subseteq D$ and*

$$\|f_n(t) - f_{n+1}(t)\|_{\mathbf{B}} \geqslant \delta,$$

for all $n \in N$ and $t \in [0, 1)$.

Consequently, a Banach space containing a bounded non-σ-dentable set is a Banach space without the Radon-Nikodým property.

Proof Let $\mathbf{B}$ be a Banach space and D be a bounded non-σ-dentable subset of $\mathbf{B}$. It is enough to construct $\{f_n\}$, a $\{\mathscr{F}_n\}$-martingale in $L^1([0, 1), \mathbf{B})$ with $f_n([0, 1)) \subseteq D$ and $\|f_n(t) - f_{n+1}(t)\|_{\mathbf{B}} \geqslant \varepsilon$ for some fixed $\varepsilon > 0$ and all $t \in [0, 1)$.

In order to construct the martingale $\{f_n\}$, by the condition on D there is a $\varepsilon > 0$ such that for each $x \in D$ there is a sequence of positive real numbers $\{\alpha_n(x)\}$ with

$\sum_{n=1}^{\infty} \alpha_n(x) = 1$ and a sequence $\{x_n(x)\}$ in D with $\|x - x_n(x)\|_{\mathbf{B}} \geq \varepsilon$ for all n such that

$$x = \sum_{n=1}^{\infty} \alpha_n(x) x_n(x). \tag{5.16}$$

Pick $\bar{x} \in D$ arbitrarily and set $f_1 = \bar{x}\chi_{[0,1)}$ and $\mathscr{F}_1 = \{\phi, [0, 1)\}$. Then we have

$$f_1 = \sum_{n=1}^{\infty} \alpha_n(\bar{x}) x_n(\bar{x}) \chi_{[0,1)}.$$

Let $\beta_m = \sum_{n=1}^{m} \alpha_n(\bar{x})$ $(\beta_0 = 0)$ and set $I_m = \left[\beta_{m-1}, \beta_m\right)$ for $m \geq 1$. Define

$$f_2 = \sum_{n=1}^{\infty} x_n(\bar{x}) \chi_{I_n}.$$

Now, taking μ the Lebesgue measure, $\mu\left(I_n\right) = \alpha_n(\bar{x})$, it follows that

$$\int_{[0,1)} f_2 d\mu = \int_{[0,1)} f_1 d\mu = \bar{x}.$$

Hence the conditional expectation $\mathbf{E}\left[f_2 \mid \mathscr{F}_1\right] = f_1$. Let $\mathscr{F}_2$ be the σ-field generated by $\{I_n\}$ and note that $\|f_2(t) - f_1(t)\|_B \geq \varepsilon$ for all $t \in [0, 1)$.

Instead of giving a formal inductive proof, we shall show how to construct f_3 and $\mathscr{F}_3$ from f_2 and $\mathscr{F}_2$. Write $\Delta_2 = \{I_n\}_{n=1}^{\infty}$, then f_2 can be written as $f_2 = \sum_{E \subset \Delta_2} x_E \chi_E$ where $x_E = x_n(\bar{x})$ when $E = I_n$. The construction of f_3 from f_2 is similar to the construction of f_2 from f_1 except that we must work inside each of the E's $\in \Delta_2$. Fix $E \in \Delta_2$ and note that by (5.16)

$$x_E = \sum_{n=1}^{\infty} \alpha_n\left(x_E\right) x_n\left(x_E\right)$$

with $\alpha_n\left(x_E\right)$ and $x_n\left(x_E\right)$ as in (5.16). Then one has

$$f_2 \chi_E = \left(\sum_{n=1}^{\infty} \alpha_n\left(x_E\right) x_n\left(x_E\right)\right) \chi_E.$$

By the usual measure theory argument, it is enough to consider E to be half-open intervals. Let $E = [a, b)$, and $\beta_n = (b - a) \sum_{m=1}^{n} \alpha_n\left(x_E\right)$ with $\beta_0 = 0$. Also for the moment, let $I_n = \left[a + \beta_{n-1}, a + \beta_n\right)$ and define f_3 on E by

$$f_3 \chi_E = \sum_{n=1}^{\infty} x_n\left(x_E\right) \chi_{I_n}.$$

Since $\mu(I_n) = \alpha_n(x_E)(b-a)$, we have

$$\int_E f_3 d\mu = (b-a)x_E = \int_E f_2 d\mu.$$

Thus, defining f_3 on each $E \in \Delta_2$ as above; it follows Δ_2 from above that

$$\mathbf{E}[f_3 \mid \mathscr{F}_2] = f_2.$$

Further from (5.16) we obtain $f_3([0,1)) \subseteq D$ and $\|f_3(t) - f_2(t)\|_B \geqq \varepsilon$ for all $t \in [0,1)$. Finally, let $\mathscr{F}_3$ be the σ-field generated by all the intervals I_n constructed above as E ranges over Δ_2. $\square$

5.3.2 Khintchine's Inequality in Banach Spaces

The *Khintchine's inequality* (1.15), discussed in detail in Chap. 1 and which, as we saw in Chap. 4, is the seminal root of Burkholder's inequality in martingale theory, also has important implications in the geometry of Banach spaces that we will review briefly.

Definition 5.12 (*Type and cotype*) Let $\mathbf{B}$ be a Banach space, let $p \in [1,2]$ and $q \in [2, \infty]$.

(i) The space $\mathbf{B}$ is said to have type p if there exists a constant $\tau \geqslant 0$ such that for all finite sequences $x_1, \ldots, x_N$ in B we have

$$\left(\mathbf{E}\left\|\sum_{n=1}^{N} x_n r_n\right\|_{\mathbf{B}}^{p}\right)^{1/p} \leqslant \tau \left(\sum_{n=1}^{N} \|x_n\|_{\mathbf{B}}^{p}\right)^{1/p}. \tag{5.17}$$

(ii) The space $\mathbf{B}$ is said to have cotype q if there exists a constant $c \geqslant 0$ such that for all finite sequences $x_1, \ldots, x_N$ in X we have

$$\left(\sum_{n=1}^{N} \|x_n\|_{\mathbf{B}}^{q}\right)^{1/q} \leqslant c \left(\mathbf{E}\left\|\sum_{n=1}^{N} x_n r_n\right\|_{\mathbf{B}}^{q}\right)^{1/q}, \tag{5.18}$$

with the usual modification for $q = \infty$.

where $\{r_n\}$ is the family of Rademacher functions considered in Chap. 1.

We denote by $\tau_{p,\mathbf{B}}$ and $c_{q,\mathbf{B}}$ the least admissible constants in (5.17) and (5.18) and refer to them as the type p constant and cotype q constant of $\mathbf{B}$, respectively.

It is easy to check that the inequalities defining type and cotype cannot be satisfied for any $p > 2$ and $q < 2$, respectively, even in one-dimensional spaces $\mathbf{B}$. This explains the restrictions imposed on these numbers.

Example 5.3 Every Hilbert space $\mathbf{H}$ has type 2 and cotype 2, with constants $\tau_{2,\mathbf{H}} = 1$ and $c_{2,\mathbf{H}} = 1$. Indeed, for all $h_1, \ldots, h_N \in H$ we have

$$\mathbf{E}\left\| \sum_{n=1}^{N} h_n r_n \right\|_{\mathbf{H}}^{2} = \mathbf{E}\left[\left\langle \sum_{n=1}^{N} h_n r_n \mid \sum_{n=1}^{N} h_n r_n \right\rangle \right] = \mathbf{E}\left[\sum_{n=1}^{N} \sum_{m=1}^{N} r_n \bar{r}_m \langle h_n \mid h_m \rangle \right]$$

$$= \sum_{n=1}^{N} \sum_{m=1}^{N} \mathbf{E}\left[r_n \bar{r}_m \right] \langle h_n \mid h_m \rangle = \sum_{n=1}^{N} \langle h_n \mid h_n \rangle = \sum_{n=1}^{N} \| h_n \|_{\mathbf{H}}^{2} .$$

This result can be considered as a generalization of the parallelogram law.

By the triangle inequality, every Banach space $\mathbf{B}$ has type 1, with constant $\tau_{1,B} = 1$. Likewise, the inequalities

$$\sup_{n \geq 1} \| x_n \| \leq \left\| \sum_{n=1}^{N} \varepsilon_n x_n \right\|_{L^1(\Omega;\mathbf{B})} \leq \left\| \sum_{n=1}^{N} \varepsilon_n x_n \right\|_{L^\infty(\Omega;\mathbf{B})},$$

show that every Banach space has cotype ∞, with constant $c_{\infty,\mathbf{B}} = 1$. We say that $\mathbf{B}$ has non-trivial type if $\mathbf{B}$ has type p for some $p \in (1, 2]$, and finite cotype if it has cotype q for some $q \in [2, \infty)$.

From the fact that the sequence space norm $\| \cdot \|_{\ell_N^t(\mathbf{B})}$ is decreasing and the probability space norm $\| \cdot \|_{L^t(\Omega;\mathbf{B})}$ is increasing in $t \in [1, \infty)$, it follows that if $\mathbf{B}$ has type p (cotype q), then it also has type u for all $u \in [1, p]$ (cotype v for all $v \in [q, \infty]$) and $\tau_{u,\mathbf{B}} \leq \tau_{p,B}$ $\left(c_{v,\mathbf{B}} \leq c_{q,\mathbf{B}} \right)$. In particular, every Hilbert space $\mathbf{H}$ is also an example of a space with type p and cotype q for every $p \in [1, 2]$ and $q \in [2, \infty]$.

On the other hand, Kwapien [110] proved that if $\mathbf{B}$ is a Banach space where the Khintchine inequality verifies for coefficients in such a space for some $p > 1$, in particular, for $p = 2$, for instance, then it is isomorphic to a Hilbert space. Additionally, another Kwapien's result says that a Banach space with Rademacher type and cotype 2 is equivalent to a Hilbert space up to isomorphism. Thus Hilbert spaces are the only Banach spaces with both type 2 and cotype 2. On the other hand, the optimality constants for the Khintchine inequality on Banach spaces were obtained by U. Haagerup, see [94].

5.3.3 Unconditional Martingale Differences (UMD) Spaces

In this final section, we will review, without proof, a very important class of Banach spaces that, in some sense, are natural for analysis on them and that can be characterized using martingale theory.

Recall, from Chap. 4, that by Theorem 4.7 if $X = \{X_n\}_n$ is a $\{\mathscr{F}_n\}$-martingale and $Y = \{Y_n\}_n$ is a multiplier sequence then if X is L^1-bounded $(Y \circ X) = \{(Y \circ X)_n\}_n$ the martingale transform of X by Y converges a.e.,

$$\text{If } \| X \|_1 < \infty \text{ then } \{(Y \circ X)_n\}_n \text{ converges a.e.} \tag{5.19}$$

On the other hand, by Theorem 4.43 we know that

$$\lambda \mathbf{P}\{(Y \circ X) > \lambda\} \leqslant c_1 \|X\|_1,$$

and from Theorem 4.19 for $1 < p < \infty$,

$$\|(Y \circ X)\|_p \leqslant c_p \|X\|_p.$$

As mentioned before, these results are analogous to the classical theorems of I. Privalov, A. Kolmogorov, and M. Riesz, respectively, which show how harmonic functions control their conjugates.

Now, a natural question is what Banach spaces $\mathbf{B}$ do (5.19) hold for all $\mathbf{B}$-valued martingales and their transforms? Or for what Banach spaces $\mathbf{B}$ does there exist a positive real number C such that inequality (4.59) holds for all $\mathbf{B}$-valued martingales and their transforms? And finally, a similar question can be raised with respect to the inequality (4.54). We shall show that all three of these questions lead to the same class of Banach spaces, the unconditional martingale differences spaces, or simply UMD spaces.

The equivalence of (5.19) and (4.59) in a Banach space setting contrasts with the nonequivalence of two results of Doob [69]: For $\mathbf{B} = \mathbb{R}$ from Theorem 3.14 we have

$$\text{if} \ \ \|X\|_1 < \infty \ \ \text{then} \ \{X_n\}_n \ \text{converges a.e.}$$

and from the maximal inequality (3.27), Theorem 3.7

$$\lambda \mathbf{P}\{X^* > \lambda\} \leqslant \|X\|_1,$$

for any $\lambda > 0$.

Chatterji in [52] proved that (5.19) holds for all martingales with values in $\mathbf{B}$ if and only if $\mathbf{B}$ has the Radon-Nikodým property. On the other hand, (3.27) holds for every Banach space since $\{|X_1|, |X_2|, \ldots\}$ is a real-valued submartingale. Therefore, (5.19) and (3.27) are not equivalent conditions on a Banach space.

Additionally, it is easy to prove, as we did in Proposition 4.3, that in the presence of the Khintchine inequality, inequalities (4.44) and (4.54) are equivalent. Then, in general, from what was said above, the class of Banach spaces where the Burkholder inequality (4.54) is verified is very restricted.

Definition 5.13 (*Burkholder*) Given a Banach space $\mathbf{B}$, it is said that $\mathbf{B}$ is an unconditional martingale difference space or simply a UMD space, if for $1 < p < \infty$ there exists a positive real number $\alpha_p(B)$ such that

$$\left\|\sum_{k=1}^{n} \epsilon_k \Delta X_k\right\|_p \leqslant \alpha_p(B) \left\|\sum_{k=1}^{n} \Delta X_k\right\|_p = C_p \|X_n\|_p,$$

for all $\mathbf{B}$-valued martingales $X = \{X_n\}$, all sequences $\{\epsilon_n\}$ with $\epsilon_n \in \{-1, 1\}$, and all $n \geqslant 1$,

It is possible to replace the $\{\epsilon_n\}$ sequence by suitable random coefficients, i.e., by multiplier sequences $Y = \{Y_n\}$.

This question and closely related questions have been of interest to Maurey [118], Pisier [139], Diestel and Uhl [68], and others. This class of spaces appears to depend on p but, in fact, it does not as was proved by Maurey [118], and we will see it in another form.

We know, as a consequence of Theorem 4.19 that (4.58) holds, which implies that $\mathbb{R}$ is a UMD space, and from this follows immediately that the Lebesgue spaces ℓ^p and $L^p(0,1)$ are UMD spaces, for $1 < p < \infty$. On the other hand, it can be proved that any UMD space is reflexive, in fact, they are super-reflexive[6] see [118] so, for example, ℓ^1, ℓ^∞ are not UMD spaces. Additionally, Pisier [139] has constructed an example showing that a super-reflexive space need not be UMD, i.e., there are super-reflexive spaces that are not UMD.

On the other hand, there is also an application to singular integrals which is a very important topic in harmonic analysis. First of all a little bit of history, a question that excited the interest of more and more mathematicians during the early part of the twentieth century, see, for instance, Zygmund [176]: How does the size of a periodic function control the size of its conjugate? Consider, for example, the trigonometric polynomial

$$f(\theta) = \frac{a_0}{2} + \sum_{k=1}^{N} (a_k \cos k\theta + b_k \sin k\theta),$$

where the coefficients $a_0, \ldots, a_N, b_1, \ldots, b_N$ are real numbers and the size of f is measured by its norm $\|f\|_p$ in $L^p(0, 2\pi)$:

$$\|f\|_p^p = \int_0^{2\pi} |f(\theta)|^p d\theta.$$

The polynomial conjugate of f, denoted by $\tilde{f}$, is given by

$$\tilde{f}(\theta) = \sum_{k=1}^{N} (a_k \sin k\theta - b_k \cos k\theta).$$

The orthogonality of the trigonometric functions implies that $\|g\|_2 \leqslant \|f\|_2$ and, for many years, the question was whether or not this inequality could be extended in some sense to L^p. In fact, as Marcel Riesz eventually proved, see [143, 144],

$$\|\tilde{f}\|_p \leqslant c_p \|f\|_p, \quad 1 < p < \infty,$$

---

[6] Suppose that $\mathbf{B}_1$ and $\mathbf{B}_2$ are Banach spaces. Then $\mathbf{B}_1$ is *finitely representable* in $\mathbf{B}_2$ if, for all finite-dimensional subspaces E of $\mathbf{B}_1$ and all $\lambda > 1$, there is a linear map $T : E \to \mathbf{B}_2$ such that, for all $x \in E$,

$$\lambda^{-1} \|x\|_{\mathbf{B}_1} \leqslant \|Tx\|_{\mathbf{B}_2} \leqslant \lambda \|x\|_{\mathbf{B}_1}.$$

A Banach space $\mathbf{B}$ is *super-reflexive* if it is reflexive and every Banach space that is finitely representable in $\mathbf{B}$ is also reflexive.

where the constant depends only on p and not on the coefficients or the positive integer N. This work, with his related work on the interpolation of operators, has had a profound influence on twentieth-century analysis.

Observe the clear analogy between this problem and the inequality (4.54) of Theorem 4.43 that establish the control of the L^p-norm of the martingale transforms with the L^p-norm of the martingale.

The extension of Riesz's inequality for the non-periodic case is then completely natural. Let $f \in L^p(\mathbb{R})$ and $\varepsilon > 0$. If $H_\epsilon f$ denotes the truncated Hilbert transform of f, that is,

$$H_\varepsilon f(x) = \frac{1}{\pi} \int_{|y|>\varepsilon} \frac{f(x-y)}{y} dy,$$

then, for $1 \leqslant p < \infty$, the limit

$$H f(x) = \lim_{\epsilon \to 0} H_\varepsilon f(x)$$

exists almost everywhere and, for $1 < p < \infty$,

$$\|H f\|_p \leq c_p \|f\|_p.$$

Additionally, $f + i H f$ is the boundary function (the non-tangential limit a.e.) of a function analytic in the upper half-plane $\mathbb{R}^{d+1}_+$.

Riesz's results have been greatly extended. A decisive step was taken by A. P. Calderón and A. Zygmund [47] who showed that for a large class of kernels $K : \mathbb{R}^n \setminus \{0\} \to \mathbb{C}$, the limit

$$T f(x) = \lim_{\varepsilon \to 0} \int_{|y|>\varepsilon} f(x-y) K(y) dy$$

exists a.e. if $f \in L^p (\mathbb{R}^n)$ for some p satisfying $1 \leqslant p < \infty$,

$$\|T f\|_p \leqslant c_p \|f\|_p,$$

if $1 < p < \infty$.

For many kernels, the Calderón-Zygmund inequality can be derived from the Riesz inequality with the use of the so-called method of rotation.

Now let $\mathbf{B}$ be a Banach space (real or complex) and let $a_0, \ldots, a_N, b_1, \ldots, b_N$ be elements of $\mathbf{B}$. Then

$$f(\theta) = \frac{a_0}{2} + \sum_{k=1}^{N} (a_k \cos k\theta + b_k \sin k\theta)$$

and

$$\tilde{f}(\theta) = \sum_{k=1}^{N} (a_k \sin k\theta - b_k \cos k\theta)$$

define elements f and $\tilde{f}$ of the Lebesgue-Bochner space $L^p((0, 2\pi) : \mathbf{B})$ and the L^p-norm of f as before in the sense of Bochner integral. Within a decade or two of Riesz's discovery its validity in the B-valued case began to be considered. For example, some work of Bochner and Taylor implies that Riesz's inequality does not hold if B is the Lebesgue sequence space ℓ^1. Even the inequality

$$\|\tilde{f}\|_2 \leqslant C\|f\|_2$$

does not hold if $\mathbf{B} = \ell^1$ or ℓ^∞. For what Banach spaces $\mathbf{B}$ does

$$\|\tilde{f}\|_p \leqslant c_p\|f\|_p, \quad 1 < p < \infty,$$

hold for $1 < p < \infty$?

Similarly, after Calderón and Zygmund had written their papers on singular integral operators, a similar question arose: What happens if the assumption that $f \in L^p(\mathbb{R}^n)$ is replaced by the assumption that $f \in L^p(\mathbb{R}^d : \mathbf{B})$. For some of the early work related to this question, see, for instance, A. Benedek, A. P. Calderón, and R. Panzone [13] or E. Stein [152].

In [35] Burkholder proved that, as in the classical case, if $\mathbf{B}$ is a UMD Banach space, i.e., α_p is finite, then, the Hilbert transform for functions with values in $\mathbf{B}$, $f \in L^p(\mathbb{R} : \mathbf{B})$ $1 < p < \infty$ exists a.e., and

$$\|\mathrm{H}f\|_p \leqslant \alpha_p(B)\|f\|_p.$$

The converse was proved by J. Bourgain [22], i.e., if the Hilbert transform H exists and is bounded in $L_B^p(\mathbb{R})$ for some $1 < p < \infty$ then $\mathbf{B}$ is a UMD space, i.e., $\alpha_p(\mathbf{B})$ is finite and is, indeed, the norm of H. It is also known from [23] that the UMD-property implies the boundedness of all invariant singular integrals or more generally standard multiplier operators with some regularity assumptions. The importance of the UMD-property, especially in connection with PDEs, is further evidenced by recent new results on operator-valued singular integrals (R-boundedness) [172]. Thus, UMD Banach spaces are the Banach spaces where the classical harmonic analysis can be extended naturally.

Finally, in [32], D. Burkholder gave a geometrical characterization of UMD spaces. First, we need some definitions.

Definition 5.14 A function $\zeta : \mathbf{B} \times \mathbf{B} \to \mathbb{R}$ is symmetric if $\zeta(x, y) = \zeta(y, x)$ and is biconvex if both $\zeta(\cdot, y)$ and $\zeta(x, \cdot)$ are convex on B for all $x, y \in B$.

Definition 5.15 A Banach space $\mathbf{B}$ is ζ-convex if there is a biconvex function $\zeta : B \times B \to \mathbb{R}$ such that $\zeta(0, 0) > 0$ and

$$\zeta(x, y) \leqslant \|x + y\|_\mathbf{B} \text{ if } \|x\|_\mathbf{B} \leqslant 1 \leqslant \|y\|_\mathbf{B}. \tag{5.20}$$

We note some properties of this condition:

(i) Any such function ζ determines the norm of **B** up to equivalence.
(ii) If B is ζ-convex, then any space isomorphic to **B** is ζ-convex.
(iii) The class of ζ-convex spaces contains ℓ^r and $L^r(0,1)$, $1 < r < \infty$, as well as all finite-dimensional spaces.
(iv) A Banach space **B** is an inner product space if and only if there is a function ζ on $\mathbf{B} \times \mathbf{B}$, as above, with $\zeta(0,0) = 1$.

For other properties and all the proofs, see [32] or [33].

In [32], D. Burkholder proved that **B** is a UMD space if and only **B** is ζ-convex, see [32, Theorem 3.1]. From this equivalence, then it is clear that the definition of UMD spaces is independent of the value of p. A comprehensive study on UMD spaces and ζ-convex spaces can be found in [131, Chap. 3].

5.4   Learning and Martingale Theory

In this section, we will give a brief review of the relation between prediction and martingale theory. In their book *Predictions, Learning, and Games* [50], N. Cesa-Bianchi and G. Lugosi introduce a clear idea of the *model of prediction with expert advice* which provides the foundations to the theory of prediction of individual sequences.

Prediction with expert advice is based on the following protocol for sequential decisions: the decision-maker is a forecaster whose goal is to predict an unknown sequence $y_1, y_2 \ldots$ of elements of an *outcome space* $\mathscr{Y}$. The forecaster's predictions $\widehat{p}_1, \widehat{p}_2 \ldots$ belong to a *decision space* $\mathscr{D}$, which we assume to be a convex subset of a vector space. In some special cases, we take $\mathscr{D} = \mathscr{Y}$, but in general $\mathscr{D}$ may be different from $\mathscr{Y}$.

The forecaster computes his predictions in a sequential fashion, and his predictive performance is compared to that of a set of reference forecasters that we call *experts*. More precisely, at each time t the forecaster has access to the set $\left\{ f_{E,t} : E \in \mathscr{E} \right\}$ of expert predictions $f_{E,t} \in \mathscr{D}$, where $\mathscr{E}$ is a fixed set of indices for the experts. On the basis of the experts' predictions, the forecaster computes his own guess $\widehat{p}_t$ for the next outcome y_t. After $\widehat{p}_t$ is computed, the true outcome y_t is revealed. The forecaster incurs loss $\ell(\widehat{p}_t, y_t)$ and each expert E incurs loss $\ell(f_{E,t}, y_t)$. The loss function is usually a bounded non-negative function, not always convex, i.e., $\mathscr{D} = \mathscr{Y} = \{0, 1\}$ with $\ell(p, y) = \chi_{\{p \neq y\}}$.

The predictions of forecaster and experts are scored using this *loss function* $\ell :$ $\mathscr{D} \times \mathscr{Y} \to \mathscr{R}$. The prediction protocol can be naturally viewed as the following repeated game between "forecaster," who makes guesses $\widehat{p}_t$, and "environment," who chooses the expert advice $\left\{ f_{E,t} : E \in \mathscr{E} \right\}$ and sets the true outcomes y_t. The forecaster's goal is to keep as small as possible the *cumulative regret* (or simply *regret*) with respect to each expert. This quantity is defined, for expert E, by the sum

$$R_{E,n} = \sum_{t=1}^{n} \left(\ell \left(\widehat{p}_t, y_t \right) - \ell \left(f_{E,t}, y_t \right) \right) = \widehat{L}_n - L_{E,n},$$

where we use $\widehat{L}_n = \sum_{t=1}^{n} \ell \left(\widehat{p}_t, y_t \right)$ to denote the *forecaster's cumulative loss* and $L_{E,n} = \sum_{t=1}^{n} \ell \left(f_{E,t}, y_t \right)$ to denote the *cumulative loss of expert E*. Hence, $R_{E,n}$ is the difference between the forecaster's total loss and that of expert E after n prediction rounds. For instance, if $E = \{1, \dots N\}$ then

$$R_{i,n} = \sum_{t=1}^{n} (\ell \left(\widehat{p}_{t,t}, y_t \right) - \ell \left(i, y_t \right)),$$

for each $i \in \{1, \dots N\}$.

We also define the *instantaneous regret* with respect to expert E at time t by $r_{E,t} = \ell \left(\widehat{p}_t, y_t \right) - \ell \left(f_{E,t}, y_t \right)$, so that $R_{E,n} = \sum_{t=1}^{n} r_{E,t}$. The quantity $r_{E,t}$ could be considered as the regret the forecaster feels of not having listened to the advice of expert E right after the t outcome y_t has been revealed.

Assume that the number of experts is finite, $\mathscr{E} = \{1, 2, \dots, N\}$, and use the index $i = 1, \dots, N$ to refer to an expert. The goal of the forecaster is to predict so that the regret is as small as possible for all sequences of outcomes. For example, the forecaster may want to have a vanishing per-round regret, that is, to achieve

$$\max_{i=1,\dots,N} R_{i,n} = o(n) \quad \text{or, equivalently,} \quad \frac{1}{n} \left(\widehat{L}_n - \min_{i=1\dots,N} L_{i,n} \right) \xrightarrow{n \to \infty} 0,$$

where the convergence is uniform over the choice of the outcome sequence and the choice of the expert advice. This ambitious goal could be by using martingale-type tools

One of the key ideas of the theory of prediction of individual sequences is that in such a situation randomization may help; in other words, we want to introduce probability techniques. We consider a version of the sequential prediction problem in which the forecaster has access, at each time instant, to an independent random variable U uniformly distributed on the interval $[0, 1]$. The scenario is conveniently described in the framework of playing repeated games.

Let us consider a game between a player (the forecaster) and the environment. At each round of the game, the player chooses an action $i \in \{1, \dots, N\}$, and the environment chooses an action $y \in \mathscr{Y}$. The player's loss $\ell(i, y)$ at time t is the value of a loss function $\ell : \{1, \dots, N\} \times \mathscr{Y} \to [0, 1]$. Now suppose that, at the tth round of the game, the player chooses a discrete probability distribution $\mathbf{p}_t = (p_{1,t}, \ldots, p_{N,t})$ over the set of actions and plays action i with probability $p_{i,t}$. Formally, the player's action I_t at time t is defined, drawing U_t uniformly in $[0, 1]$, by

$$I_t = i \quad \text{if and only if} \quad U_t \in \left[\sum_{j=1}^{i-1} p_{j,t}, \sum_{j=1}^{i} p_{j,t} \right),$$

so that

$$\mathbf{P}\{I_t = i \mid U_1, \ldots, U_{t-1}\} = p_{i,t}, \quad i = 1, \ldots, N,$$

where $U_1, U_2, \ldots$ are independent random variables, uniformly distributed in the interval $[0, 1]$. Suppose the environment chooses action $y_t \in \mathscr{Y}$. The loss of the player is then $\ell(I_t, y_t)$.

If $I_1, I_2, \ldots$ is the sequence of forecaster's actions and $y_1, y_2, \ldots$ is the sequence of outcomes chosen by the environment, the player's goal is to minimize the cumulative regret

$$\widehat{L}_n - \min_{i=1,\ldots,N} L_{i,n} = \sum_{t=1}^{n} \ell(I_t, y_t) - \min_{i=1,\ldots,N} \sum_{t=1}^{n} \ell(i, y_t),$$

that is, the realized difference between the cumulative loss and the loss of the best pure strategy.

Formally, an oblivious opponent is defined by a fixed sequence $y_1, y_2, \ldots$ of outcomes, whereas a non-oblivious opponent is defined by a sequence of functions $g_1, g_2, \ldots$, with $g_t : \{1, \ldots, N\}^{t-1} \to \mathscr{Y}$, and each outcome Y_t is given by $Y_t = g_t(I_1, \ldots, I_{t-1})$. Thus Y_t is measurable with respect to the σ-algebra generated by the random variables $U_1, \ldots, U_{t-1}$.

Interestingly, any forecaster that is guaranteed to work well against an oblivious opponent also works well in the general model against any strategy of a non-oblivious opponent, in a certain sense. To formulate this fact, we introduce the expected loss

$$\bar{\ell}(\mathbf{p}_t, Y_t) := \mathbf{E}\left[\ell(I_t, Y_t) \mid U_1, \ldots, U_{t-1}\right] = \sum_{i=1}^{N} \ell(i, Y_t) \, p_{i,t}.$$

An important property of the forecasters is that the probability distribution $\mathbf{p}_t$ of play is fully determined by the past outcomes $Y_1, \ldots, Y_{t-1}$; that is, it does not explicitly depend on the past plays $I_1, \ldots, I_{t-1}$. Formally, $\mathbf{p}_t : \mathscr{Y}^{t-1} \to \mathscr{D}$ is a function taking values in the simplex $\mathscr{D}$ of probability distributions over N actions.

Note that the next result, which assumes that the forecaster is of the above form, is in terms of the cumulative expected loss $\sum_{t=1}^{n} \bar{\ell}(\mathbf{p}_t, Y_t)$, and not in terms of its expected value

$$\mathbf{E}\left[\sum_{t=1}^{n} \bar{\ell}(\mathbf{p}_t, Y_t)\right] = \mathbf{E}\left[\sum_{t=1}^{n} \ell(I_t, Y_t)\right].$$

Consider a randomized forecaster such that, for every $t = 1, \ldots, n$, $\mathbf{p}_t$ is fully determined by the past outcomes $y_1, \ldots, y_{t-1}$. Assume the forecaster's expected regret against an oblivious opponent satisfies

$$\sup_{y^n \in \mathcal{Y}^n} \mathbf{E}\left[\sum_{t=1}^{n} \ell\,(I_t, y_t) - C \sum_{t=1}^{n} \ell\,(i, y_t) \right] \le B, \quad i = 1, \ldots, N$$

for some fixed B. If the same forecaster is used against a non-oblivious opponent, then it can be proved that

$$\sum_{t=1}^{n} \bar{\ell}\,(\mathbf{p}_t, Y_t) - C \sum_{t=1}^{n} \ell\,(i, Y_t) \le B, \quad i = 1, \ldots, N$$

holds. Moreover, for all $\delta > 0$, with probability at least $1 - \delta$ the actual cumulative loss satisfies, for any (non-oblivious) strategy of the opponent,

$$\sum_{t=1}^{n} \ell\,(I_t, Y_t) - C \min_{i=1,\ldots,N} \sum_{t=1}^{n} \ell\,(i, Y_t) \le B + \sqrt{\frac{n}{2} \ln \frac{1}{\delta}}, \tag{5.21}$$

for some $C > 0$.

A key ingredient for the proof of this result is that the random variables $\ell\,(I_t, Y_t) - \bar{\ell}\,(\mathbf{p}_t, Y_t)$, $t = 1, 2, \ldots$ form a sequence of bounded martingale differences (with respect to the sequence $U_1, U_2, \ldots$ of randomizing variables). Here is where it starts the connection between prediction and martingale theory. For instance, applying the Azuma-Hoeffding inequality for martingales, inequality (5.21) can be obtained. For details of this result and further discussions, we refer to the book of Cesa-Bianchi and Lugosi [50].

Related with this kind of applications of martingale theory, B. Bercu and T. Touati [15] in the context of online statistical learning improved in an important way the former result using exponential inequalities and the so-called self-normalized martingales and concentration inequalities.

For further results with subject related to concentration inequalities and martingales theory we refer to S. Boucheron et al. [25] and B. Bercu et al. [14].

5.5 Exercises

(1) Let $\{\xi_n, n \ge 1\}$ be independent random variables with $\mathbf{E}[\xi_n] = 0$, $\mathbf{E}[\xi_n^2] = 1$ for $n \ge 1$. If

$$\tau = \inf \left\{ n \ge 1 : \sum_{1}^{n} \xi_j > 0 \right\},$$

then

$$\mathbf{P}\{\tau < \infty\} = 1 \text{ and } \mathbf{E}[\tau^{1/2}] = \infty.$$

(2) Let $\{\xi_n\}_n$ a sequence of independent random variables with Bernoulli distribution,

$$\mathbf{P}\{\xi_k = 1\} = p, \mathbf{P}\{\xi_k = -1\} = 1 - p; \ p \in (0, 1),$$

and let $S_0 = 0$ and $S_n = \sum_{k=1}^{n} \xi_k$ for any $n \in \mathbb{N}$ and set $\mathscr{F}_n^{\xi} = \sigma(\{\xi_1, \xi_2, \ldots, \xi_n\})$. Given a, b integers such that $a < 0 < b$, let τ be a $\mathscr{F}_n^{\xi}$-stopping time defined as

$$\tau = \inf \{n : S_n = a \text{ or } S_n = b\}.$$

Thus, τ is the exit time of $\{S_n\}$ from (a, b). Prove that

$$\mathbf{P}\{S_\tau = a\} = p_a = \begin{cases} \dfrac{1 - \left(\frac{q}{p}\right)^b}{\left(\frac{q}{p}\right)^a - \left(\frac{q}{p}\right)^b} & \text{if } p \neq q, \\[2ex] \dfrac{b}{b-a} & \text{if } p = q = \frac{1}{2}, \end{cases}$$

and then $\mathbf{P}\{S_\tau = b\} = p_b = 1 - p_a$. Moreover,

$$\mathbf{E}[\tau] = \begin{cases} \dfrac{ap_a + bp_b}{p-q} & \text{if } p \neq q, \\[1ex] |ab| & \text{if } p = q = \frac{1}{2}. \end{cases}$$

[HINT: Prove first that $\mathbf{P}\{\tau < \infty\} = 1$ and $\mathbf{E}[\tau] < \infty$. For the case $p = q = \frac{1}{2}$ use that $S_n = \sum_{k=1}^{n} \xi_k$ and $Y_n = S_n^2 - n$ are $\left\{\mathscr{F}_n^{\xi}\right\}$-martingales and use Doob's optional stopping time, Theorem 3.1. Finally, if $p \neq q$ consider $\left\{X_n = \left(\frac{q}{p}\right)^{S_n}\right\}_n$, prove that $\{X_n\}$ is a $\left\{\mathscr{F}_n^{\xi}\right\}$-martingale and then use Doob's optional stopping time, Theorem 3.1, again; see [148, Chap. 1].]

(3) The precedent of Bernstein inequality (4.27), discussed in Chap. 4, is a version for random variables due to Azema, see [50],

   (i) For any $a \in \mathbb{R}$ and $x \in [-1, 1]$, show that

$$e^{ax} \leq \frac{1+x}{2}e^a + \frac{1-x}{2}e^{-a} = \cosh a + x \sinh a.$$

   (ii) Suppose that $\{\xi_1, \ldots, \xi_n\}$ are $[-1, 1]$-valued random variables on the probability space $(\Omega, \mathscr{F}, \mathbf{P})$ with the property that, for each $1 \leq m \leq n$,

$$\mathbf{E}\left[\xi_{j_1} \cdots \xi_{j_m}\right] = 0 \quad \text{for all } 1 \leq j_1 < \cdots < j_m \leq n.$$

Show that, for any $\{a_j\}_{j=1}^n \subseteq \mathbb{R}$,

$$\mathbf{E}\left[\exp\left(\sum_{j=1}^n a_j \xi_j\right)\right] \le \prod_{j=1}^n \cosh a_j \le \exp\left(\frac{1}{2}\sum_{j=1}^n a_j^2\right),$$

and conclude that

$$\mathbf{P}\left\{\sum_{j=1}^n a_j \xi_j \ge R\right\} \le \exp\left(-\frac{R^2}{2\sum_{j=1}^n a_j^2}\right), \quad R \in [0,\infty).$$

(4) Prove that every Hilbert space $\mathbf{H}$ is also an example of a space with type p and cotype q for every $p \in [1,2]$ and $q \in [2,\infty]$.

(5) The *Rademacher maximal function of a martingale* $\{X_n\}_{n\in\mathbb{Z}}$ is defined as

$$M_{\mathrm{Rad}}X := \sup\left\{\left\|\sum_{k\in\mathbb{Z}} r_k \lambda_k X_k\right\|_{L^2(\Omega;\mathbf{B})} : \lambda_k \in \bar{B}_{\ell^2(\mathbb{Z})} \text{ finitely non-zero}\right\},$$

here $\bar{B}_{\ell^2(\mathbb{Z})}$ denotes the closed unit ball of $\ell^2(\mathbb{Z})$ and $\{r_k\}$ the Rademacher functions, see [100] and [101, Chap. 3].

Now, a Banach space $\mathbf{B}$ has the RMF property if

$$\|M_{\mathrm{Rad}}f\|_p \le C_p \|X\|_{L^p(\Omega;\mathbf{B})}$$

for all $p \in (1,\infty)$ and all L^p-martingales $\{X_k\}_{k\in\mathbb{Z}}$ with values in $\mathbf{B}$.

 (i) Prove that if a Banach space $\mathbf{B}$ has type 2 and $\{X_k\}_k$ is a martingale with values in $\mathbf{B}$, then

$$M_{\mathrm{Rad}} X \le \tau_{2,\mathbf{B}} X^*.$$

Then conclude that $\mathbf{B}$ has the RMF property.
 (ii) Prove that L^p-spaces with $2 \le p < \infty$ has the RMF property.
(iii) Prove that if $\mathbf{B}$ is a UMD Banach space it has the RMF property.
    [Hint: Use Rubio de Francia's extension of the Fefferman-Stein theorem.]

(5) It is well known that the space $L^2(\Omega, \mathscr{F}, \mathbf{P})$ is a Hilbert space (under equivalent relation), with inner product $\langle X, Y \rangle = \mathbf{E}[XY]$. For any $X = \{X_n\}$ sequence in this space, let $\mathscr{M}^n$ be the closed subspace obtaining by closure of the set of all finite real-valued linear combinations of elements of $\{X_k; 1 \le k \le n\}$. Let's call

$\mathscr{P}_n$ the orthogonal projection onto $\mathscr{M}^n$. Given $\{Z_k\}$ is a centered non-correlated L^2-process, usually called *white noise*, define the *linear processes* $\{X_k; k \in \mathbb{Z}\}$

$$X_k = \sum_{j=0}^{\infty} a_j Z_{k-j},$$

where $a_0 \neq 0$, $\sum_{j=0}^{\infty} a_j^2 < \infty$. Show that

$$\mathbf{E}[X_{n+1} | X_k, -\infty < k \leqslant n] = \mathscr{P}_n(X_{n+1}),$$

if and only if the process $\{Z_k\}$ is a martingale difference sequence, see [27].

Glossary of Symbols

- C will always denote a constant, not necessarily the same in each occurrence.
- $\delta_{i,j}$ is the Dirac delta,

$$\delta_{i,j} = \begin{cases} 1, & \text{if } i = j, \\ 0, & \text{if } i \neq j. \end{cases}$$

- $\operatorname{sgn} a$ is the sign of $a \in \mathbb{R}$

$$\operatorname{sgn} a = \begin{cases} 1, & \text{if } a \geq 0, \\ -1, & \text{if } a < 0. \end{cases}$$

- $a \wedge b = \min\{a, b\}$.
- $a \vee b = \max\{a, b\}$.
- $a \approx b$ means that there exists constants A, B such that $Aa \leq b \leq Ba$.
- $A \subset B$ means A is a subset of B.
- A^c denotes the complement of A

$$A^c = \{x : x \notin A\}.$$

- $A \cup B$ is the union of A and B,

$$A \cup B = \{x : x \in A \text{ or } x \in B\}.$$

- $A \cap B$ is the intersection of A and B,

$$A \cap B = \{x : x \in A \text{ and } x \in B\}.$$

- $A \backslash B$ is the set difference,

$$A \backslash B = \{x : x \in A \text{ and } x \notin B\}.$$

- χ_E denotes the characteristic function of the set E,

$$\chi_E(x) = \begin{cases} 1 & \text{if } x \in E, \\ 0 & \text{otherwise.} \end{cases}$$

- $\mathbb{N}$ is the set of natural numbers

$$\mathbb{N} = \{1, 2, 3, \ldots\}.$$

- $\mathbb{N}_0 = \mathbb{N} \cup \{0\}$ is the set of non-negative integers.
- $\mathbb{Z}$ is the set of integers

$$\mathbb{Z} = \{\ldots, -3, -2, -1, 0, 1, 2, 3, \ldots\}.$$

- $\mathbb{Z}^-$ is the set of negative integers

$$\mathbb{Z} = \{\ldots, -3, -2, -1\}.$$

- $\mathbb{R}^d$ is the d-dimensional real space,

$$\mathbb{R}^d = \left\{ x = (x_1, \ldots, x_d) : x_i \in \mathbb{R}, \ i = 1, \ldots, d \right\}.$$

- Given $x = (x_1, \ldots, x_d) \in \mathbb{R}^d$,

$$|x| = \left(\sum_{i=1}^d x_i^2 \right)^{1/2}$$

  is its Euclidean norm in $\mathbb{R}^d$, $d \geq 1$.
- Given $x, y \in \mathbb{R}^d$, $d \geq 1$ $\langle x, y \rangle$ is the inner product in $\mathbb{R}^d$,

$$\langle x, y \rangle = \left(\sum_{i=1}^d x_i y_i \right)^{1/2}.$$

- $(\mathbf{H}, \| \cdot \|_{\mathbf{H}})$ a real Hilbert space.
- $(\mathbf{B}, \| \cdot \|_{\mathbf{B}})$ a real Banach space.
- $m(A)$ is the Lebesgue measure of the Borel set A in $\mathbb{R}^d$, $A \in \mathscr{B}(\mathbb{R}^d)$, $d \geq 1$.
- $(\Omega, \mathscr{F}, \mathbf{P})$ a probability space, where Ω is an arbitrary set, $\mathscr{F}$ is a σ-algebra of sets of Ω and $\mathbf{P}$ is a probability measure defined on $\mathscr{F}$.

- $([0, 1], \mathscr{B}([0, 1]), m)$ is the Lebesgue probability space, $\mathscr{B}([0, 1])$ the σ-algebra of Borel sets on $[0, 1]$ and m the Lebesgue measure defined on $\mathscr{B}([0, 1])$.
- $\{\chi_k^{(n)}\}$ the Haar functions.
- $\{r_n\}$ the Rademacher functions.
- $\{\psi_k^{(n)}\}$ the Walsh functions.
- $\{\mathscr{F}_n\}$ is a filtration on the σ-algebra $\mathscr{F}$, i.e., an increasing family of sub-σ-algebras of $\mathscr{F}$, $\mathscr{F}_n \subset \mathscr{F}$, and $\mathscr{F}_n \subset \mathscr{F}_{n+1}$, $n \geq 1$.
- ξ, ζ, η random variables defined on the probability space $(\Omega, \mathscr{F}, \mathbf{P})$.
- $\mathbf{E}[\xi] = \int_\Omega \xi d\mathbf{P}$ the expected value of the random variable ξ.
- $\mathbf{E}[\xi \mid \mathscr{G}]$ the conditional expected value of the random variable ξ with respect to the σ-algebra $\mathscr{G}$.
- $\{X_n\}, \{Y_n\}, \{Z_n\}, \{W_n\}, \{U_n\}, \{V_n\}$ stochastic processes defined on the probability space $(\Omega, \mathscr{F}, \mathbf{P})$.
- Given $X = \{X_n\}$ a $\{\mathscr{F}_n\}$-martingale, $\Delta X_1 = X_1$ and for $n > 1$, $\Delta X_n = X_n - X_{n-1}$. $\Delta X = \{\Delta X_n\}$ is the sequence of differences of the process X. It is clear that

$$X_n = \sum_{k=1}^{n} \Delta X_k.$$

- Given $X = \{X_n\}$ a $\{\mathscr{F}_n\}$-martingale and $Y = \{Y_n\}_n$ be a multiplier sequence, then $(Y \circ X) = \{(Y \circ X)_n\}_n$, is the transform of X by Y,

$$(Y \circ X)_n = \sum_{k=1}^{n} Y_k \Delta X_k = \sum_{k=1}^{n} Y_n (X_k - X_{k-1})$$

  for any $n \in \mathbb{N}$.
- Given $X = \{X_n\}$ a $\{\mathscr{F}_n\}$-martingale, X_n^* is the maximal function of order n

$$X_n^* = \max_{0 \leq k \leq n} |X_k|$$

  and X^* is the maximal function of de X.

$$X^* = \sup_k |X_k|.$$

- For any $0 < p < \infty$ for $X = \{X_n\}$ a $\{\mathscr{F}_n\}$-martingale,

$$\|X_n\|_p = \left(\mathbf{E}\left[|X_n|^p\right]\right)^{1/p}$$

  for all n and

$$\|X\|_p = \sup_n \|X_n\|_p.$$

- Given $X = \{X_n\}$ a $\{\mathscr{F}_n\}$-martingale, $S_n(X)$ is the quadratic function of X of order n

$$S_n(X) = \left(\sum_{k=1}^{n} (\Delta X)_k^2 \right)^{1/2},$$

and $S(X)$ is the quadratic function

$$S(X) = \left(\sum_{k=1}^{\infty} (\Delta X)_k^2 \right)^{1/2},$$

if the series converges, otherwise it would be infinite.

- Given $X = \{X_n\}$ a $\{\mathscr{F}_n\}$-martingale, $\langle X \rangle_n$ is the quadratic variation of X of order n

$$\langle X \rangle_n = \sum_{i=1}^{n} \left(\Delta X_j \right)^2$$

and $\langle X \rangle$ is the variation of X

$$\langle X \rangle = \sum_{j=1}^{\infty} \left(\Delta X_j \right)^2,$$

if the series converges, otherwise it would be infinite.

- Given $X = \{X_n\}$ a $\{\mathscr{F}_n\}$-martingale, $s(X)$ is the conditional quadratic function.

$$s(X) = \left(\sum_{k=1}^{\infty} \mathbf{E}\left[(\Delta X_k)^2 \mid \mathscr{F}_{k-1} \right] \right)^{1/2}.$$

- Given $X = \{X_n\}_{0 \leqslant n \leqslant N}$ and $Y = \{Y_n\}_{0 \leqslant n \leqslant N}$ two sequences of random variables on $(\Omega, \mathscr{F}, \mathbf{P})$, $X_0 = Y_0 = 0$, the quadratic covariation of X and Y, $\langle X, Y \rangle = \{\langle X, Y \rangle_n\}_{0 \leqslant n \leqslant N}$, is defined for $n \geqslant 1$, as

$$\langle X, Y \rangle_n = \sum_{k=1}^{n} \Delta X_k \Delta Y_k.$$

Appendix B

In this appendix, in order to make the book as much as self-contained as possible, we present some basic probability theory results widely used throughout the text and some crucial limit results. Our main reference is the book of Richard Durret [74].

B.1 Probability Space and Events

- Ever since A. Kolmogorov first set down a clear mathematical foundation for prob- ability theory in 1933, the tradition has been for probability texts from the most elementary to the most advanced to begin their exposition by defining a probability space $(\Omega, \mathscr{F}, \mathbf{P})$.

- A probability space is a triplet $(\Omega, \mathscr{F}, \mathbf{P})$, where Ω is an arbitrary set, called events space, or sample space $\mathscr{F}$ is a σ-algebra of subsets of Ω which means that:

(S1) $\Omega \in \mathscr{F}$
(S2) $A^c \in \mathscr{F}$ if $A \in \mathscr{F}$
(S3) If $\{A_n\}_{n=1}^{\infty} \in \mathscr{F}$ then $\bigcup_{n=1}^{\infty} A_n \in \mathscr{F}$.

and $\mathbf{P}$ is a probability measure, which means that:

(P1) $\mathbf{P} : \mathscr{F} \to [0, 1]$
(P2) $\mathbf{P}(\Omega) = 1$

W. Urbina-Romero and R. Rios, *An Introduction to the Modern Martingale Theory and Applications*, Texts in Applied Mathematics 81, https://doi.org/10.1007/978-3-031-88903-5

(P3) $\mathbf{P}$ is σ additive: if $\{A_n\}_{n=1}^{\infty} \in \mathscr{F}$ with $A_i \cap A_j = \emptyset$, $i \neq j$ then

$$\mathbf{P}\left(\bigcup_{n=1}^{\infty} A_n\right) = \sum_{n=1}^{\infty} \mathbf{P}(A_n).$$

The smallest σ-algebra is the the so-called *trivial* σ-algebra,

$$\mathscr{F} = \{\emptyset, \Omega\}.$$

- The most basic example of a probability space is

$$\Omega = \{0, 1\}, \mathscr{F} = \{\emptyset, \{0\}, \{1\}, \Omega\}$$

  and

$$\mathbf{P}(\{0\}) = \mathbf{P}(\{1\}) = 1/2.$$

  This is the classical model for coin tossing.
- If $\mathscr{C}$ is a class of subsets of Ω, the σ-algebra generated by $\mathscr{C}$, $\sigma(\mathscr{C})$ is the intersection of all the σ-algebras that contains the class $\mathscr{C}$.
- If $\mathscr{T}$ is a topology on Ω, the σ-algebra de Borel on Ω, or Borel sets of Ω is define as $\sigma(\mathscr{T})$.
- If $\Omega = [0, 1]$ and $\mathscr{B}([0, 1])$ the Borel σ-algebra of Borel set in $[0, 1]$ (i.e. the smallest topology containing the open subintervals of the interval $[0, 1]$) the pair $([0, 1], \mathscr{B}([0, 1]))$ is known as the Borel measurable space.
- In $([0, 1], \mathscr{B}([0, 1]))$, using Cartheodory extension theorem (see Durret [74] Appendix A.1, Theorem 1.3), there exists a probability measure λ such that $\lambda(a, b) = b - a$ for any $0 \leq a < b \leq 1$ known as the (classical) Lebesgue measure; also known as uniform distribution on $[0, 1]$. This is, probably the most important probability space, or at least one of the most used.
- Two sets A and B in $\mathscr{F}$ are independent if

$$\mathbf{P}(A \cap B) = \mathbf{P}(A)\mathbf{P}(B).$$

- Any finite collection of measurable sets $\{A_1, \ldots A_n, \cdot\}$ is said to be independent if, for any finite subset $I \subset \{1, \ldots, n, \ldots\}$, we have

$$\mathbf{P}\left(\bigcap_{i \in I} A_i\right) = \prod_{i \in I} \mathbf{P}(A_i)$$

B.2   Random Variables and Limit Theorems

- A random variable ξ is a measurable function from an abstract probability space $(\Omega, \mathcal{F}, \mathbf{P})$ and the Borel space over $\mathbb{R}$ $(\mathbb{R}, \mathcal{B}(\mathbb{R}))$ i.e. $\xi : \Omega \to \mathbb{R}$ such that such that $\xi^{-1}(B) \in \mathcal{F}$ for all $B \in \mathcal{B}(\mathbb{R})$.
- Given a random variable ξ and a function $\Psi : \mathbb{R} \to \mathbb{R}$ Borel measurable, then $\eta = \Psi(\xi)$ is also a random variable.
- The distribution of a random variable ξ is the induced measure on $\mathbb{R}$, defined as

$$\mathbf{P}_\xi(B) = \mathbf{P} \circ \xi^{-1}(B) = \mathbf{P}\{\xi \in B\},$$

  all $B \in \mathcal{B}(\mathbb{R})$. The distribution of ξ describes its random behavior.
- The (cumulative) distribution function of X is defined as

$$F_\xi(x) = \mathbf{P}\{\xi \leqslant x\}, \quad -\infty < x < \infty.$$

  F_ξ is a monotone non-decreasing right continuous function with left limits, such that

$$\mathbf{P}_\xi(B) = \int_B dF_\xi(x),$$

  where the integral is in the Lebesge-Stieltjes sense. In particular,

$$\mathbf{P}\{a < \xi \leq b\} = \int_{(a,b]} dF(x) = F(b) - F(a),$$

  for any pair of real numbers $a < b$.
- *Change of variables* For any measurable function $f : \mathbb{R} \to \mathbb{R}$

$$\int_\Omega f(\xi)\, d\mathbf{P} = \int_{-\infty}^\infty f(x)\, dF_\xi(x),$$

  where the integral is in the Lebesge-Stieltjes sense.
- The expected value of a random variable ξ is defined as

$$\mathbf{E}[\xi] = \int_\Omega \xi\, d\mathbf{P} = \int_\infty x\, dF_\xi(x),$$

  which is well-defined if and only if $\int_\Omega |\xi|\, d\mathbf{P} < \infty$.
- If X is a non-negative random variable

$$\mathbf{E}[\xi] = \int_0^\infty (1 - F_\xi(\lambda))d\lambda = \int_0^\infty \mathbf{P}\{\xi > \lambda\}d\lambda.$$

- Two random variables ξ, η are independent if and only if

$$\mathbf{E}[\Phi(\xi)\Psi(\eta)] = \mathbf{E}[\Phi(\xi)]\mathbf{E}[\Psi(\eta)],$$

  for any Φ, Ψ bounded Borel functions.
- The independence notion can be extended to any collection of random variables. A family of random variables $\{\xi_k\}_k$, is said to be independent if for any finite subfamily of subindexes $1 \leqslant n_1 < n_2 < \cdots < n_l$,

$$\mathbf{E}[\Phi_1(\xi_{n_1})\Phi_2(\xi_{n_2}) \cdots \Phi_l(\xi_{n_l})] = \mathbf{E}[\Phi_1(\xi_{n_1})]\mathbf{E}[\Phi_2(\xi_{n_2})] \cdots \mathbf{E}[\Phi_l(\xi_{n_l})],$$

  for any $\Phi_1, \Phi_2, \ldots, \Phi_l$ bounded Borel functions.

- *Hölder's inequality* If $p, q \in [1, \infty]$ with $1/p + 1/q = 1$, then

$$\mathbf{E}[|\xi\eta|] \leq \|\xi\|_p \|\eta\|_q.$$

- *Minkowski's Inequality.* If $\mathbf{E}|\xi|^p < \infty$, $\mathbf{E}|\eta|^p < \infty$, $1 \leq p < \infty$, then we have

$$\left(\mathbf{E}|\xi + \eta|^p\right)^{1/p} \leq \left(\mathbf{E}|\xi|^p\right)^{1/p} + \left(\mathbf{E}|\eta|^p\right)^{1/p}.$$

- *Tchebyshev's inequality* If φ is a nondecreasing nonnegative function, ξ is a random variable, and $\lambda > 0$ such that $\varphi(\lambda) > 0$, then

$$\mathbf{P}\{\varphi(\xi) \geq \lambda\} \leqslant \frac{\mathbf{E}[\varphi(\xi)]}{\varphi(\lambda)}.$$

- *Markov's inequalityt* Let ξ is a random variable, and $\lambda > 0$ then

$$\mathbf{P}\{|\xi| \geq \lambda\} \leqslant \frac{\mathbf{E}[|\xi|]}{\lambda}.$$

- Finally, we have several very important limit results for sums of independent random variables.

Theorem B.1 (weak law of large numbers) *Given a sequence of independent random variables $\{\xi_n\}_{n=1}^{\infty}$, with $_{\centerdot}[X_n] = \mu$ and $\mathrm{Var}(\xi_n) \leq C < \infty$ for all $n \geq 1$, then $\frac{1}{N}[\sum_{n=1}^{N} \xi_n]$ converges to μ in probability and in L^2, i.e.,*

$$\lim_{N \to \infty} \mathbf{P}\left\{\left|\frac{\sum_{n=1}^{N} \xi_n}{N} - \mu\right| > \varepsilon\right\} = 0, \tag{B.1}$$

and

$$\lim_{N \to \infty} \int_{\Omega} \left|\frac{\sum_{n=1}^{N} \xi_n}{N} - \mu\right|^2 d\mathbf{P} = 0. \tag{B.2}$$

Theorem B.2 (Kolmogorov's three-series theorem) *Given a sequence of independent random variables* $\{\xi_n\}_{n=1}^{\infty}$. *Let* $\lambda > 0$, *then* $\sum_{n=1}^{\infty} \xi_n$ *converges a.s. if and only if*

(i) $\sum_{n=1}^{\infty} \mathbf{P}\{|\xi_n| > \lambda\} < \infty,$
(ii) $\sum_{n=1}^{\infty} \mathbf{E}[\xi_n \chi_{\{|\xi_n| \leq \lambda\}}]$ *converges, and*
(iii) $\sum_{n=1}^{\infty} \mathrm{Var}\left(\xi_n \chi_{\{|\xi_n| \leq \lambda\}}\right) < \infty.$

Theorem B.3 (Strong Law of Large Numbers) *Given a sequence of independent and identically distributed random variables* $\{\xi_n\}_{n=1}^{\infty}$, *with* $\mathbf{E}[|\xi_n|] < \infty$ *and* $\mathbf{E}[\xi_n] = \mu$ *for all* $n \geq 1$,

$$\lim_{k \to \infty} \frac{\sum_{k=1}^{n} \xi_k}{n} \to \mu, \quad a.e.\ \mathbf{P} \tag{B.3}$$

We have the *Central Limit Theorem*. First, for sequence of random variables,

Theorem B.4 (Central Limit Theorem CLT) *Given a sequence of independent and identically distributed random variables* $\{\xi_n\}_{n=1}^{\infty}$, *with* $\mathbf{E}[\xi_n] = \mu$, $\mathrm{Var}\,(\xi_n) = \sigma^2 > 0$, *then*

$$\frac{\sum_{k=1}^{n}(\xi_k - \mu)}{\sigma n^{1/2}}$$

converges in distribution to the standard normal distribution, as $n \to \infty$, *i.e.,*

$$\int_{\Omega} f\left(\frac{\sum_{k=1}^{n}(\xi_k - \mu)}{\sigma n^{1/2}}\right) d\mathbf{P} \to \frac{1}{\sqrt{2\pi}} \int_{-\infty}^{\infty} f(x) e^{\frac{x^2}{2}}\, dx,$$

as $n \to \infty$, *for any bounded continuous function* f.

Second, for triangle arrays of random variables,

Theorem B.5 (Lindeberg-Lévy CLT) *Suppose that for each* n, $\{\xi_{n,k}\}_{1 \leq k \leq m_n}$, $m_n \to \infty$, *are independent centered random variables such that*

$$\lim_{n \to \infty} \sum_{k=1}^{m_n} \mathbf{E}\left[\xi_{n,k}^2\right] = \sigma^2 > 0,$$

and

$$\lim_{n \to \infty} \sum_{k=1}^{m_n} \mathbf{E}\left[\xi_{n,k}^2 \chi_{\{|\xi_{n,k}| > \varepsilon\}}\right] = 0,$$

for all $\varepsilon > 0$. *Then* $\sum_{k=1}^{m_n} \xi_{n,k}$ *converges in distribution to a normal distribution* $N\left(0, \sigma^2\right)$ *as* $k_n \to \infty$, *i.e.,*

$$\int_\Omega f\left(\frac{\sum_{k=1}^{n}(X_k - \mu)}{\sigma n^{1/2}}\right) d\mathbf{P} \to \frac{1}{\sqrt{2\pi}\sigma} \int_{-\infty}^{\infty} f(x) e^{\frac{x^2}{2\sigma}} \, dx,$$

as $n \to \infty$, *for any bounded continuous function* f.

References

1. Abramowitz, M. and Stegun, I.A. editors, *Handbook of Mathematical Functions with Formulas, Graphs and Mathematical Tables*, Reprint of the 1972 edition, Dover Publications, Inc., New York, (1992).
2. Aldous, D. J. *Unconditional bases and martingales in $L^p(F)$*, Math. Proc. Cambridge Phil, Soc, 85 (1979), 117–123,
3. Austin, D. G. *A sample function property of martingale* Ann.Math. Stat. 37 (1966) 1866–1867.
4. Báez Duarte, L. *Another look at the Martingale Theorem*. J. of Math. Anal. and Applications. Vol 23 #3 (1968) 551–557.
5. Bañuelos, R. & Davis, B. *Donald Burkholder's work in Martingales and Analysis* in Selected Works of Donald l. Burhkolder. B. Davis and R. Song editors. Springer New York (2010). 1–22.
6. Bañuelos, R. & Davis, B. *Introduction to memorial issue for Donald Burkholder (1927–2013)* Ann. Prob., 45(1) (2017) 1–3.
7. Bañuelos, R. & Osękowski, A. *Burkholder inequalities for submartingales, Bessel processes and conformal martingales.* Amer. J. Math., 135 (2013), 1675–1698.
8. Bañuelos, R. & Osękowski, A. *On the Bellman function of Nazarov, Treil and Volberg.* Zbl 1311.42065 Math. Z. 278, No. 1-2, (2014) 385–399.
9. Bañuelos, R. & Osękowski, A. *Sharp maximal L_p-estimates for martingales* Illinois J. Math., 58 (2014), 149–165.
10. Bañuelos, R. & Osękowski, A. *Sharp Martingale Inequalities and Applications to Riesz Transforms on Manifolds, Lie Groups and Gauss Space.* J. Funct. Anal. 269, No. 6, (2015) 1652–1713.
11. Bañuelos, R. & Osękowski, A. *Stability in Burkholder's differentially subordinate martingales inequalities and applications to Fourier multipliers.* J. Math. Pures Appl. (9) 119 (2018), 1–44.
12. Bañuelos, R, & Osękowski, A. *A weighted maximal inequality for differentially subordinate martingales.* Proc. Am. Math. Soc. 146, No. 5, (2018) 2263–2275.
13. Benedek, A., Calderón, A. P. & Panzone, R., *Convolution operators on Banach space valued functions*, Proc, Nat. Acad. Sci. 48 (1962) 356–365.
14. Bercu, B., Delyon B., & Rio, E. *Concentration Inequalities for Sums and Martingales* Springer-Briefs in Mathematics. Springer Cham Heidelberg (2015).
15. Bercu, B. & Touati, T. *New insights on concentration inequalities for self-normalized martingales.* Electron. Commun. Probab. 24 (2019), no. 63, 1–12.

W. Urbina-Romero and R. Rios, *An Introduction to the Modern Martingale Theory and Applications*, Texts in Applied Mathematics 81, https://doi.org/10.1007/978-3-031-88903-5

16. Betz, C. *Introducción a la Teoría de la Medida e Integración* Universidad Central de Venezuela (UCV) Facultad de Ciencias. Caracas (1992).

17. Beznosova, O. V. *Linear bound for the dyadic paraproduct on weighted Lebesgue space $L_2(\omega)$* J. Funct. Anal 255 (2008) 994–1007.

18. Billingsley, P. *Probability and Measure.* Wiley, N.Y. (1986).

19. Bollobás, B. *Martingale inequalities.* Math. Proc. Cam. Phi.Soc. Vol 87 (1980), 377–382.

20. Bogachev, V. I. *Gaussian measures* Vol. 62 of Mathematical Surveys and Monographs. American Mathematical Society, Providence, RI, (1998).

21. Bose, A, Chakrabarty, A. & Subhra Hazra, R. *A Little Book of Martingales* Texts and Readings in Math. Springer Nature Singapore. (2024).

22. Bourgain, J. *Some remarks on Banach spaces in which martingale difference sequences are unconditional,* Ark. Mat. 21 (1983), 163–168.

23. Bourgain, J. *Vector-valued singular integrals and the H^1-BM0 duality* in: Probability theory and harmonic analysis (Cleveland, Ohio, 1983), 1–19, Monogr. Textbooks Pure Appl. Math., 98, Dekker, New York, (1986).

24. Brzozowski, M & Osękowski, A. *Weighted Maximal Inequalities for Martingale Transforms.* Probability and Mathematical Statistics Vol. 41, Fasc. 1 (2021), pp. 89–114.

25. Boucheron, S., Lugosi, G., Massart, P. *Concentration Inequalities. A Nonasymptotic Theory of Independence* Oxford Univ. Press (2013).

26. Breiman, L. *Probability.* Addison-Wesley, Reading Ma. (1968).

27. Brockwell, P. & Davis, R. *Times series: theory and methods* Springer Verlag New York (1991).

28. Burkholder, D. L. *Maximal inequalities as necessary conditions for almost everywhere convergence.* Z.W. 3 (1964) 75–88.

29. Burkholder, D.L. *Martingale Transform.* Ann. Math. Stat. 37 (1966) 1494–1504.

30. Burkholder, D.L. *Distribution function inequalities for martingales.* Ann. Prob.1 (1973) 19–42.

31. Burkholder, D. L. *A sharp inequality for martingale transforms,* Ann. Prob. 7 (1979), 858–863.

32. Burkholder, D. L. *A Geometrical Characterization of Banach Spaces in which Martingale Difference Sequences are Unconditional.* Ann. Prob. Vol. 9, No. 6, (1981) 997–1011.

33. Burkholder, D. L. *Martingale transforms and the geometry of Banach spaces.* Proc. of the Third International Conference on Probability in Banach Spaces, Tufts Univ., 1980. Lecture Notes in Math. 860 (1981): 35–50.

34. Burkholder, D. L. *A nonlinear partial differential equation and the unconditional constant of the Haar system in L^p.* Bull. Amer. Math. Soc. (N.S.), 7(3) (1982) 591–595.

35. Burkholder, D. L. *A geometric condition that implies the existence of certain singular integrals of Banach space-valued functions.* in Conference on Harmonic Analysis in honor of Antoni Zygmund, Vol I, II. W. Beckner, A.P. Calderón, R. Fefferman and P.W. Jones (Eds.) Wadsworth Math. Ser, Belmont, CA.(1983) 270–286.

36. Burkholder, D.L. *Boundary value problems and sharp inequalities for martingale transform* Ann. Prob.12 (1984) 647–702.

37. Burkholder, D.L. *Elementary proof of an inequality of A. E. A Paley.* Bull. London Math. Soc. 17 (1985) 474–478.

38. Burkholder., D. L. *An extension of a classical martingale inequality* In Probability theory and harmonic analysis (Cleveland, Ohio, 1983), vol. 98 Monogr. Textbooks Pure Appl. Math. Dekker, N.Y. (1986) 21–30.

39. Burkholder, D. L. *Martingales and Fourier analysis in Banach spaces.* In Probability and Analysis (Varenna,1985), vol. 1206 Lecture Notes in Math., Springer, Berlin, (1986) 61–108.

40. Burkholder, D. L. *A proof of Pelczynski's conjecture for the Haar system,* Studia Math. 91 (1988) 268–273.

41. Burkholder, D. L. *Sharp inequalities for martingales and stochastic integrals.* Colloque Paul Levy sur les processus stochastiques. Asterisque **157-158** (1988), 75–94.

42. Burkholder, D. L. *Explorations in martingale theory and its applications,* E'cole d'Ete de Probabilités de Saint-Flour XIX 1989, Lecture Notes in Math.,1464, Springer, Berlin, (1991) 1–66.

43. Burkholder, D. L. *Strong Differential Subordination and Stochastic Integration.* Ann. Prob. Vol. 22, No. 2, (1994) 995–1025.

44. Burkholder, D. L. & Gundy, R. F. *Extrapolation and interpolation and quasilinear operators on martingales.* Acta Math. 124 (1970) 249–304.

45. Burkholder, D.L., Davis, B.L. & Gundy, R. F. *Integral inequalities for convex functions of operators on martingales* Sixth Berkeley Symposium Math. Statist. Prob.2 (1972). 223–240.

46. Burkholder, D. L., Gundy, R. F. & Silverstein, M. L. *A maximal function characterization of the class H^p,* Trans. Amer. Math. Soc. 157 (1971), 137–153.

47. Calderón A. P. & Zygmund, A., *On the existence of certain singular integrals,* Acta Math. 88 (1952), 85–139.

48. Calderón A. P. & Zygmund, A. *On singular integrals,* Amer. J. of Math. 78 (1956), 289-309.

49. Calderón, A. P. & Zygmund, A. *A note on the interpolation of sublinear operators.* Amer. J.Math, 78, (1956), 282–288.

50. Cesa-Bianchi, N. & Lugosi, G. *Prediction, learning, and games.* Cambridge University Press, Cambridge, (2006).

51. Chatterji, S. D. *A note on the convergence of Banach-space valued martingales.* Math. Ann., 153, (1964) 142–149.

52. Chatterji, S. D. *Martingale convergence and the Radon-Nikodým theorem in Banach Spaces.* Math. Scand. 22 (1968) 21–41.

53. Chow, Y. S. *Local convergence of martingales and the law of large numbers.* Ann. Math. Statist. 36, (1965), 552–558.

54. Chow, Y. S. *Convergence of sum of squares of martingale differences.* Ann. Math. Statist, 39, (1968), 123–133.

55. Chow, Y. S. *Martingale extensions of a theorem of Marcinkiewicz and Zygmund.* Ann. Math. Statist. 40 (1969) 427–433.

56. Chow, Y. S. & Teicher, H. *Probability Theory. Independence, Interchangeability, Martingales.* Springer-Verlag, N.Y. (1978).

57. Chung, K. L. *A graduate course in Probability.* 2 ed. Acad Press. N.Y. (1974).

58. Coifman, R. R. & Fefferman, C. *Weighted norm inequalities for maximal functions and singular integrals,* Studia Math. 51, (1974) 241–250.

59. Cox, D. C. *The best constant in Burkholder's weak-L^1 inequality for the martingale square function,* Proc. Amer. Math. Soc. 85 (1982), 427–433.

60. Courant, R., Friedrichs, K., & Lewy, H. *Uber die partiellen Dijferenzengleichungen der mathematischen.* Physik, Mathematische Annalen, 100 (1928), 32–74.

61. Dellacherie, C. & Meyer, P.A. *Probabilité et Potential.* Hermann, Paris. (1980).

62. Davis, B. *A comparison test for martingale inequalities.* Ann. Math. Statist. 40 (1969), 505–508.

63. Davis, B. *Divergence Properties of some Martingales Transform* Annals of Math. Stat. Vol 40, No. 5 (1969) 1832–1834.

64. Davis, B. *On the integrability of the martingale square function* Israel. J. Math. 8 (1970) 187–190.

65. Davis, B & Song, R. *Selected Works of Donald l. Burhkolder* Springer New York (2010).

66. Dellacherie, C., Meyer, P.A.*Probabilities and potential A.* North-Holland Math. Studies 29, North-Holland (1978).

67. Dellacherie, C., Meyer, P. A. *Probabilities and potential B.* North-Holland Math. Studies 72, North-Holland (1982).

68. Diestel, J. & Uhl Jr., J. J. *Vector Measures.* Math Surveys #15 Amer. Math. Soc., Providence R.I. (1977).

69. Doob, J. L. *Stochastic Processes.* Wiley, N.Y. (1953).

70. Dragicevic, O & Volberg, A. *Bellman function and dimensionless estimates of classical and Ornstein-Uhlenbeck Riesz transforms,* J. of Oper. Theory 56 (2006),167–198.

71. Duoandikoetxea, J. *Fourier Analysis.* Graduate Studies in Math vol 29. AMS. Providence (2001).

72. Duflo, M. *Random Iterative Model* Springer Verlag, Berlin (1997).

73. Durret, R. *Brownian Motion and martingales in Analysis*. Wadsworth. Belmont, (1984).

74. Durret, R. *Probability Theory and Examples*. Fifth Ed. Cambridge Univ. Press (2019).

75. Enflo, P. *Banach spaces which can be given an equivalent uniformly convex norm*. Israel J. Math. 13 (1972): 281–288.

76. Fefferman, C. *Characterization of bounded mean oscillation*. American Math. Soc. Bulletin 77 (1971) 587–588.

77. Fefferman, C & Stein, E. *H^p spaces of several variables*. Acta Math. 129 (1972): 137–193.

78. Feller, W. *An Introduction to Probability Theory and its Applications*. Vol 2, 2 Ed., Wiley, N.Y. (1971).

79. Fujita, T., Kawanishi, Y. *A proof of Itô's formula using a discrete Itô's formula* Studia Sci. Math. Hungarica, Vol.45, No.1, (2008), 125–134.

80. Garcia Cuerva, J. & Rubio de Francia, J. L. *Weighted norm inequalities and related topics*. North-Holland. Amsterdam (1970).

81. Garsia, A. M. *On a convex function inequality for martingales*. Ann.Prob.1 (1973) 171–174.

82. Garsia, A. M.*Martingale inequalities*, Seminar Notes on Recent Progress. Math. Lecture notes series. New York: Benjamin Inc. (1973).

83. Garsia, A. M. *The Burgess Davis inequalities via Fefferman's inequality*, Ark. Mat.11 (1973), 229–237.

84. Geiss, S., Montgomery-Smith, G., & Saksman, E. *On singular integral and martingale transforms*, Trans. Amer. Math. Soc., 362(2), (2010) 553–575.

85. Gundy, R. F. *Martingale theory and pointwise convergence of certain orthogonal series*. Trans. Am. Math. Soc. 124 (1966) 228–248.

86. Gundy, R. F. *On a class of martingale series*. Proc. of the Conference on Orthogonal expansions and their continuous analogs. Southern IL. Univ. Edwardsville (1967). 99 –101.

87. Gundy, R. F. *A decomposition for a L^1-bounded martingales*. Ann. Math. Statist. 39 (1968) 134–138.

88. Gundy, R. F. *On the class L log L martingales and singular integrals*. Studia Math. 33 (1969) 109–118.

89. Gundy, R. F. *The martingale version of a theorem of Marcinkiewicz and Zygmund*. Ann. Math. Statist, 38 (1967) 725–734.

90. Gundy, R. F. *Orthogonal expansions and their continuous analogs*. Proc. Conference Southern Illinois Univ. (1968) 99–102.

91. Gundy, R. F. & Varopoulos, N. Th. *Les transformations de Riesz et les integrales stochastiques*, C. R. Acad. Sci. Paris Ser. A-B 289 (1979), A13–A16.

92. Gundy, R. F. & Kazarian, K. *Stopping times and local convergence for spline wavelets expansions*. Rutgers Univ. Technical report, #97 − 002, (1997).

93. Halmos, P. *Invariants of certain stochastic transformations: The mathematical theory of gambling systems*. Duke Math. J. 5 (1939). 461–478.

94. Haagerup, U. *The best constants in the Khintchine inequality*, Studia Math. 70 no. 3 (1982), 231–283.

95. Hall, P. & Heyde,C.C.*Martingale Limit Theory and its Application* Academic, Harcourt Brace Jovanovich, Publishers New York-London (1980).

96. Herz, C. *Bounded mean oscillation and regulated martingales*. Trans. Amer. Math. Soc. 193, 199-215 (1974)

97. Herz, C.: *H_p-spaces of martingales, $0<p \le 1$*. Z. Wahrscheinlichkeitstheorie Verw. Geb. 28, (1974) 189–205.

98. Hunt, R. *Almost everywhere convergence of Walsh-Fourier series of L^2 functions*. Actes. Inter. Congress Math. 2 (1970) 655–661.

99. Hunt, R. A., Muckenhoupt, B. & Wheeden, R. L. *Weighted norm inequalities for the conjugate function and Hilbert transform*, Trans.Amer. Math. Soc. 176, (1973) 227–251.

100. Hytönen, T. P., A. McIntosh, and P. Portal. *Kato's square root problem in Banach spaces*. J. Funct. Anal., 254 (3) (2008) 675–726.

101. Hytönen, T., van Neerven, J., Veraar, M. & Weis, L. *Analysis in Banach Spaces Volume I & II* Ergebnisse der Mathematik und ihrer Grenzgebiet Springer Verlag, N.Y. (2016 & 2017).

102. Izumisawa M. & Kazamaki, N. *Weighted Norm Inequalities for for Martingales.* Tohoku Math. Journ. 29 (1977), 115–124.

103. Kac, M. *Statistical independence in Probability, Analysis, and Number Theory.* Carus Math. Monogr. #12 MAA.

104. Kac, M. & Steinhaus, H. *Sur les fonctions indépendantes II.* Studia Math. 6 (1936) 59–66.

105. Karatzas, I. & Shreve, E. *Brownian Motion and Stochastic Calculus* 2 ed. Springer Verlag, N.Y. (1991).

106. Kaczmarz, S. & Steinhaus, H *Le systeme orthogonal de M. Rademacher.* Studia Mathematica 2 (1930) 231–247.

107. Khinchine, A. Y. A. & Kolmogorov, A. N. *Uber Konvergenz von Reihen deren Glieder durch den Zufall bestimmt werden.* Rec. Math. (Mat. Sbornik) 32 (1925) 668–677.

108. Koosis, P. *Introduction to Hp spaces.* London Math. Soc. Lecture note series. Cambridge University Press. Cambridge (1980).

109. Krasnoselskii, M & Rutickii, Ya. B *Convex Functions and Orlicz Spaces* (trans. Leo F. Boron). Noordhoff LTD, Groningen. (1961).

110. Kwapien, S. *Isomorphic characterization of inner spaces by orthogonal series with vector-valued coefficients.* Studia Math. 44 (1972) 583–595.

111. Lindestrauss, J. & Tzafriri,L. *Classical Banach space.* Springer Verlag. Berlin (1979).

112. Lai, T.L. & Wei, C. Z *A note on martingale difference sequences satisfying the local Marcinkiewicz-Zygmund condition.* Bull of Inst. of Math Acad. Sinica Vol. 11, No. 1 (1983) 1–13.

113. Liptser, R.S. & Shiryayev, A.N. *Statistics of Random Processes I* Springer Verlag. Berlin (2000).

114. Marmolejo, M & Valencia, E *Martingalas discretas. Aplicaciones.* Revista Integración. Universidad Industrial de Santander Vol. 27, No. 2 (2009) 135–171.

115. Marcinkiewicz, J. *Quelques theorémes sur les séries orthogonales.* Ann. Soc. Polon. Math. 16 (1937) 84–96.

116. Marcinkiewicz, J. & Zygmund, A. *Sur les fontions independantes.* Fund. Mat., 29, (1937), 60–90.

117. Marcinkiewicz, J. & Zygmund,A. *Quelques théoremes sur les fonctions independentes.* Studia Math. 7(1938) 104–120.

118. Maurey, B. *Systeme de Haar* Seminaire Maurey-Schwartz, 1974-1975, Ecole Polytechnique, Paris. (1975).

119. Maurey, B. & Pisier, G. *Séries de variables aléatoires vectorielles indépendantes et propriétés géométriques des espaces de Banach.* Studia Math. LVIII. (1976) 45–90.

120. McConnell,T. *On Fourier multiplier transformations of Banach-valued functions,* Trans. Amer. Math. Soc. 285 (1984), 739–757.

121. Muckenhoupt, B. *Weighted norm inequalities for the Hardy maximal function.* Trans. Amer. Math. Soc. 165 (1972), 207–226.

122. Muckenhoupt, B. and Wheeden, R. *Weighted norm inequalities for fractional integrals* Trans. Amer. Math. Soc., 192 (1974), 261–274.

123. Narváez, M. & Urbina, W. *Desigualdades de Burkholder-Davis-Gundy generalizadas y desigualdades del buen λ.* Acta Científica Venezolana, 4(48), (1997), 211–215.

124. Narváez, M. & Urbina, W. *Local properties of martingales: some new proofs.* Acta Científica Venezolana. Suplemento 2 vol 52. (2001) 39–44.

125. Nazarov, F. L. & Treil, S. R. *The hunt for a Bellman function: applications to estimates for singular integral operators and to other classical problems of harmonic analysis,* St. Petersburg Math. J. 8 (1997) 721–824.

126. Nazarov, F. L. & Volberg, A. *Heating of the Ahlfors-Beurling operator and estimates of its norm,* St. Petersburg Math. J. 15 (2004), pp. 563–573.

127. Nazarov, F. L. Reznikov, A., Vasyunin, V., & Volberg, A. *On weak weighted Estimates of Martingale Transform and Dyadic Shift.* Analysis and PDE Vol. 11, No. 8, (2018).

128. Nazarov, F. L., Treil, S. R.& Volberg, A. *The Bellman functions and two-weight inequalities for Haar multipliers,* J. Amer. Math. Soc., 12 (1999) 909–928.

129. Neveu, J. *Mathematical Foundations of the Calculus of Probability*. Holden-Day, San Francisco (1965).

130. Neveu, J. *Discrete paramenter martingales*. North Holand Math. Lib. 10. North Holland. Amsterdan, (1975).

131. Osękowski, A. *Sharp martingale and semimartingale inequalities*. Monografie Matematyczne 72, Birkhaüser Basel, (2012).

132. Osękowski, A. *Weighted Inequalities for Martingale Transforms and Stochastic Integrals* Mathematika 63 (2017) 433–450.

133. Osękowski, A. *Weighted inequalities for the martingale square and maximal functions.*Stat. Probab. Lett. 120, (2017) 95–100.

134. Paley, R.E.A.C. *A remarkable series of orthogonal functions I*. Proc. London Math. Soc. 34 (1932) 241–264.

135. Paley, R. E. A. C. & Zygmund, A. *On some series of functions III*. Proc. Camb. Phil. Soc, 28, (1932) 190–205.

136. Panzone, R. *Alternative proofs for certain upcrossing inequalities*. Ann. Math. Stat. 38 (1967) 735–741.

137. Pereyra, M.C. *Lecture notes on dyadic harmonic analysis*. Second Summer school in analysis and mathematical physics. Topics in analysis: harmonic, complex, nonlinear and quantization. Cuernavaca Morelos, Mexico, June 12-22, 2000. S. Pérez-Esteva, C.Villegas eds. Contemporary Mathematics 289 AMS, (2001) 1–61.

138. Pereyra, M. C. *Dyadic harmonic analysis and weighted inequalities: the sparse revolution*. Aldroubi A., Cabrelli C., Jaffard S., Molter U. (eds) New Trends in Applied Harmonic Analysis, Volume 2. Applied and Numerical Harmonic Analysis. Birkhauser, Cham (2019) 259–239.

139. Pisier, G. *Un exemple concernant la super-reflexivite*. Seminaire Maurey-Schwartz, 1974-1975, Ecole Polytechnique, Paris (1975).

140. Pisier, G. *Martingales with values in uniformly convex spaces*. Israel J. Math. 20 (1975) 326–350.

141. Pisier, G. *Martingales in Banach spaces*, Cambridge Studies in Advanced Mathematics, vol. 155, Cambridge University Press, Cambridge, (2016).

142. Rademacher, H. *Einige Satze ueber Reihen von allgemeinenen Orthogonal funktionen*. Math. Ann. vol 87 (1922) 112–138.

143. Riesz, M. *Les fonctions conjuguees et les series de Fourier* C. R. Acad. Sci. Paris, 178 (1924), 1464–1467.

144. Riesz, M. *Sur les fonctions conjugue'es*. Math. Z. 27 (1927), 218–244.

145. Rosenthal, H.P. On the subspaces of L^p ($p>2$) spanned by sequences of independent random variables. Israel J. Math. 8, (1970) 273–303.

146. Rota, G. C. *An "Alternierende Verfahren" for general positive operators* Bull. Amer. Math. Soc. 68 (1962) 95–102.

147. Schipp, F., Wade, W. & Simon, P. *m Walsh Series: An introduction to dyadic Harmonic Analysis*. Akadémiai Kiadó, Budpest (1990).

148. Shiryayev, A.N. *Probability*. 3 ed. Springer Verlag, N.Y. (2019).

149. Schauder J. *Eine Eigenschaft des Haarschen Orthogonalsystems*, Math. Z. 28 (1928) 317–320.

150. Steele, J. M. *Stochastic Calculus and Financial Applications* Applications of Math. 45 Springer Verlag, NY (2001).

151. Stein, E. M. *The development of square functions in the work of A. Zygmund*. Conference on Harmonic Analysis in Honor of Antoni Zygmund, W. Beckner, A.P. Calderón, R. Fefferman and P.W. Jones (Eds.) Wadsworth, Belmont, Calif.(1982).

152. Stein, E. M. *Singular Integrals and Differentiability Properties of Functions*. Princeton Univ. Press. Princeton. (1970).

153. Stein, E. M. *Topics in Harmonic Analysis related to the Littlewood-Paley theory*. Princeton Univ. Press. Princeton. (1970).

154. Steckin, S. B. *On the best lacunary system of functions* (in Russian), Izv. Acad. Nauk. SSSR, Ser. Mat. 25, pp. 357–366. (1961).

155. Stroock, D. W. *Probability Theory. An analytic view*. 2 ed. Cambridge Univ. Press NY. (2011).

156. Szarek, S. J. *On the best constants in the Khintchine inequality.* Studia Math.LVIII (1976) 197–208.
157. Szabados, T *A Discrete Itô's formula.* Limit Theorems in Probability and Statistics, Pécs, 1989. Colloq. Math. Soc. János Bolyai, vol. 57, pp. 491–502. North-Holland, Amsterdam (1990).
158. Szabados, T & Székely, B. *Stochastic Integration Based on Simple, Symmetric Random Walks* J. Theor Probab 22: (2009) 203–219.
159. Taylor, S. J. *Introduction to Measure and Integration* Cambridge Univ. Press (1966).
160. Torchinsky, A. *Real Variable Methods in Harmonic Analysis.* Academic Press San Diego (1986).
161. Urbina, W. *Operators Semigroups associated to Classical Orthogonal Polynomials and Functional Inequalities.* in *Orthogonal Families and Semigroups in Analysis And Probability - CIMPA Workshop Mérida, Venezuela, 2006.* French Mathematical Society (SMF). Séminaires et Congrès 25 (2012).
162. Urbina-Romero, W. *Gaussian Harmonic Analysis.* Springer Monograph in Mathematics, Springer-Nature (2019)
163. Vasyunin, V. & Volberg, A. *Burkholder's function via Monge-Ampère equation.* Ilinois. J. Math. 54, No. 4, 1393-1428 (2010).
164. Vasyunin, V. & Volberg, A. *The Bellman Function Technique in Harmonic Analysis* Cambridge studies in advanced mathematics 186, Cambridge Univ. Press N.Y. (2020).
165. Vershynin, R. *High-Dimensional Probability. An Introduction with Applications in Data Science* Cambridge University Press (2018).
166. Vilenkin, N. J. *On a class of complete orthonormal systems.* Izv. Akad. Nauk. SSSR, Ser. Math. 11, 363–400 (1947)
167. Wald, A. *On cumulative sums of random variables.* Ann. Math. Stat. 15(3), (1944) 283–296.
168. Wald, A. *Some generalizations of the theory of cumulative sums of random variables.* Ann. Math. Stat. 16(3), (1945) 287–293.
169. Walsh, J. L. *A closed set of normal, orthogonal functions,* Amer. J. Math. 55 (1923) 5–24.
170. Wheeden, R. & Zygmund, A. *Measure and integral. An introduction to Real Analysis.* Marcel Dekker, Inc New York (1977).
171. Wiener, N. *Fourier Integral and Certain of its Applications* Cambridge Univ. Press, N.Y. (1988).
172. Weis, L. *Operator-valued Fourier multiplier theorems and maximal L^p regularity* Math Ann. 319 (2001) 735–758.
173. Weisz, F. *Martingale Hardy Spaces and their Applications in Fourier Analysis* Lecture Notes in Math. 1568 (1994).
174. Williams, D. *Probability with Martingales* Cambridge Mathematical Textbooks, Cambridge Univ. Press, N.Y.(1991).
175. Young, R. M. G. *On the best constants in the Khintchine inequality,* J. London Math. Soc. 14 (1976) 496–504.
176. Zygmund, A. *Trigonometric Series.* 2 ed. Cambridge Univ. Press, N.Y. (1959).

Index

The manufacturer's authorised representative in the EU is Springer
Nature Customer Service Centre GmbH, Europaplatz 3, 69115 Heidelberg,
Germany. If you have any concerns regarding our products, please
contact ProductSafety@springernature.com

Printed and bound by CPI Group (UK) Ltd, Croydon, CR0 4YY
20/05/2026
02114595-0002